Ob

To Irene – K.B.
To Salwa – J.S.

Object-Oriented Design with UML and Java

K. Barclay
J. Savage

ELSEVIER
BUTTERWORTH
HEINEMANN

Amsterdam • Boston • Heidelberg • London • New York • Oxford • Paris
• San Diego • San Francisco • Singapore • Sydney • Tokyo

Elsevier Butterworth-Heinemann
Linacre House, Jordan Hill, Oxford OX2 8DP
200 Wheeler Road, Burlington, MA 01803

First published 2004

British Library Cataloguing in Publication Data
A catalogue record for this book is available from the British Library

ISBN 0 7506 6098 8

For information on all Elsevier Butterworth-Heinemann
publications visit our website at http://books.elsevier.com

Typeset by Charon Tec Pvt. Ltd, Chennai
Printed and bound in Great Britain

Contents

Preface

This book presents an introduction to Object-Oriented Design with the Unified Modelling Language and the Java Programming Language. The target audience for the book is undergraduate students in computing, computing science and software engineering. However, practising software professionals who wish to update their knowledge in this important area will also find it of value.

The text assumes no prior knowledge of object orientation on the part of the reader. However, some experience of the Java programming language is expected. The book is a basis for an academic course in which object-oriented design and the Java language are jointly presented. Appendices F and G can be used as indicators of the level of Java programming skills required. They can also be used to add to the reader's knowledge of Java programming.

Considerable effort has been taken to present the topics in a clear and orderly manner. Each chapter participates in the development of the object concept from simple objects and classes through to abstract classes, specialization, dynamic binding and polymorphic behaviour. A number of illustrative object models are developed, specifications for the classes presented, and Java implementations are programmed. All are reproduced from the computer and should execute correctly on any machine supported by a standard Java environment.

A particular feature of the book is the inclusion of case studies. They are used to illustrate various aspects of analysis and design and the Java language. They provide fundamental and systematic software development in a relevant context not otherwise achievable with small examples. Where appropriate, each chapter includes the analysis and design accompanied with the program listings. The case study is first introduced as a relatively simple application with a text-based user interface. It is then developed over a number of versions and given various makeovers. From its humble beginnings it is transformed to better accommodate revision and change, and given a contemporary graphical user interface. It is further enhanced by applying design patterns to its architecture and completed by refactoring some of its parts to obtain more leverage from our work.

A short list of the major issues raised in each chapter is repeated as a summary at the chapter end. All chapters include a set of exercises. They include the construction of new designs as well as modifications and extensions to the given illustrations. Selected solutions to these exercises are available from the book's website.

The Object Management Group (OMG) has adopted the Unified Modelling Language (UML) as a standard for presenting object-oriented designs. The UML is a visual modelling language that is both simple and extensible, and integrates the best software engineering practices. It is used to visualize, specify, document and construct the artefacts of an object-oriented analysis and design.

The emergence of object-oriented (OO) methods offers the opportunity to introduce object-oriented modelling and design into the development process from the beginning, ahead of any implementation concerns. Development methods or processes are both cultural and organizational. UML does not mandate one method; rather it aims to be applicable to a variety of methods. In this book we introduce a lightweight process that guides the development lifecycle and provides the reader with an accessible introduction to object-oriented modelling.

Many practising professionals consider Java as the implementation language of choice for object-oriented development. The adoption of Java as a programming language presents a number of major challenges, not the least of which is to adopt and fully embrace the object-oriented paradigm. This book explores some of these difficulties and their resolution, by imposing on the language the use of an object model. It is expressed as a UML design, which, in addition to constraining the form of the model, also offers guidance in the production of the final Java code.

The UML reinforces the holistic nature of the object model where there is less division between the phases of the software lifecycle. It also provides the balance between the power and flexibility of the Java programming language and the control required in its usage. In this way the UML reduces the risk involved in the adoption of Java as an object-oriented implementation language.

The UML permits developers to capture the detail of their design without the need to embrace the Java language. It has sufficient expressive power to record the classes, their basic attributes, relationships with other classes, and the behaviour of their operations. The benefit from the approach is that there is a significant shift of emphasis away from detailed programming difficulties on to the higher ground of analysing the meaning and accuracy of the model. Further, as the mappings from the model to the implementation are developed, we recognize emerging and repeated patterns. As a consequence much of the programming collapses into a coding chore.

Organization

The book is presented in three parts. The first is covered by chapters 1 to 4 inclusive. In these chapters we introduce object-oriented modelling, the UML, and the realization of these models as Java programs. At this stage we do not exploit the full object-oriented paradigm, but seek to provide a firm foundation with object-based solutions to applications.

The second part is presented in chapters 5 and 6 where we embrace specialization and its various aspects such as substitution and polymorphism. Specialization is one of the distinguishing features of the object-oriented approach.

In these two parts we aim to provide the reader with a solid understanding of object-oriented development. We have deliberately put the paradigm to the fore at the expense of providing a sophisticated user interface. In these early chapters a simple text-based interface is used in the illustrative applications.

In the third part of the book we consider further aspects of object orientation. With a solid understanding of object-oriented principles from the first two parts we are able to easily replace the text-based interface with a contemporary graphical interface. This is the subject of chapter 7. In the final two chapters we then investigate more advanced object-oriented issues such as design patterns and refactoring.

Chapter 1: Object Technology — gives an account of the fundamental concepts of object orientation and presents an introduction to topics discussed in more detail later in the book. It describes software development as a modelling activity with objects as the primary building blocks. It outlines the attributes of objects, encapsulation, information hiding, abstraction, the class, message passing, inheritance and polymorphism.

Chapter 2: Object-Oriented Analysis and Design — gives an account of the fundamental features of an object-oriented analysis and design (OOAD). The chapter introduces the UML as a notation for capturing and presenting the results of an OOAD. To obtain the full leverage of the UML we must superimpose on it a lifecycle development process. Here we introduce a lightweight process, one that is not overly bureaucratic, to drive the development lifecycle. We then continue to follow this process throughout subsequent chapters.

At this point the reader may care to review appendices A and B for installation and user instructions for the ROME tool presented with the book. ROME may be used to create and maintain these object models.

Chapter 3: Implementing Objects with Java — presents a detailed account of the implementation in Java of simple object-oriented systems. Great care is taken to use best practice so as to avoid opaque or confusing code. The bulk of the code developed can be traced back to the class and other UML analysis diagrams of the preceding chapter. Java concerns such as class declaration, attribute declaration and method definition, message sending, parameter passing, scope and duration of objects and control flow are now readily accessible through their early presentation in the object model.

Chapter 3 further develops Java implementations of object models by considering the implementation of architectural relationships. The examples from this chapter draw upon the collection classes included in the Java application programming interface (API). Incremental development through a series of versions is offered as part of the overall development method to reduce complexity and risk.

Chapter 4: Case Study: A Library Application — is a case study intended to consolidate the issues raised in the preceding chapters. A relatively simple lending library is modelled and detailed with the UML. Emphasis is placed upon using the principles of good software engineering within the context of an OOAD. The case study presents a strategy for delivering an OOAD, through a number of controlled iterations driven by use-cases. Each iteration is realized as successive increments to the developed code.

The three versions of the case study developed in this chapter follow the lightweight process introduced in chapter 2. Further, we introduce how to include testing into an OO development by considering use-cases that act as the basis for test-cases.

The final version considers the issue of progressing from a simple menu-driven, text-oriented interface to a graphical user interface. The principle issue addressed here is the separation of any input and output from the application domain model. This readily permits the substitution of one user interface by another, and is the subject of chapter 7.

Chapter 5: Specialization — completes the discussion of OOAD with the specialization of classes, the redefinition of methods and the use of the polymorphic effect.

The notion of base classes, interfaces and deferred methods is presented. Other issues including public and protected attributes, as well as private operations, are also addressed. The realization of these features in Java is demonstrated.

Chapter 6: The Library Application Revisited — is a case study that revisits the library application of chapter 4. As before it consolidates the most important issues raised in the preceding chapter. Specialization of classes and the use of the polymorphic effect are highlighted in the context of this application. Further, a series of iterations of this library application ultimately delivers a reusable framework suitable for this and other similar applications.

Chapter 7: Graphical User Interfaces — evolves the library case study by incorporating a graphical user interface on to the application domain model developed in chapters 4 and 6. The user interface is constructed from the Swing class library. Throughout the chapter we continue to highlight the designs in the application as well as the object-oriented models found throughout the Swing classes.

Chapter 8: Design Patterns — further extends our study of OOAD by considering design patterns that communicate best practice solutions to common problems. In this chapter we investigate and demonstrate a range of design patterns, captured through UML diagrams and implemented in Java. We demonstrate, with a number of illustrations, how these design patterns can be exploited.

Chapter 9: Case Study: A Final Review — completes our study of OOAD by demonstrating how refactoring can enhance the quality of our work. Once again we emphasize the importance of controlled development through relatively small iterations. The changes are supported by making testing a critical element in this activity.

Tool support

The UML has been incorporated into a graphical design tool called ROME by one of the authors (K Barclay). This object modelling environment permits the user to prepare and edit various UML diagrams. Further, the class diagrammer in ROME is used to generate the Java code. The user creates a class diagram with ROME, generates the Java program from this diagram, compiles the program, and then executes it. If the program fails to compile correctly or produces the wrong output, the user makes the necessary changes in ROME and repeats the cycle.

The ROME tool is offered on an as-is basis to the reader without any implied warranty. It is a fully functional UML modelling tool and Java code generator, by which the reader can follow the presentation of this book. It has been used extensively by the authors for the past six years in their teaching programmes. The authors have taken every precaution to ensure the robustness of this distribution. The product is offered in good faith to the readers of the book. However, the authors cannot accept any liability for any computer system fault resulting while using the software. Equally, the authors imply no warranties of merchantability, or fitness for a particular purpose. The authors shall not be liable for any damages suffered by a user when using any generated code from the ROME software.

Acknowledgements

The authors are deeply grateful for the encouragement and stimulation to this project given by the Head of School, Professor Jon Kerridge (School of Computing, Napier University, Edinburgh). We also thank the students who actively participated in the development of ROME, both as guinea pigs while it was being formulated, and as contributors as it evolved into its present form. More than 1000 undergraduate and postgraduate students of the School have had direct experience of ROME. Most acted as unofficial beta-test sites for the tool. The authors also greatly benefited from the experiences and insights offered by the practitioners who attended their professional development programmes.

Finally, the authors wish to extend their sincere thanks to the team at Butterworth-Heinemann. In particular we thank Alfred Waller for having faith in the book and Jodi Burton in her editorial role.

Software distribution

The authors have prepared a supporting website (case sensitive URL):

http://www.dcs.napier.ac.uk/~kab/jeRome/jeRome.html

It contains the ROME software, designs and working programs for all the illustrations and case studies presented in the book. Answers to selected exercises are also provided. As new features are regularly included the reader is advised to consult this site for up-to-date information.

October 2003

K A Barclay
W J Savage

1

Object Technology

This book is concerned with Object-Oriented Design, the Unified Modelling Language (UML) and the Java Programming Language. It seeks to demonstrate that a Java application, no matter how small, can benefit from some design during its construction. Various aspects of that design are captured and documented with the UML.

In this book we are primarily concerned with the middle ground between object-oriented design and implementation. Many textbooks exist that are solely concerned with the Java programming language (see the bibliography). These books give little or no explicit consideration to the question of design. A much smaller number of analysis and design books have been published. They often say little on the matter of realizing their designs. Here, we seek to offer a bridge between the two.

The benefit from this approach is that there is a significant shift of emphasis away from detailed programming issues on to the higher ground of analysing the meaning and accuracy of the design. Further, as mappings from the design to the implementation language are established, we recognize emerging and repeated patterns and consequently much of the programming activity often collapses into a coding chore.

This introductory chapter offers a roadmap into the remainder of the book. Here, we present the essence of object-oriented computing, and provide the necessary introductory background. We examine the fundamentals of object-oriented computing including modelling, analysis, design and implementation. The concepts are framed around everyday illustrations in which we concentrate on introducing the vocabulary of object-oriented systems.

Computer technology has developed extremely quickly since its inception. Today, computer-based systems impact on much of our lives in many spheres including banking, medical, flight reservation, educational and military applications. They are distinguished by having large amounts of software at their core. The capabilities of these systems are derived from the complex computer programs that control them.

Although we better understand the process of developing computer software, we frequently deliver it late and over budget. Often the software fails to do what the user requires and is difficult to maintain and modify. These remarks have always applied to the computer software industry, and while we have improved the technologies to support the development process, they have not fully matched the size and complexity of contemporary systems. Object technology is considered the best to deliver on these challenges, offering us the means to improve application development, reliability and scalability.

1

1.1   Background

During the 1990s object technology entered the mainstream computing landscape. The two primary fronts were in programming languages and in the introduction of object-oriented methods. The development of object-oriented methods was the subject of much research by both organizations and individuals. Notable leaders include Grady Booch (Booch 1991) and Jim Rumbaugh (Rumbaugh 1991). The various approaches promoted by these and others each had some merit but also had the effect of fragmenting the industry.

In the mid-1990s Booch and then Rumbaugh, and later Ivor Jacobson, formed the Rational organization (http://www.Rational.com). The aim was to combine their individual approaches and the contributions of others. Their efforts were offered for public scrutiny, thereby obtaining industry acceptance for a single unified object-oriented (OO) notation. It offers a means for capturing and recording the various elements of an object-oriented analysis and design (OOAD). During the late 1990s Rational published a number of versions of the UML. Subsequently in 1997 the Object Management Group (OMG) approved the UML as a vendor-neutral standard.

1.1.1   Modelling

In the same manner that an architect's blueprint presents design details for a building, the UML allows software models to be constructed, viewed and manipulated during analysis and design. *Modelling* is a proven technique used in a variety of disciplines. For example, engineering models are used in the design and development of motor cars (automobile engineering), aeroplanes (aero-engineering) and bridges (civil engineering). Similarly, meteorologists have developed mathematical models to predict weather patterns.

Models are central to many human activities. They provide a blueprint for some artefact we wish to manufacture or understand. A blueprint offers a measure of repeatability ensuring standardization of the product. This is as equally important to the software customer as it is to a customer purchasing, say, household goods such as a television or a washing machine.

Through a model we aim to provide a better understanding of the system under development. Models aid our understanding of especially complex systems and help ensure we have correctly interpreted the system under development. For example, automobile engineers use models to design new motor cars that meet a number of criteria including their aerodynamics. The cars can be prototyped in fibreglass and tested in wind tunnels. A similar argument can be applied to software development whereby a model can be used to ensure all its requirements are met.

A model also permits us to evaluate our design against criteria such as safety or flexibility. Similarly, UML models permit an application's design to be evaluated and critiqued before implementation. Changes are much easier and less expensive to make when they are made in the early phases of the *software lifecycle*.

Models help us capture and record our software design decisions as we progress toward an implementation. This proves to be an important communications vehicle between the development team members as well as between them and the customer. Development team members can discuss their designs with the client to ensure they have fully understood his needs.

Models can be layered and hence provide varying levels of detail. Abstract software models omit large amounts of fine-grained detail and permit us to gain a high-level view of the system and its architecture. These views permit us to focus on various parts of the system without recourse to the details of program code. Repeated refinement of these models can be used to progress them toward the final code.

An architect when designing a building often constructs a number of diagrams that present it from a variety of perspectives. One view is, of course, the structure of the building and is vital to the construction company. This same view is also important to the customer since it reveals the details of the accommodation, its layout and the access to stairs, elevators, etc. However, electrical (or plumbing) contractors are more interested in the run of electrical circuits (water supplies and drainage pipes) and their supply to the building from the utility companies. Hence an architect would construct a number of blueprints highlighting these various facets.

In a similar manner, the software architect can offer a range of UML diagrams that view the system from different perspectives. Some give a static view of the application with the architectural configuration of the objects as the primary focus. Other UML models emphasize the dynamic behaviour of the objects and their interactions.

1.1.2   UML

The UML defines a diagrammatic notation for describing the artefacts of an OOAD. Through the UML we can visualize, specify, construct and document our software application. As our software systems become ever larger and ever more complex we need to manage that complexity and, in a sense, simplify it so we have a better understanding of it. Often, visualizing the software graphically is more appropriate than struggling to understand it in program code.

By inspecting our models we can identify deficiencies in our designs as well as opportunities to enhance them. The UML acts as a specification language in which we can precisely and unambiguously capture our design decisions.

Finally, from our UML diagrams we can derive programming language code. This is referred to as *forward engineering* — the generation of code from UML models. This is an approach we advocate through this textbook. The models are at the core of our designs. The code is an outcome of that modelling activity and is itself a design document. The models dictate the code that we ultimately produce.

1.1.3   Analysis and design models

An *analysis model* used in software development aims to document various facets of the real world problem that we are modelling. In an object-oriented system development this would typically involve identifying the significant application objects and the application processing to be performed.

Development methods that pre-dated object technology often deliberately delineated between the analysis phase and the design stage. Commonly, they would also use a linear or *waterfall model* for the development process in which the design stage only follows after all the analysis has been completed. Further, these separate analysis and design stages often resulted in a conflict between them, especially where different models and notations were deployed.

Object-oriented methods are characterized by using the same modelling concepts throughout the software lifecycle. This way, the solution that emerges during analysis is carried through into design and finally to code. Objects identified during analysis should also be present in the final code. This offers a seamless integration of the stages, not otherwise found with other approaches. The design models augment the analysis models with additional detail and, perhaps, introduce further low-level system objects required in the implementation.

1.1.4   Development process

The UML is a modelling language. It has no notion of a development process, which must accompany a method. The dictionary defines a method as a systematic or orderly procedure. The authors of the UML understood this distinction and deliberately sought to separate the language used to document a software design from the process used to develop it. They recognized that processes are influenced by many considerations such as the nature of projects and the culture of organizations.

Object-oriented design is usually conducted within an *iterative process*. This is vital to ensure that we can revisit earlier decisions when corrections or modifications are necessary. This is not unreasonable. After all, initial design decisions may require revision, especially in new projects or in those that are less well understood by the development team. The iterative process continues until the full system is developed.

In common with many object-oriented developers, an iterative process is also accompanied by an *incremental* style of development. Each increment introduces some additional functionality on to the previous stage. Often, the new increment only adds a small feature so that we can fully test it and its effect on the existing system and its architecture.

Each iteration needs to be accompanied by an objective that can be checked. Otherwise, there is a danger of the process degenerating into *undisciplined hacking*. An iterative approach is further enhanced when the customer is closely involved in the system development. Each new iteration can be presented to the customer to obtain feedback and to ensure his active participation throughout the project.

1.2   Using the UML

The next chapter presents a detailed discussion of the UML. Here, we simply consolidate our earlier discussions by considering the more important elements of an object-oriented software system. This will act as an introduction to the more detailed discussions that follow.

1.2.1 Objects: combined services and data

An object-oriented system comprises a number of software objects that interact to achieve the system objective. The software objects usually mimic the real-world objects of the application domain. The real-world objects may have a physical presence or may represent some well-understood conceptual entity in the application. For example, in a university application we might have software objects that represent students. Equally, we may have software objects representing programmes of study at a university even though they have no physical existence.

Objects are characterized by having both *state* and *behaviour*. The state of an object is the information an object has about itself. For example, a student object may have a name, a date of birth and a university matriculation number. Equally, a programme of study object might have the name of the programme, its duration and the name of the programme leader. The behaviour of an object describes the actions the object is prepared to engage in. For example, we might ask a programme of study object for its duration. We might ask a student object for its age. This would involve the student object performing a calculation based on its date of birth and today's date.

The behaviour of an object is described by the set of *operations* it is prepared to perform. One object interacts with another by asking it to perform one of its advertised operations. This interaction is achieved by one object sending a *message* to another. The first object is known as the *sender object* and the second object is known as the receiver or *recipient object*. The only messages an object can receive belong to the set of operations it can accept. In the UML, objects and message passing are usually captured by a *sequence diagram* as shown in figure 1.1. Here a university object is shown sending the message getAge to a student object.

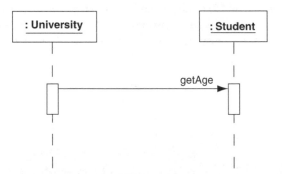

Figure 1.1 *Message passing in a sequence diagram*

When an object receives a message it performs some action. This action is described by a *method*. A method is the processes the receiving object follows when servicing the message. For example, if a university object sends a message to a student object asking for its name, then the student object simply replies to the sender with one part of its state, namely, the name. However, if a university object sends a student object the message asking for its age, then the method that student object must follow is more elaborate. First, it must obtain today's date. This might be achieved by the student object sending a message to some calendar object. The student object then has to perform some complex date arithmetic on its date of birth and today's date to determine its age. In the sequence diagram of figure 1.1 this processing is shown as an *activation*, the rectangle adjacent to the message arrow.

Figure 1.1 also implies *message propagation*. When one object receives a message it often sends a cascade of other messages to other objects. The university object sends a message to the student object asking for its age. In turn, it sends a message to some calendar object requesting today's date. This example also demonstrates that an OO system is a mix of objects interacting to achieve the required objective.

A university would typically have a large number of students. Unlike real students, all student objects exhibit the same behaviour and carry the same knowledge about themselves. We might model a student object with a name, date of birth and matriculation number. The actual state values for two student objects are presumably different since university matriculation numbers are unique. With a large university population we might, however, expect two or more students with the same name or two or more with the same date of birth.

They are, however, all subject to the same behaviours. If one student can be asked for their age by sending some suitable message, then all students can be sent this message. How is this determined? All of our student objects support a single abstraction that we may choose to call **Student**. Other abstractions from this problem domain might include **University**, **ProgrammeOfStudy** and **Tutor**. We refer to the abstraction as the *class* of the object.

A class is effectively a blueprint or template that fully describes the abstraction. The **Student** class describes any number of student objects. The **Tutor** class describes any number of tutor objects. The class describes the information an object holds to represent its state. The items of information are called *attributes* (sometimes also called *properties*). The class also defines the behaviours of such objects, listing the operations they can perform, i.e. the messages they can receive. The effect of these operations is described by its method. Figure 1.2 shows a simplified *class diagram* for a **Student** class.

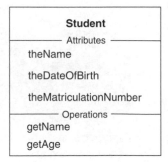

Figure 1.2   *UML class diagram (simplified) for a Student class*

In this figure we have a class **Student** with two operations and three attributes. Any student object we create from this template will have a state comprising three values for the attributes, i.e. **theName**, **theDateOfBirth** and **theMatriculationNumber**. Further, any

Student object can be sent messages to obtain their name or their age, i.e. **getName** and **getAge**.

Figure 1.3 shows how in the UML we present a particular example of an object from some named class. This we refer to as an *object instance* or simply an object. The upper part of the figure names the class to which the object instance belongs, here **Student**. It also labels the object with some identifier by which we can refer to that object (**s1**). The lower part presents the attribute values maintained by the instance and represents its state. Here, for example, this particular **Student** instance has **theName** attribute with the value **Ken Barclay** as part of its state. Such a diagram element may be part of a much larger *object diagram* (or *collaboration diagram*) that we describe with the UML (see chapter 2).

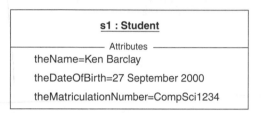

Figure 1.3   *An object instance*

As a further example, consider how we might model a bank account. An account can exhibit a variety of behaviours such as debiting or crediting some monetary amount, or requesting the account's current balance. These behaviours give rise to some of the likely account operations. Debit and credit transactions document the amount involved in the transaction, changing the balance for an account. The balance, along with the account number must be maintained by each bank account instance. Every example of a bank account carries its own data values for these attributes (see figure 1.4).

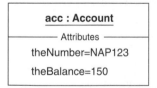

Figure 1.4   *A bank account instance*

To model an account as an object we describe its behaviours as operations and its state with attributes. During the execution of a system, an account object is requested to carry out its various operations, changing its attribute values as needed to reflect the effect of its actions. For example, in figure 1.4, a **debit** operation applied to such an account object results in a change to the value of **theBalance** attribute.

Some operations are used to get information about an object, while others have some effect on an object's state. The operations that only give information about an object are referred to as *enquiry operations*. They enquire about some state information held by the object. The operation to obtain the value of an account's balance is of this type.

The operation that performs a debit transaction on a bank account object changes the current balance it holds. This category of operation is described as a *transformer operation*. A transformer operation changes one or more of the object instance attribute values. Both operations refer to the values of the object's attributes, collectively the *state* of the object. Transformer operations result in a state change, while enquiry operations do not usually affect the state of the instance. In the class diagram of figure 1.5 we recognize the operation getBalance as an enquiry operation, while debit and credit are transformer operations.

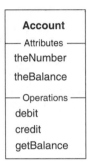

Figure 1.5   *Class diagram with Account class*

1.2.2   *Objects make excellent software modules*

The concept of an object is both simple and yet extremely powerful. Objects make ideal software modules. Each object instance forms a self-contained entity. Everything an object knows is expressed in terms of its attributes and everything it can perform is expressed by its list of operations. For this reason, objects are described as *highly cohesive*. All the characteristics of an object provide some well-bounded behaviour for the particular abstraction they represent (*encapsulation*).

Consider a motor car. A car has various controls that are used to control and operate it. The gear shift, for example, is used to change gear on a manually operated vehicle or to select the drive on one that is an automatic. Cars usually have tachometers to show the speed of the vehicle. Other controls include the accelerator and the brake.

The internal components of the car are implemented by the many mechanical and electronic devices contained within the body of the car (usually under the engine bonnet). This metal carcass isolates the driver from the internals. Since the driver has no direct contact with these components there is no likelihood that he will damage himself or the car. Consider the position where, instead of an accelerator pedal, the driver controls the speed by using a screwdriver on some internal control screw. The accelerator would usually have some restricted amount of movement, limiting the driver to a certain maximum speed. Without this restriction it is possible that adjustments made directly to the control screw may set the speed above its maximum safe working level and consequently damage vital components.

Equally, by isolating the driver from the internals, and only permitting a trained engineer access to the components, a second benefit is obtained. If the car develops a fault, then an engineer may replace the faulty component. The change has, however, been effected without any operational change. The driver still uses the car in exactly the same way. Only the repair of the faulty component has changed.

These ideas are also mirrored in object technology. They present the same benefits to software as they do to cars. In a software object the attributes are hidden from a user but may be replaced without affecting user software that relies on that object abstraction (*information hiding*). In a software object behaviour is defined by its operations and not by its private representation. The operations govern what we can ask a software object to perform. If we supply an unacceptable value when invoking an operation, then the effect can be denied, ensuring the correct state of the object. For example, a debit operation applied to a bank account might only be permitted if sufficient funds are available.

Since object attributes and operations only define their abstraction and no other, they usually exhibit *loose coupling* with other objects. This is highly desirable because *strong coupling* makes software components harder to understand, change or correct. For example, when defining an object's operation we need not concern ourselves with the requester (client) of that operation. Only the effect of that operation on the receiving object's attributes need be considered. Equally, the client requesting a service need not be concerned with how that request is achieved by the recipient object.

1.2.3   Object interaction is expressed as messages

Figure 1.1 demonstrated that objects interact with each other by sending a message from the sender requesting the recipient object to carry out one of its advertised operations. In figure 1.1 a University object requests the age of a Student object. Equally, a bank object may send one of its account objects a credit operation. Here, the bank is the sending object and the account is the receiving object. A message identifies the recipient object and the name for the operation to be performed. The message name represents one of the operations of the class to which the recipient belongs. If the message requires any further details they are given as the *message parameters*.

To request a bank account object to engage in a transaction to debit it by some monetary amount, some sending object, such as an automated teller machine (ATM), might send the account object the message:

acc debit 50

Here, acc is the identifier of the receiver object which is some bank account object, debit is the operation it is being asked to execute, and 50 is the actual parameter informing the receiver account of the amount involved in the transaction. The UML *collaboration diagram* in figure 1.6 portrays this message passing between objects. In the diagram we have two objects: an Account object with identifier acc and an anonymous (no identifier) ATM object. The latter sends a message to the Account object to perform the operation debit with the actual parameter value of 50.

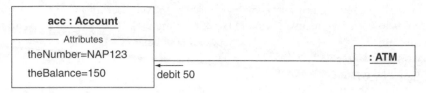

Figure 1.6 *Message from an ATM to an account object*

When a message is received by some recipient object then an action is performed. This action usually involves some or all of the values of the attributes representing the state of the receiving object. The action will also use any message parameters. The logic associated with this action is described by the method for the operation. The *method* refers to the *algorithm* that is applied when an operation is executed. The method for the debit operation applied to an Account object involves reducing the value of theBalance attribute by the actual message parameter value of 50.

We have noted how transformer operations usually result in a state change to the receiving object, while enquiry operations merely request information from it. The only means of communication is a message sent from a sender to a receiver. In the case of an enquiry operation a secondary information flow is observed. Here, the sender is expecting a response from the receiver in the form of a *return value*. Occasionally, transformer operations also supply return values, say, to report to the sender that the designated task has been completed successfully. Be sure to recognize that a return value is not another message.

Observe also the asymmetry of the messaging concept. The recipient object, when defining its operation's logic, does not concern itself with the object that is sending the message. Equally, the sender need not be concerned with how the operation is implemented by the recipient. As noted in the preceding section this greatly assists with the production of high quality software systems by the separation of these two concerns.

1.3   Classes: sets of similar objects

An object-oriented system is characterized as a set of interacting objects. It is therefore common to have more than one object of any given kind. For example, a bank will certainly have a number of customer accounts each of which carries out the same actions and maintains the same kind of information. The single class Account (such as figure 1.5) could represent the entire collection of account objects (such as in figure 1.4). The class contains the specification and definition of its operations (methods) and its attributes. The actual accounts are represented by instances of this class, each with its own unique identifier (say, acc1, acc2, ...). Each instance contains data that represents its own particular state. When an account receives a message to carry out one of its operations, it uses the method definition for the operation given in its class and applies it to its own attribute values.

Figure 1.7 shows the Account class and two instances of that class. The instances have identifiers acc1 and acc2. The Account class has three services provided by the operations debit, credit and getBalance. The attributes maintained by every instance of this class have their own values for theNumber and theBalance. For example, in the instance with identifier acc1 these attributes are respectively DEF456 and 1200. The message:

acc1 debit 50

results in the execution of the method for the debit operation applied to the Account object acc1. This operation might be defined in terms of subtracting the value of the message parameter 50 from the value of the attribute theBalance presently held by the Account object acc1. The effect of this transformer operation produces a state change in the object acc1, reducing theBalance to 1150.

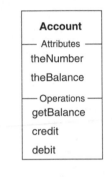

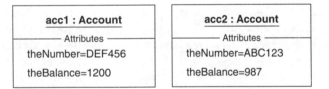

Figure 1.7   *The Account class and two instances*

Normally objects do not exist in isolation. They form relationships with each other and related objects engage in message passing. In the object diagram of figure 1.8 we show two Account objects that are related to the same Bank object. The Account objects are not related to each other. Thus the Bank object can send messages to either or both Account objects and the latter can send messages to the Bank object. Significantly, since there is no relation between the Account objects therefore they cannot engage in message passing.

The classes and the relationships between them are modelled in the class diagram in figure 1.9. The annotation 0..* indicates that one Bank object can be related to none (0) or more (*) Account objects. We shall have much more to say on these diagrams in the next chapter. The object diagram of figure 1.8 is a particular example of a configuration of objects described by this class diagram. In principle, there are an arbitrary number of object diagrams that are based on a single class diagram. Thus a class diagram is a abstract description of all possible configurations of objects.

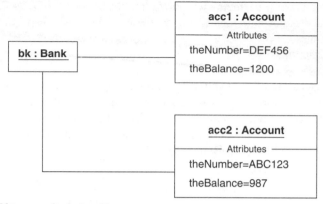

Figure 1.8   *Objects and relationships*

Figure 1.9   *Class diagram with a relationship*

1.3.1   Specialization

Object-oriented models can give rise to many classes of objects. Such complex models may be rationalized and simplified by arranging the classes into hierarchies. The hierarchies assist in the categorization of the types of object instances represented by the many classes. This categorization of knowledge is found in many science and engineering disciplines and greatly assists in simplifying complex systems.

Many people, whatever their particular kind, share common characteristics. These characteristics include both state information and behaviours. For example, name, home address and social security number are data shared by all people. Equally, all share the same behaviour of being able to ask a person for their age. People may be further specialized as a particular kind of person, e.g. student. In addition to all the qualities associated with people they also have additional characteristics peculiar to being a student, such as a matriculation number. Everything that applies to an ordinary person also applies to students and every instance of a student is implicitly an instance of a person.

The **Student** class is described as a *specialization* of the **Person** class. Conversely, the class **Person** is a *generalization* of the class **Student**. The specialized class **Student** is said to *inherit* all the features of its generalization class **Person**. Thus if any person has a name, then so do students. If we can ask a person for their age then we can do the same with a student. In fact, any operation that may be applied to a **Person** instance may also be applied to a **Student** instance. The converse, however, is not true. A **Student** object may have attributes and operations peculiar to it, such as a matriculation number. Since a **Person** is a generalization of a **Student**, only those common characteristics are applicable to **Persons**. Hence we are not permitted to ask a person for their matriculation number since this is only applicable to students.

Specialization is the mechanism by which one class is defined as a special case of another class. The specialized class includes all the operations and attributes of the general class. The specialized class is said to inherit all the features (or characteristics) of the general class. The specialized class may introduce additional operations and attributes peculiar to it. In addition to the operations inherited, the specialized class may choose to *redefine* the behaviour of any one of these. This, as we shall see shortly, is used when the specialized class has a more specific way of defining that behaviour. The specialized class is commonly known as the *subclass* and the general class its *superclass*.

Specialization is often described as *programming by difference*. Since the Student class inherits from the class Person, then the Student class need only implement those differences between itself and the generalized Person class. So, for example, the Student class need only introduce the additional attributes peculiar to students, say, a matriculation number. Further, through inheritance, the specialized Student class need not reprogram the inherited operations. The difference is any additional operations and any redefined operations. The subclass then benefits from a significant amount of *code reuse*.

In our banking illustration, the class Account could be specialized into two subclasses CurrentAccount and DepositAccount. The specialization relation is illustrated as a directed arrow from the subclass to the superclass as shown in figure 1.10. Each inherits the general characteristics of its common superclass. Either subclass may then add to the set of operations and attributes of the superclass, or *redefine* the behaviour of one or more inherited operations.

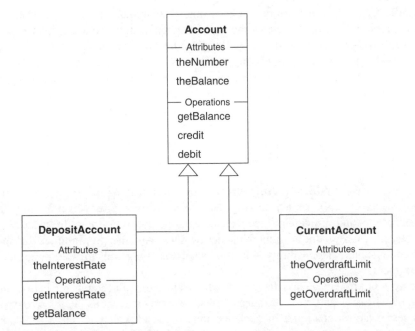

Figure 1.10 *Two subclasses of bank account*

In Figure 1.10 any instance of the class CurrentAccount has the attributes theNumber, theBalance and theOverdraftLimit. The latter attribute is introduced in the CurrentAccount class itself, while the other attributes are inherited from the Account superclass. Equally, any instance of CurrentAccount can respond to the messages debit, credit, getBalance and getOverdraftLimit. Again, the first three operations are inherited from the superclass and the last is defined for the CurrentAccount class. A similar arrangement applies to the DepositAccount class in which the attributes of an instance are theNumber, theBalance (both inherited) and theInterestRate (declared by the class). The operations are debit, credit (both inherited), getInterestRate (defined by the class) and getBalance (redefined).

Redefining the definition of an operation in a subclass permits a specialized implementation of the method. For example, the operation getBalance in the Account class may be defined to simply return the present value of the attribute theBalance. In the subclass DepositAccount the reappearance of the operation getBalance indicates a redefinition that might, for example, deal with any interest accruing.

1.3.2   Polymorphism

Through inheritance, one class is formed as a specialization of an existing class, inheriting all the features of that existing class. This proves to be a particularly important concept that supports re-usability of existing code. Inheritance also gives rise to the notions of *polymorphism* and *dynamic binding*. These additional concepts provide support by which software systems may be modified to accommodate changes to its specification.

The dictionary definition for polymorphism is "having many forms". The class definition for DepositAccount declares the explicit specialization from class Account and reveals that a DepositAccount is an Account with additional attributes, operations and redefined operations. Hence an instance of DepositAccount may be substituted for an Account instance. This is permitted since an instance of the class DepositAccount can be sent the same messages as an instance of the Account class. Equally, an instance of CurrentAccount may also be used where an Account instance is expected. This means, for example, a Bank object may be introduced with a number of Account objects associated with it. The Bank does not need to concern itself with whether they are CurrentAccount objects or DepositAccount objects. They are all some kind of Account.

This approach is radically different from that which is employed in conventional systems where it is necessary to populate code with complex selection statements to identify the kind of account then execute some appropriate logic. In these systems the determination of the account kind lies wholly with the programmer. In object-oriented systems responsibility for this selection is given to the programming environment.

Figure 1.11 presents a class diagram in which a number of accounts are held or maintained by a bank. A particular instance of the class Bank is responsible for zero or more instances of the various kinds of bank accounts, some of which are DepositAccounts and some CurrentAccounts. The labelled line is an *association* demonstrating a one-to-many relationship between the Bank class and the general Account class. This kind of object model is the subject of this book and is formally introduced in the next chapter.

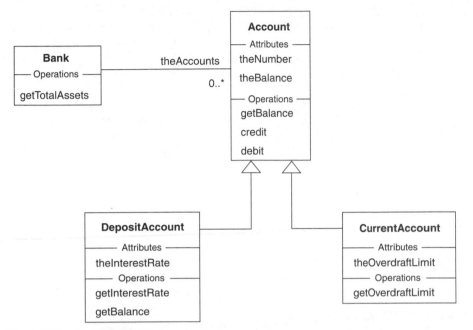

Figure 1.11 *Bank/Account class diagram*

When a Bank object sends the message getBalance to each of the many Account objects it holds, then the definition for this operation in the Account class is used. Some accounts will, of course, be interest-bearing DepositAccounts that, through redefinition of the getBalance method, have a different way of computing the amount of interest. To obtain the correct selection of method we defer the choice of method to execute until *run-time* using a mechanism known as *dynamic binding*. This is done by recording that the *polymorphic effect* is required on the getBalance operation. If the operation getBalance in the class Account is polymorphic, then when the message is sent to each account instance the appropriate definition of the operation is executed according to the class of the receiving object. Effectively, the receiving object knows to which class it belongs and executes the appropriate method.

Thus when the getBalance message is received by a DepositAccount instance, then the redefined version of the method from that class is executed. When a CurrentAccount object receives the same message, then since that class does not redefine the operation, the method executed is that defined and inherited from the superclass Account. The UML collaboration diagram in figure 1.12 demonstrates how the Bank object determines the value of its total assets. Each account is either an instance of a DepositAccount or an instance of a CurrentAccount. The Bank is unaware of this fact and simply sends the message getBalance to each. When the single CurrentAccount object receives the message getBalance it simply executes the method inherited from

the Account superclass. The two DepositAccount objects use the redefined method in
their subclass.

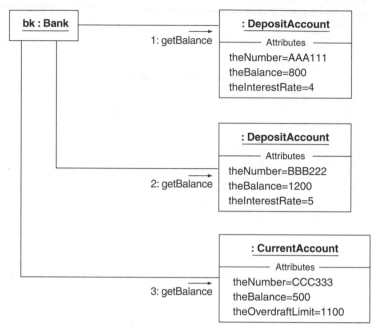

Figure 1.12   *The dynamic binding effect*

Polymorphism contributes to program maintenance. If we were to add a new kind of
account to the system, say the class SavingAccount as a specialization of the Account
class, it might implement its own getBalance operation. The Bank class does not need
to know about this enhancement. Again, it sends the getBalance message to every
account. If a particular instance happens to be from this new class it will choose its own
definition for this operation.

1.4   Tools

Today, modern software development normally takes place with the support of software
development tools. They are generally described as *Computer Aided Software Engineering* (CASE) tools. Many commercial CASE tools are large and complex software
systems that support many of the stages of the software development process. They are
often known as upper-case tools because of their support for most aspects of a process.
By comparison, lower-case tools provide less support but usually require a much shorter
learning curve.

To support the reader throughout the remainder of this textbook, the authors have made
available a lower-case tool ROME (see appendices A and B). That way, the reader can
follow the book's content and have access to CASE tool support for various parts of the
discussion. Whereas upper-case tools strive to automate activities, ensure consistency

across the models, provide various management reporting, etc., ROME does not include these features to ensure we have a much simpler tool to operate. At the end of this study however, we expect that the reader should be capable of progressing to the more advanced commercial tools. For example, consult the website http://www.rational.com, http://www.togethersoft.com.

The ROME modelling tool has been used throughout this book to create the many UML diagrams shown. The current version of ROME supports the majority of diagrams described by the UML. Further, the class diagrammer in ROME is used to generate the Java code.

See the section entitled "Software distribution" in the Preface for details of how to obtain and install the supplied software.

1.5 Summary

1. The Unified Modelling Language, UML, is an internationally agreed notation for recording the various elements of an object-oriented analysis and design. The UML defines a number of views of a system through various diagrams such as class and collaboration diagrams.
2. The UML must be augmented with a process to guide the development of the software. In an OOAD the same modelling concepts are used throughout the software development process.
3. An object-oriented system is characterized as a set of communicating objects.
4. An object is a set of operations together with a state that the object retains between invocations of any of its operations. Transformer operations result in a change to that state while enquiry operations report on the state.
5. An object instance is a particular example of an object from some named class and can be shown in a UML object diagram. A class is a blueprint or template describing an arbitrary number of such instances and is presented in a class diagram.
6. Objects interact through message passing shown in either UML collaboration or sequence diagrams. One object sends another object a message that invokes one of the recipient's operations. The message is bound to the operation's definition given in its class or its superclass.
7. Classes may be classified into a hierarchy starting from the general and leading to the more specific. A subclass is a specialization of its immediate superclass. A subclass inherits all the features of its superclass. It may add further features and redefined operations. An instance of a subclass is an instance of its superclass and may substitute at any time for an instance of the latter.
8. Inheritance also gives rise to the notions of polymorphism and dynamic binding. A dynamically bound, polymorphic message sent to an object binds to the operation definition in the class to which the object belongs.

1.6 Exercises

1. Explain why modelling is central to many human activities. Other than those already described, give one further example for the use of models.

2. What do you understand by the term process? Object-oriented development processes employ iterations and increments. Outline what you understand by these terms.

3. What role does the UML play in an OO development process? Is the UML the process itself?

4. Objects are said to have both state and behaviour. Explain what is meant by these terms?

5. Carefully distinguish between the terms "operation" and "message". What is the relationship between the terms "message" and "method"?

6. What do you understand by the term "class"? What are the two principal features or characteristics of a class? How do they relate to the notions of state and behaviour?

7. Give a UML class diagram in the manner of figure 1.2 for the class **Person** with **theName** and **theDateOfBirth** attributes as well as the operations **getName**, **getAge** and **changeName**.
   Show two instances of class **Person** in an object diagram representing yourself and someone in your family.
   Suggest other plausible attributes and operations for the **Person** class.

8. The **Account** class in figure 1.5 shows two attributes and three operations. The enquiry operation **getBalance** interrogates an **Account** object for part of its state. Which attribute value is returned by this operation? How is this value used when implementing the additional operation **isOverdrawn** that indicates the state of an **Account** object?
   When we **debit** or **credit** an **Account**, which attribute value is affected? By what amount is it changed? How is this change value introduced with the **debit** and **credit** operations?

9. How is an operation invoked? What are the three principal components? Which component is optional? When missing, what category of operation do we typically then have? What is a method in this context?

10. Figure 1.10 shows three kinds of bank account. The operations **debit**, **credit** and **getBalance** of the **Account** class operate on which attributes? What attribute does operation **getOverdraftLimit** in the **CurrentAccount** class use? How will the redefined operation **getBalance** of class **DepositAccount** operate? What attributes are involved and to which class do they belong?

11. Suppose the **CurrentAccount** class of figure 1.11 redefined the **getBalance** operation, and a **Bank** instance is related to a number of **DepositAccount** objects or **CurrentAccount** objects (as in figure 1.11). If the **Bank** sends the **getBalance** message to all, which versions of this operation are executed? How is this selection determined?

2

Object-Oriented Analysis and Design

The previous chapter described an object-oriented system as one composed of software objects collaborating with each other to achieve some common goal. Clearly it is of paramount importance that the objects and their relationships are correctly identified and recorded.

Unfortunately, it is difficult to give hard and fast rules as to how this can be accomplished in all cases. Part of the reason is that problem solving is very much a human activity. However, over the past few years several *object-oriented analysis and design* methods have become popular (Jacobsen 1999, Priestley 2000). Although they differ in some important respects each has at its core the belief that we naturally think in terms of objects and that they make excellent software components.

We must appreciate that the UML is a notation and not a method. It has no notion of a process that has to be followed as part of a method. Its purpose is to allow developers to capture and present the results of an object-oriented analysis and design. In this chapter we present a process that can be followed in which the UML is used to capture elements of the system development.

In the following sections we introduce some of the more important UML diagrams. We shall first illustrate them by considering a small example that continues into the next chapter when we consider the implementation for our design in Java. Sections 2.2 and 2.3 introduce applications of the UML to this small demonstration. Later sections (2.4 and 2.5) revisit other aspects of the diagrams not covered by the illustration.

The reader is invited to review the materials in appendix D, which summarize most of the UML features we use throughout this textbook. It is also recommended that specialist textbooks for the UML be consulted (Priestley 2000, Larman 2002).

2.1 Fundamentals of an OOAD

The UML does not mandate a particular software development lifecycle. In fact, its aim is to act as a modelling language and to be process independent. To gain the full leverage of the UML we must impose on it a method. The original authors of the UML have proposed the Rational Unified Process (Jacobsen 1999), a heavyweight, industrial-strength method. Unfortunately, for many projects it is overly elaborate and complex.

In this textbook we offer a lightweight process as a vehicle for presenting an introduction to the subject. This is also in keeping with recent developments in *agile systems development* (Beck 1999) and *extreme programming* (Cockburn 2001) which seek to simplify and reduce the bureaucracy of the heavyweight methods. We shall, however, retain the accepted guiding principles that an OOAD process should be:

- use-case driven
- architecture centric
- iterative and incremental

A *use-case* is a typical interaction between a user and the system under development. It is used to capture some functionality to be provided by the software system. For example, in a banking application a use-case may document that a requirement of the system is to support transactions on bank accounts, e.g. make a deposit. Similarly, in a word-processor application changing the font of some text might be presented as a use-case.

Developing use-cases is a significant activity in the OOAD process. They can be used to communicate with clients, as statements of intent for developers and as specifications for the testing that should be applied to the system when it is under development.

A process that is *architecture centric* makes the system's architecture the primary focus during development. A model, presented as various UML diagrams, gives different perspectives of the system. Some diagrams emphasize the static structure of the system's architecture while others capture some of its dynamic behaviours.

Any software system is actually a model of a problem that exists in the real (or imagined) world. Therefore the more closely a software model corresponds to the actual problem then the more effective it will be. OOAD methods recognize this fact and use key abstractions (objects) taken from the problem domain as the fundamental building blocks for the software system.

With this approach there is only one model of the system no matter what stage of development it is at. System development is a process that progressively adds more detail to the model until such time that it can be executed on a computer. Further, there can be different views (perspectives) of the model at any given point in its development. Each view has a specific purpose and uses only part of the information held in the full model. In this way multiple and easily understood views of a complex system can be presented.

An *iterative* process aims to release a series of versions of the software. Each version augments its predecessor with some additional functionality. We need to ensure that each iteration has a clearly defined aim to avoid undisciplined development. Some iterations may not involve new designs but are concerned with refining or *refactoring* the model to enhance its quality or its usefulness. Within an iteration we conduct an incremental style of development in which we introduce small changes as the software is developed with the aim of minimizing any risk.

2.1.1 A lightweight process

Our lightweight process employs use-cases to drive the development process. Based on the use-cases identified we create a series of views of the model that describe various perspectives of the system. They present both static and dynamic views. Arising from these views we derive a design for our model to realize them. The design need not be fully elaborated. Rather, we are interested in an outline design which we are confident is sufficiently developed that the implementation can be achieved. It is not intended that it is a blueprint that cannot be revised. The design sketches a solution that may be subject to revision as the system develops. Using the design as a guideline, a number of iterations are used to grow our solution.

The elements of our process are captured by the *activity diagram* shown in figure 2.1. The diagram shows the order in which each activity is undertaken. The process is heavily influenced by a forward-engineering approach in which the design ultimately determines the final code. If we wish to make revisions to the code, they are captured in changes to the design and subsequently to the revised code. The code and the design are inextricably woven together since they form different aspects of the same model.

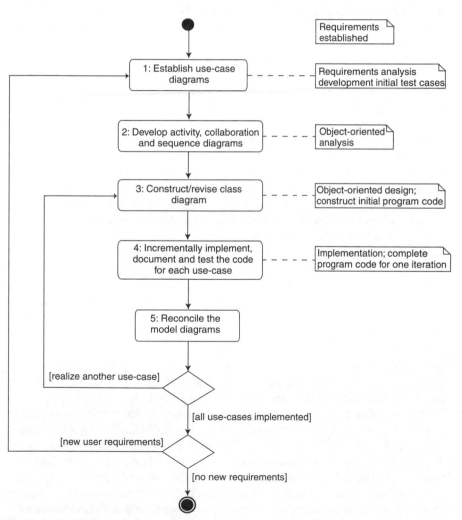

Figure 2.1   *A lightweight process*

The software lifecycle usually begins with the requirements capture in which we seek to determine the purpose of the software. Determining the requirements is usually performed in conjunction with the customer. The outcome from this activity might be a short report or voluminous tomes depending on the scale and complexity of the system.

Frequently the *requirement documents* are subject to interpretation. They can be incomplete and ambiguous, so that further work is necessary to clarify these anomalies. Notwithstanding these failings, requirements introduce the vocabulary of the problem domain and allow us to set its boundaries.

When the requirements have been specified then a number of activities are undertaken. Some are repeated as we progress through analysis, design and implementation. The steps in our lightweight process are:

1. Develop a set of use-cases that describe the capabilities we expect from the system. In many cases we will initially fail to gather all of the use-cases but it is vital to try to capture the most important. They provide a point of communication between the customer and the developer. They can also help in correcting the ambiguities and omissions in the requirements documentation. Use-cases can also operate as the basis for test-cases since they define the functionality sought from the software.

2. Begin the analysis phase by postulating what objects are present in the problem domain, how they interact to achieve some part of the system objective, and how they form relationships so that communication may take place. This work consists of establishing a number of sequence, object, collaboration and activity diagrams that demonstrate how each use-case is realized. The activity diagrams reveal some of the processes within the system. The object and collaboration diagrams highlight the principle objects drawn from the problem domain and their architecture. Collaboration and sequence diagrams help identify and document the behaviours expected of the objects in the problem domain.

3. These analysis views reveal how a set of interacting objects deliver the functionality described by the use-cases. We can then formulate a class diagram for the system's architecture. A fully decorated class diagram records the features of the class including the state and behaviours we can expect of objects of each class. The class diagram and the analysis diagrams should be considered jointly to ensure we have consistency between the views of the model. For example, if a collaboration diagram shows objects that are related, then that same knowledge must be present in the class diagram. From the class diagram we can generate the initial program code. This will produce the Java classes populated with class attributes, class operations and the architectural relations that must exist between them.

4. We can augment the initial code with method bodies and start the process of implementing and testing the development. Here it is best to conduct this work incrementally. Often it is possible to fully develop one class in isolation. When we are satisfied with it we can progress to another, possibly related, class. Its implementation can proceed safe in the knowledge that the first class is relatively stable. Each increment should proceed with the aim of ultimately delivering one of the use-cases. Throughout, we often have the customer closely associated with developments so we have his active involvement.

5. Implementation may reveal some changes required to the UML diagrams. It may be a simple augmentation to the model that does not invalidate the overall system architecture. However, we should be prepared to make revisions to the class diagram where it offers some significant improvement. *Refactoring* (Fowler 1999) a class diagram to obtain a more elegant solution often produces cleaner code that is simpler to enhance and maintain. Any changes made here need to be reconciled with the model produced during analysis. After all, if new requirements are subsequently introduced into the project it is important that these documents are consistent with the existing code.

2.2 Illustration

Consider the outline for the following scenario:

The development of a computer system is required by a community bank. The community bank is a new venture to introduce banking services to a local community that do not normally use the facilities of the national banks. A system is required whereby customers may open accounts and perform the usual transactions on these accounts (credit the account, debit the account, and obtain the current balance). The bank is also required to provide to government the value for its total assets.

Typically, such a statement of requirements would need further elaboration. For our example we may need to know who actually opens an account. Is it performed directly by the customer using an automated teller system or online system, or is it performed indirectly through a bank clerk? *Requirement documents* for large projects might fill a number of ponderously large tomes. The information presented in requirement documents is often assembled by working with the customer and their domain experts to obtain a statement of their needs. Since this documentation is approved by the customer it is also known as the *negotiated statement of requirements*.

However, we adapt the use of a requirement document by our iterative approach to system development. We initially identify and document the system requirements with some simple UML diagrams. From these emerge the design model that operates as a blueprint to guide our future developments. Incrementally we can grow our application until we satisfy all the requirements.

Our analysis begins by first identifying the functionality the system must provide to the users. Here, we seek the tasks the user will perform with the aid of the system. As we noted earlier, object-oriented systems are described as use-case driven. Therefore, the formation of the system is determined by considering a number of use-cases. A *use-case* is a single task that a user needs the system to perform. The totality of these individual use-cases should then describe the full system functionality. Our specification, for example, suggests a use-case in which a customer should be able to perform a debit transaction on the customer's account.

We typically have a variety of users interacting with the system. Clearly there are the customers, but we might also have to consider bank clerks and bank managers. The customer may be the user in one use-case who instructs the system to make a deposit into that customer's account. The bank manager may be another user who has the necessary permissions to obtain the total bank assets, and this is shown in another use-case.

In a use-case, a user is described as an actor. An *actor* represents a user having a particular *role* when interacting with the system. We show a customer as an actor since he represents the source of the stimulus to the system when making a transaction. The manager is also an actor when obtaining the value of the bank's assets. In some examples the same actor may have different roles. For example, our bank manager may also be a customer of his own bank.

Notice, that in some systems actors may not be people. They might, for example, be other computer systems. We might eventually introduce an automated teller machine (ATM) to give customers 24-hour access, and the ATM is effectively an actor in an appropriate use-case. Sometimes actors are other physical devices, such as health monitors in medical applications or radar systems in military applications.

A sample use-case drawn from our banking application illustration is shown in figure 2.2. The diagram records that a bank clerk interacts with the system to open a new customer account. Here the clerk is the actor and is presented as an iconic stick man. The use-case is shown as an ellipse, briefly describing the task. Where an actor participates in a use-case, then this relationship is shown as an *interaction* line between the two.

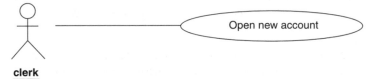

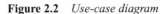

Figure 2.2   *Use-case diagram*

We might augment the use-case with appropriate documentation. Our documentation should aim to further elaborate some detail of the use-case. For example, figure 2.3 includes a UML note in the diagram attached to the use-case.

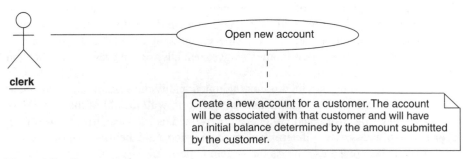

Figure 2.3   *Documented use-case*

Generally, documentation for a use-case describes its normal behaviour. It may be that the customer is already known to the bank and has an unacceptable financial record. Perhaps our customer already has a number of overdrawn accounts and is a possible financial risk. In that case no new account would be opened, but we do not capture this in our use-case. We may, however, record this knowledge in our documentation so that we may incorporate this behaviour into our final system. Later, we show how to capture this exceptional behaviour.

Figure 2.4 is a full set of the use-cases from our banking system application. Our clerk is involved in one use-case "Open new account", the bank manager in another "Determine total bank assets", and the customer in the remaining three use-cases. Notice that a customer makes three kinds of transactions: debit, credit and obtain account balance. We will limit this system by assuming that a customer has at most one account with the bank.

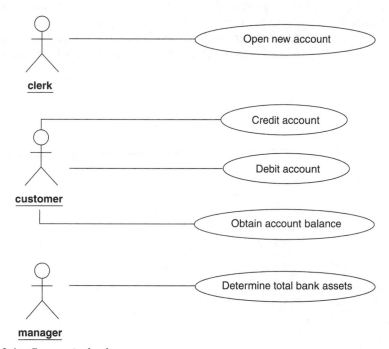

Figure 2.4   *Community bank use-cases*

2.2.1   *Realizing the use-cases*

The next stage in modelling our application is to consider how each use-case can be realized. In effect we are moving from an external view of the system as perceived by a user, to an internal perspective that reveals how each use-case will be achieved. The work also reveals some of the objects in our system, the behaviours they are responsible for, and their relationships with each other.

This activity involves capturing some of the dynamic behaviour of the system as well as some architectural aspects. The UML has a range of diagrams for this purpose. In this section we use activity diagrams, sequence diagrams, collaboration diagrams and object diagrams. These *analysis diagrams* can be considered as "for-instances" in which a particular configuration of objects is considered and represented by a UML diagram. By choosing concrete examples we seek to gain an understanding of the objects, their configuration and how their interaction achieves the required functionality of the system. During the design activity we aim to move from these examples to a single (class) diagram that describes all possible arrangements.

An *activity diagram* shows the flow of control through each separate activity. It emphasizes the flow of control among the activities, whereas an interaction diagram emphasizes the flow of control among the objects (see later in this section). Activity diagrams can be presented at various levels of granularity. At the lowest levels we can use them to model the flow of control within a method (see next chapter). At this stage in the development lifecycle we use activity diagrams to elaborate use-cases.

Figure 2.5(a) is an activity diagram demonstrating how we describe the "Open new account" use-case. We start from the initial state symbol (solid circle), flow through the two activities (soft boxes) and end with the final state symbol (bull's eye). The first activity we carry out is to establish the new account and the second is to register it with the bank.

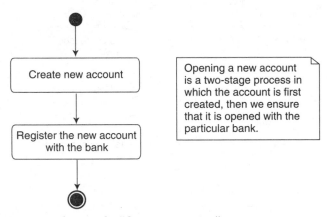

Figure 2.5(a)   *Activity diagram for "Open new account"*

Two *activities* are shown in this diagram: "Create new account" and "Register new account with the bank". The first connects to the second with a *transition* shown by the directed arrow. The activity diagram captures that this particular transition only occurs after completion of the first activity.

A more elaborate version of this activity diagram is given in figure 2.5(b). Here, a *decision* symbol (diamond) is accompanied with two *guard expressions* on each outgoing transition. The guard [good customer] shows how the activity proceeds if the customer has a good credit rating. For a good customer the new account is created and registered with the bank. For a bad customer (one with a poor credit rating, perhaps) no actions are shown. In reality, some activity associated with this exceptional behaviour would normally be described. Chapter 6 considers this further.

The UML supports two kinds of *interaction diagrams*. They are referred to as *sequence diagrams* and *collaboration diagrams*. Essentially they convey the same information. However, each has a different focus. A sequence diagram shows the sequence of messages that take place between interacting objects, with an emphasis on their time ordering. A collaboration diagram captures this same information, although less

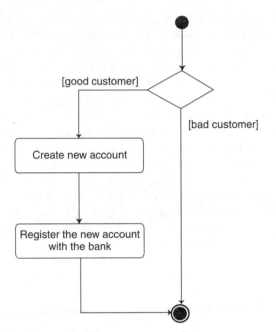

Figure 2.5(b)   *Activity diagram for "Open new account"*

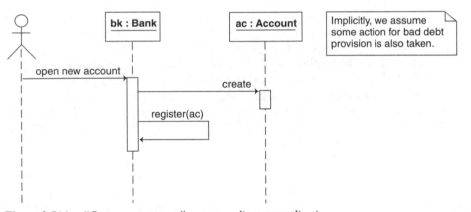

Figure 2.5(c)   *"Open new account" sequence diagram realization*

conveniently, but it additionally reveals the architectural relationships that exist between the objects. The latter is absent from a sequence diagram, though is implicit in the messages sent between objects.

Consider the "Open new account" use-case. The result of this use-case is that there must be a new Account object established that is initialized with the customer details. They might include the customer name, the account number and the initial balance, and perhaps some unique customer identification number. Such an account object then needs to be opened with the bank and it becomes one of the many accounts that belong with this bank. Figure 2.5(c) presents a possible sequence diagram for how this use-case could be realized.

A sequence diagram presents the objects that we are interested in when describing the use-case. The diagrams can include the actors involved in the use-case, though we understand that they are outside the system we implement and do not exist as application objects. Their role is to stimulate the system to perform some functionality. In this sequence diagram we have a clerk, bank and account objects. The objects are shown as rectangular figures labelled with the object type and the object's identifier. For example, we have an Account object with *object identifier* ac, and a Bank object with identifier bk. Since all objects are shown as rectangles then the names Account and Bank distinguish the kind of object they represent. The object identifier is required where, for example, we have a number of objects of the same class and there is a need to distinguish between them.

Object identity is an important notion. It is perfectly possible that two objects of the same class have the same state information. We cannot, therefore, distinguish one from the other using this state information, yet they are unmistakably different objects. This we can record by giving each object a distinct identifier. If in our example accounts always have a unique account number, then this problem will not occur.

We present the objects at the top of the sequence diagram. Each object has an associated *lifeline* extending below it. The lifeline is the period of time during which the object plays its role in the sequence diagram. Time is understood to pass as we move vertically down through the diagram. The lifelines are decorated with *activations* (thin rectangular figures) that highlight where the object is actively involved in some processing, i.e. executing a method.

Between two activations we can show a message. A *message* is a request sent from a sender object to some recipient object to perform some processing action. In figure 2.5(c) we see the Bank object first creating the Account object then registering that new account with the Bank object itself. When a recipient object has completed its processing, control is assumed to pass back to the sender. It is perfectly possible to show the return from a message in a sequence diagram (a dashed arrow from the recipient back to the sender), but it tends to introduce clutter, so we normally make the return implicit. From figure 2.5(c) we then understand that the registering of the Account with the message register sent to the Bank object will only occur after the Account object has been successfully created.

This same knowledge can also be conveyed in a collaboration diagram. It presents a number of object instances, the relationships that exist between them, and the messages that are sent. An object instance has the same form as that in a sequence diagram. Where an object is related to another, then we show this as a *link* connecting them. The link can be decorated with messages and an arrow to identify the recipient of the message. The collaboration diagram is equivalent to the sequence diagram is shown in figure 2.6(a).

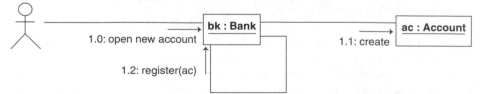

Figure 2.6(a)   *"Open new account" use-case realization*

Again we see the same two objects. Here, the messages and their ordering are shown by a simple numbering scheme. Only after the Account is created (message number 1.1) is it registered with the Bank (message 1.2).

The outcome of the messages shown in figure 2.6(a) is presented by the *object diagrams* in figures 2.6(b), 2.6(c) and 2.6(d). Figure 2.6(b) is an object diagram in which a single Bank object and two Account objects already exist. These are examples of anonymous objects that have no identifier and are shown to illustrate a possible initial configuration of objects. They have no identifier since they play no further part in the discussion. Further, the diagram shows that the Account objects are associated with this Bank object as a consequence of having been opened with this bank.

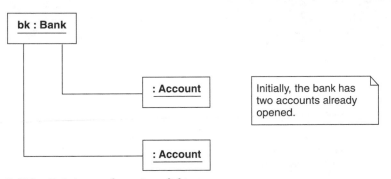

Figure 2.6(b)   *Existing configuration of objects*

Following the message 1.1: create, a newly established Account object is produced (figure 2.6(c)). It is initialized with suitable values for its attributes. At this instant this Account object has not established any relation with any other object.

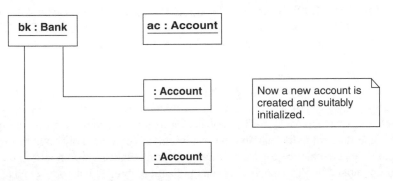

Figure 2.6(c)   *Create new Account object*

Finally, when we register this Account object with the Bank object then a link connecting them is formed. That way the bank object knows of an additional account that has been opened. This configuration of objects is shown in the object diagram of figure 2.6(d).

Figure 2.7 shows the collaboration diagram for the customer making a credit transaction on an Account. This figure reveals that the Bank and the Account objects need

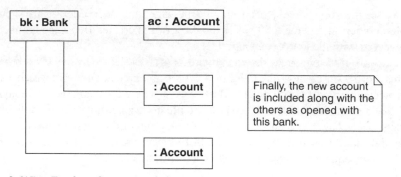

Figure 2.6(d)   *Final configuration of objects*

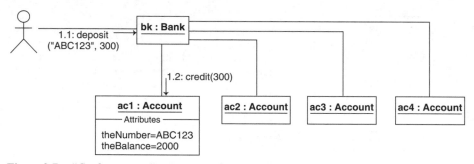

Figure 2.7   *"Credit account" use-case realization*

to be related in some way so that they may engage in message passing. We have exposed the relevant Account object to reveal some of its state information. Here, the account's number is the text (or string) ABC123 and the current balance is 2000 pounds. The customer requests that 300 pounds be deposited in this account. The message deposit("ABC123", 300) is received by the Bank object. This instructs the Bank object to deposit 300 pounds in the Account object with account number ABC123. The Bank object then sends the message credit(300) to the correct Account object, which updates its balance to 2300 pounds.

A major weakness of showing exploded instances in this way is that it presupposes how the objects will be implemented. Whilst we think of objects as having state (and behaviour), during analysis we should avoid consideration of how that state will be realized. For example, when we ask a Bank object for its total assets we should not think in terms of some state value for that amount. We might, after all, actually obtain this by an accumulation of the balance for each account. During analysis, we should make state information implicit by the behaviour of our application objects. Thus we perceive these objects in terms of the behaviours they present rather than their internal representation.

Figures 2.8(a) and 2.8(b) show a sequence diagram and corresponding collaboration diagram for a different scenario for the "Credit account" use-case shown in figure 2.7. This time we see the need for the Bank object first to identify the Account object with the required account number ABC123. Then, the credit message is sent to this Account object. Finally, the activity is concluded and no further interaction with other Account objects takes place.

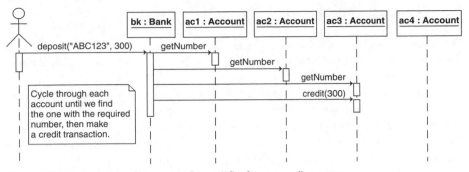

Figure 2.8(a)   *Sequence diagram realizing "Credit account" use-case*

We see in figure 2.8(a) that the customer stimulates the bank to request that 300 pounds be credited to the account with number ABC123. The Bank object interrogates each Account object in turn requesting their account number. The diagram reveals that the first two accounts are only asked for their account numbers and since neither has the required value then no further processing of them is required. If the third Account object has account number ABC123 then it is sent the additional message to credit it with 300 pounds. At this point the task is now complete and no other Account objects are involved.

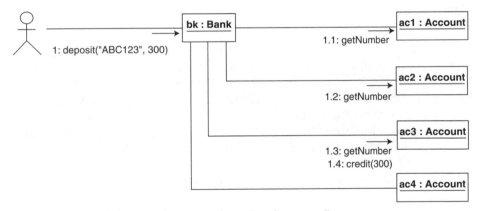

Figure 2.8(b)   *Collaboration diagram realizing "credit account" use-case*

We see from these two example diagrams how sequence and collaboration diagrams convey essentially the same information. The sequence diagram places the timing of messages at its focus. The collaboration diagram, while able to show messages and their ordering, is better at revealing the architectural relationships between objects.

We normally apply one or more of these analysis diagrams to each use-case to ensure that a configuration of objects and their interactions can support the requirement described by the use-case. We should also seek to ensure we have consistency between diagrams. For example, the sequence diagram in figure 2.8(a) implies an iteration through each Account object in which they are asked for their account number. This same knowledge is present in figure 2.8(b) while additionally revealing the architecture of the objects involved.

2.3   Toward design

The preceding sections showed an external view of the system presented by use-case diagrams, and an indication of how they might be realized with activity, sequence, object and collaboration diagrams.

We can move our analysis forward toward design with a diagram that is an abstraction for these specific diagrams. The UML *class diagram* captures the class of objects in the problem domain, the attributes and behaviours of such objects, as well as the relationships that exist between them. Development of the class diagram is a major milestone and acts as the main architectural element that guides further developments.

Figure 2.9(a) is a class diagram for our banking application. Each rectangle represents a class for some kind of object identified in the problem. Here we have classes Bank and Account. The class is a blueprint describing all the features of objects of that class. It defines the attributes that constitute the state of such an object and its behaviour when it receives some message.

Figure 2.9(a)   *Initial class diagram*

If we then further elaborate the class diagram of figure 2.9(a) based on our previous analysis, then we show the Account class with *attributes* for the account number and the balance. For an object of the Account class, the value of these attributes represents the state information for that object. Additionally, the class diagram has been augmented to show the messages that objects of each class can receive. Again, our earlier work determines that we need (among other things) to send a Bank object the messages to register some new account and to obtain the total assets of the bank. Equally, the Account class introduces the operations to get the balance, make a credit and a debit. Revising the class diagram in this manner gives figure 2.9(b).

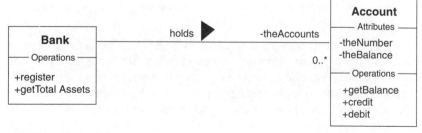

Figure 2.9(b)   *Revised class diagram*

The class diagram also captures a relationship that must exist between a Bank object and its many Account objects. In the diagram the relationship is presented as a line between the two classes. The relationship is decorated to show the role each object plays. The many Account objects are referred to by the *role name* theAccounts as shown in the figure. Further, this role is given the *multiplicity* 0..*, meaning none or more objects of the Account class are related to a Bank object. One Bank object and many Account objects participate in a holds relationship shown in the diagram. The directed arrowhead indicates the direction in which the relation is to be read. Here, it is the Bank object that holds the Account objects.

Strictly, this UML class diagram shows an *association* relationship. This is often described as a loose form of *coupling* between the objects. The objects involved (Bank and Account) are considered as peers with equal standing. Further consideration of our design might suggest that there is a much closer relationship, with the Account objects subordinate to the Bank. A class diagram can capture this relationship with a *composite aggregation* relationship (see figure 2.9(c)).

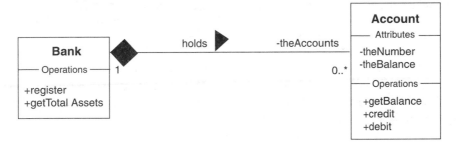

Figure 2.9(c)   *Composite aggregation relationship*

Aggregation implies a whole/part relationship between the composite and its part(s). This is a much stronger coupling than that of association. If the composite object no longer exists, then by implication, neither do the parts. For our illustration, if the bank were to cease trading, then there would be no corresponding bank accounts.

2.4   UML diagrams

In the preceding sections we have introduced various UML diagrams in the context of our banking example. In this section we revisit the diagrams and present further aspects not otherwise covered.

2.4.1   Use-case diagrams

Figures 2.2, 2.3 and 2.4 show the essentials of use-case diagrams. They are composed of actors and use-cases. Actors are related to use-cases by interactions. An actor symbol is rendered as a stick man, and a use-case as an ellipse. An interaction between an actor and a use-case is presented as an unadorned line between the two.

An actor provides the external stimuli for a use-case. It can be involved in one or more use-cases (see figure 2.4). Equally, a use-case may involve any number of actors (not shown in any of our sample use-case diagrams). While actors are given human characteristics in the form of the stick man, this need not always be the case. Often, actors that stimulate a system are other computer systems. In some cases they may be other digital devices such as computer printers, medical monitoring equipment in a hospital, or navigation systems in an aeroplane or in a ship. In our banking application ATM machines may act as actors. Equally our central banking system may play the role of an actor when initializing one of its ATMs.

2.4.2   Interaction diagrams

In the preceding section we noted that there are two types of interaction diagrams, namely, sequence diagrams and collaboration diagrams. The illustrative examples also highlighted the respective strengths of each. A sequence diagram is useful for emphasizing the time order of messages between objects. Collaboration diagrams reveal the relationships that must exist between objects so that they may engage in message passing.

The aim of an interaction diagram is to describe how groups of objects collaborate to achieve some behaviour. Usually, we employ an interaction diagram to capture the behaviour for a single use-case. Both sequence and collaboration diagrams present a number of objects and the messages that are passed between them. We can consider the example objects as representative of the totality of actual objects in the system.

2.4.3   Sequence diagrams

In a sequence diagram the objects are presented as rectangular figures. In figure 2.5(c), for example, we show two object instances: a Bank object and an Account object. The object instances are decorated with a label of the form bk : Bank, where bk is the object's identifier and Bank is the class (or type) of the object. We have already discussed the role of object identity in such diagrams. Where we have no interest in the identity, for example when we have only one such instance, then the identifier can be omitted as in :Bank. Generally, unless the identity name chosen also reveals the implied type (such as bank1), we do not omit the type name.

We are also permitted to include actors in our sequence diagrams. They need not represent software objects that we implement in our system. However, they identify the user role that has stimulated the system to behave in the way described by the diagram. The actors can be decorated in the same manner as the instances.

Each actor and object instance has a vertical lifeline, representing the object's life during the time of the interaction. When the object is participating in some activity this is shown on the lifeline as an activation, presented as a thin, vertical, rectangular figure positioned along it.

A message is represented by a directed arrow between two activations. The direction of the arrow indicates which object is the receiver and which is the sender. Generally we try to construct a sequence diagram so that the messages are shown with the recipient on the right. Since time proceeds down the diagram, then one message shown above another in the diagram precedes that message in time order. We have examples where it is valid and useful to show where an object can send a message to itself. This is presented as the arrow connecting an activation with itself (see figure 2.5(c)).

As noted earlier we seek to reduce clutter in these diagrams by not showing the return from a message when the recipient has completed its task. Where it improves the clarity of some interaction, then it can be shown as a dashed arrow back from the recipient to the sender. Be aware that these are not further messages. Such return arrows are precisely what their name suggests.

Where information is conveyed along with the message, then this can be shown as *message parameters*. For example, in figure 2.5(c) the Bank object is sent the message register(ac), where the message register is accompanied by the Account object identifier ac.

Messages can also be augmented with two additional types of information. For example, if some condition has to be met before the message is dispatched to the recipient, then we can capture this with a *condition*. This is shown as a prefix to the message, enclosed in [and]. Figure 2.10 shows that we must first check the financial stability of the customer, or some other aspect of the customer, before creating a new account for that customer.

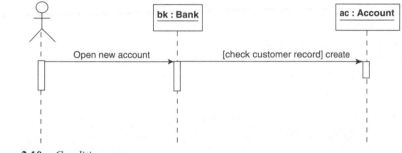

Figure 2.10   *Condition on a message*

Another decoration for a message is used to indicate where the same message is sent, in turn, to a number of similar objects. For example, where the bank is required to produce a listing of the details of every account. The message in this case is prefixed with the asterisk symbol (*) to indicate repetition (or *iteration*). In that case we can also show a group or collection of such objects as in figure 2.11. Here, the interest lies with the collection (shown as a *multi-object*) rather than with an individual member.

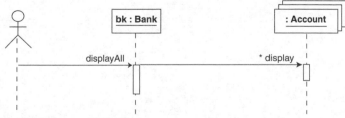

Figure 2.11   *Multi-object and repetition*

In this figure the message displayAll is required to produce a listing of the details for all the accounts held by the bank.

2.4.4   Collaboration diagrams

Collaboration diagrams use the same mix of actors and instance symbols as found in sequence diagrams. A *link* between two object instances is shown by a connecting line. The link is usually decorated with numbered messages sent between the instances. An example of a collaboration diagram appeared in figure 2.8(b). The relationship shows whether the instances are associates or are aggregates (see the discussion on class diagrams later). Like sequence diagrams we may also annotate messages in a collaboration diagram with conditions and iterations (see figures 2.10 and 2.11).

As noted earlier, a particular quality of collaboration diagrams is that they reveal or give an insight into the links that must exist between object instances. Without these links the objects cannot participate in message passing. In the first version we might simply show these as associations. Further analysis might indicate the stronger coupling of aggregation. This is a detail we can usually leave until the class diagram is constructed.

A collaboration diagram shows a representative collection of objects and the links between them. Where no messages pass between the instances, they are referred to as *object diagrams* in the UML. Their value is they make the architecture of the objects the primary focus. Ultimately, they form the basis for the important class diagram.

2.4.5   Activity diagrams

An activity diagram is often used when modelling various dynamic aspects of a system. At its simplest, an activity diagram presents the flow of control between activities. Activity diagrams focus on the separate activities within a system, whereas interaction diagrams are concerned with the execution flow through the network of objects. Activity diagrams can be used to document use-cases as well as the behaviour of methods.

The essence of an activity diagram is an *action* that represents some executable computation. They are called *action states* and are represented by soft boxes (see figure 2.12(a)) populated with text describing the action performed. Upon completion of one action, control automatically proceeds to the next along a directed transition.

Figure 2.12(a)   *Actions and transitions*

The initial and final state for an activity is shown with start and stop state symbols as shown in figure 2.12(b).

Figure 2.12(b)   *Start and stop states*

We can include branching logic in activity diagrams. They are shown as diamond-shaped figures. A *branch* has one incoming transition and two or more outgoing transitions. Figure 2.12(c) shows such a branch. Each outgoing transition is decorated with a *guard* presented as mutually exclusive boolean expressions. Here we either have a good customer or a bad customer, perhaps one that is a possible financial risk.

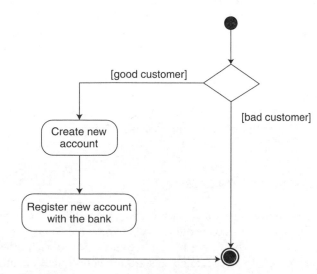

Figure 2.12(c)   *Branching*

2.5   Class diagrams

The *class diagram* is the principal diagram that we construct in an object-oriented design. Its importance lies in the fact that its content delivers the primary elements in our program code, namely the Java classes. A class diagram describes the types of the objects in the system and the relationships that exist between them. A class diagram is an abstraction for all the possible object diagrams we might construct. Class diagrams and collaboration/object diagrams need to be consistent. For example, if a possible configuration of objects can exist, then the class diagram needs to capture this information. If the class diagram shows that two object types are unrelated then no link should be shown between corresponding instances in a collaboration or object diagram. It also means that we cannot show a message from one object to another in a sequence diagram.

A class is also documented with its set of attributes and operations. The attributes represent the set of values each instance maintains as the object's state. The set of operations is the messages an object of the class may receive. Some of these operations can be identified from the various interaction diagrams we might construct. Further, these latter diagrams can give some insight into the behaviour for these operations, perhaps by sending messages to other objects in the system.

2.5.1   Representation of objects

For an object to support the operations available on its interface it must maintain an internal state. Typically an object's internal state represents the private data elements encapsulated by the object. Collectively we refer to the private data elements as the attributes of the object. Their identification is usually quite straightforward as they are normally constructed from standard classes for commonly occurring objects such as strings (a sequence of characters) and numbers.

In general, objects are significant coarse-grained elements within the context of the problem under consideration, while their attributes are much finer-grained and therefore do not merit an independent object existence. For example, the class **Employee** might have attributes that represent the name, address, age, sex, employee number and salary of each employee in a company. Considered on its own each attribute has no relevance. It is only of importance if it is seen as part of the representation of something larger, i.e. an **Employee**. This is an important point to understand, as it is common for novice designers to identify too many objects in a system.

2.5.2   Association of objects

As objects usually need to co-operate in order to achieve their effect, they commonly enter into relationships with each other. One of the most important relationships is *association*. Objects that are associates co-operate by sending messages to each other. In general association should be used where two objects are not conceptually related but within the context of the problem need to make use of each other's services. For example, an interaction in which a single **Employee** object is employed by a single **Company** object can be imagined. They are associates in the sense that the **Company** object adopts the role of the employer while the **Employee** object adopts the role of the employee. Typically as part of the run-time behaviour of the system, the employee could

Figure 2.13   *Association*

request the name of the employer while the employer could request the job title or salary of the employee. Figure 2.13 is a *class diagram* describing this relationship.

The line that connects class symbols represents an optionally named association relationship between them. Each class (and therefore each instance) may adopt a named *role* in the association. In this example we show that a relationship exists between a **Company** class, an instance of which adopts the role of the employer, and an **Employee** class, an instance of which adopts the role of the employee. The association implies that the **Employee** object and the **Company** object that participate in the relationship may make use of each other's services by sending appropriate messages to each other. In other words the association can be navigated in either direction: from the employer to the employee or from the employee to the employer.

The class diagram has the integer 1 adjacent to both the **Company** and **Employee** classes. This is their *multiplicity*. It is used to indicate that a single **Company** object and a single **Employee** object are associates. In this example the multiplicity is one in each case but we will discover below that other possibilities are common. An important point to note is that if there is no multiplicity value then it is considered to be unspecified, i.e. no decision has been made as to its value. Crucially there is no default, e.g. a value of one cannot be assumed. This is useful as it avoids any possible ambiguity in the class diagram and allows the designer more flexibility.

Some of the previous class diagrams were decorated with the operations and attributes of each class. However, when several classes are incorporated into a single class diagram it can become rather cluttered if these details are shown. In practice for all but the simplest of models the unnecessary complexity they introduce can easily obscure much more important information. For this reason it is permissible to suppress operations and/or attributes in a class diagram. Similarly, in order that explanatory text is more readable we often refer to objects by their class name. For example, we might refer to an **Employee**, or an **Employee** object. In either case the context should make it clear that we mean an object and not the class to which it belongs.

Obviously, a **Company** may employ more than one **Employee**. This leads to the **Employee** class having a multiplicity of more than one. Frequently occurring possibilities are shown in figure 2.14 in which the multiplicity denoted by the symbol * represents none or more objects.

One-to-one association: one person owns one car.

One-to-many association: one company employs many (none or more) employees.

```
 ┌─────────┐  -theTutors              -theStudents  ┌─────────┐
 │  Tutor  │─────────────────────────────────────│ Student │
 └─────────┘     *                        0..*      └─────────┘
```

Many-to-many association: may tutors teach many students;
  many students are taught by many tutors.

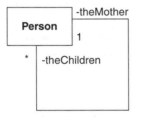

Recursive association: many child persons have one mother person.

Figure 2.14 *Associations and multiplicities*

Note that even though role names can be omitted, they are very important as they provide the means whereby an instance of one class can refer to an instance of the associated class. For example, in the second illustration of figure 2.14 an instance of the Employee class would refer to the Company object with which it is employed through the role name theEmployer. This is a point that is expanded in chapter 3 when the detailed interactions between objects are implemented as Java code.

The third illustration of figure 2.14 shows a many-to-many association. Many Tutors teach many Students and many Students are taught by many Tutors.

It is also possible to have an association between objects of the same class. For example, every person is the child of some other person, their mother. This is expressed in the fourth illustration of figure 2.14. Such an association is described as a *recursive* or *reflexive* association. Observe how for this example the role names are vital so that we can clarify the usage of the class Person.

Before leaving the association relationship between classes, it is important to understand that a class diagram does not describe a particular configuration. In effect, it describes all possible configurations. It is a general description of the relationship between two classes.

Figure 2.15   *Class diagram*

Figure 2.15 shows that one Company object acts as the employer for many Employee objects that act as the employees. It most certainly does not specify which Company object or which Employee object will enter into this association. This is the point about a class diagram that must be understood.

In order to show the architecture of a real system with real objects, and not just descriptions of objects, instances of the classes must be established with links formed between them. Just as an object is an instance of a class, a link is an instance of an association. There must be a link between each object taking part in the association. The purpose of an object diagram is to capture this information.

Figure 2.16 is an example of an object diagram. It shows a Company object identified as c1 with links to three Employee objects identified by e1, e2 and e3. As with the previous class diagrams the object diagram is simplified by suppressing irrelevant detail: in this case the attribute values of each instance.

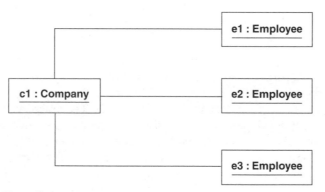

Figure 2.16   *Object diagram*

The important point to understand about this object diagram is that it represents a concrete architecture and is not abstract in the manner of a class diagram. It represents a particular configuration of objects that conforms to the general description embodied in the corresponding class diagram of figure 2.15. Clearly this is just one of many possible object diagrams that can conform to this class diagram. For example, there might be ten Employee objects as associates.

It is also clear from the object diagram that the Employee objects identified by e1, e2 and e3 are not linked to each other. This means of course that they cannot send messages to each other. Although the class diagram of figure 2.16 does specify that a single Company and many Employee objects can be linked and that an Employee object cannot be linked to another Employee object because the object diagram is concrete it does reinforce the point.

Note that there may be some confusion between the use of the phrase *the name of an object* and *the identifier for an object*. An object only has a name if that is one of its attributes and its name has nothing to do with its identifier. They are not the same. For example, an Employee object could have an identifier e1 and a name Ken. Strictly the phrase *the identifier for an object* should always be used if we are referring to object identity as opposed to an object with a name attribute. In a similar vein the phrase *the object e1* is not strictly correct. It should actually be *the object with an identifier of e1*. However, to enhance readability we normally do not make this distinction.

2.5.3   *Composite aggregation of objects*

A commonly occurring kind of association relationship between objects is *composite aggregation*. Sometimes it is also referred to as the *has-a* or *part-of* relationship as it indicates that one object (the whole) is composed of other objects (the parts). With composite aggregation the relationship between the objects is much stronger than with association in that the whole cannot exist without its parts and the parts cannot exist without the whole. Several important points are implied by this fact. They are as follows:

• deletion of the whole implies deletion of the parts
• there is only ever one whole, i.e. parts are not shared with other wholes
• parts cannot be accessed "outside" the whole, i.e. they are private to the whole and
• a message destined for a part must be sent to the whole and relayed by it to the part

This means that composite aggregation should really only be used where an object is considered to be a part of another object and not just a casual associate with an independent existence and visibility.

Clearly aggregate components are very like attributes. The difference is that the former are important enough in the context of the problem under consideration to be considered as objects in their own right, and not just ordinary attribute items. For example, a bank can be considered as having several accounts inside it. A typical example is that a message to display all the account items in the bank is received by the Bank object. In response to this message the Bank object sends a message to each Account object to display itself. This kind of execution behaviour is common in object-oriented systems and is referred to as *message propagation*.

The class diagram in figure 2.17 describes a composite aggregation class Bank and its Account parts. A line terminating with a filled diamond indicates the composite aggregation relationship. The filled diamond symbol is placed against the whole in the whole/part relationship. To avoid any possible confusion the whole is given an explicit multiplicity of 1 even though only one whole is possible. As with normal association the parts may be adorned with 1 or *, where * means none or more. Similarly, if no value is given then the multiplicity is considered to be unspecified.

As with a class diagram showing a normal association, a class diagram showing composite aggregation is completely general but the corresponding object diagram is specific. An object diagram for a specific Bank, bk with three Account items ac1, ac2 and ac3 in it is shown in figure 2.18. The filled diamond is used again to show the whole and a decorated line a link to one of its parts.

Figure 2.17 *Composite aggregation*

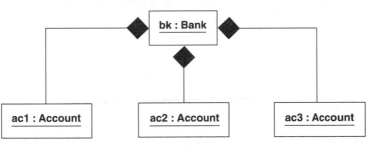

Figure 2.18 Bank *object diagram*

2.5.4 Shared aggregation

Composite aggregation requires that there is a whole/part relationship between the objects concerned. It is implicit in the relationship that a client has access only to the public interface of the whole and cannot access any of the parts directly. For example, a car engine could have a composite aggregation relationship with parts such as several cylinders and a fuel injection system. This means that a user of the engine can only gain indirect access to the engine parts through a limited set of operations such as start and stop advertised by the whole, i.e. the engine. The client is probably blissfully unaware of the existence of any of the parts as they are under the exclusive control of the whole and the client just communicates with the whole and not the parts.

By way of contrast association is a much looser form of dependency or coupling. The objects concerned are casual associates co-operating to achieve some overall effect. If a car engine is modelled as an association between its parts then a user has direct access to each part's public interface. This means that a user could adjust the settings of the fuel injection system as well as starting and stopping the engine.

Although these comparisons of composite aggregation and association are accurate, there is a widely used form of aggregation that significantly changes some of the conclusions made. It is called *shared aggregation* and in some ways it is a compromise between normal association and composite aggregation.

With shared aggregation there is still the design intention that there is a whole/part relationship but now the parts are shareable with another whole. As with composite aggregation a message destined for the part must be sent through the whole. However, deletion of the whole does not imply deletion of the parts. In effect the tightness of the coupling between the whole and its part has been loosened.

If we revisit the company/employee example shown earlier then we can model the relationship between the Company and the Employee items as one of shared aggregation. Figure 2.19 is a modified class diagram. Note that shared aggregation is shown as an open diamond to signify that the part is shareable.

Figure 2.19   *A modified class diagram using shared aggregation*

The true impact of this decision becomes apparent when a corresponding object diagram is drawn as shown in figure 2.20.

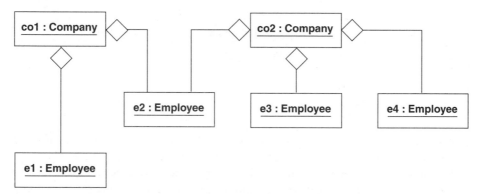

Figure 2.20   *A modified object diagram using shared aggregation*

Now we can clearly see that the Company object co1 and the Company object co2 can legally access the same Employee object e2. This way we are able to successfully model someone that works for two employers. For this and other reasons the shared aggregation relationship is commonly used as a weaker form of composite aggregation.

2.6   Summary

1. The UML is a notation for capturing and presenting the results of an OOAD. To obtain the full leverage of the UML we must superimpose on it a method.
2. A guiding principle is that an OOAD process should be use-case driven, architecture centric, iterative and incremental.
3. A use-case diagram describes a single task that a system needs to perform. The totality of the individual use-cases should describe the full system requirements.

4. Interaction diagrams present a dynamic view of the object instances. They offer a run-time view of the behaviour for all or part of the system. Two kinds of diagram document an interaction: an annotated collaboration diagram and a sequence diagram.

5. An annotated collaboration diagram highlights object structure but can also give the sequence of messages between them. A sequence diagram highlights the time-ordered sequence of messages but does not give information on object structure.

6. An object diagram presents the architectural relationship between objects. It is the same as a collaboration diagram without any messages.

7. An activity diagram is used to show the flow of control among the activities. Activity diagrams can be presented at various levels: as an analysis diagram or as method logic.

8. A class diagram records the classes identified in the problem domain together with the architectural relationships that exist between them. A class symbol is decorated with the class name, the class attributes and the class operations.

9. Relationships between classes include association and composite aggregation. Association should be used where two classes are not conceptually related but within the context of the problem need to make use of each other's services. With composite aggregation, the coupling between the classes is much stronger since the parts cannot exist without their whole. Association and aggregation may have multiplicities recorded in the class diagram. Both may also be labelled with role names.

2.7 Exercises

1. Present a clear distinction between the terms *heavyweight process* and *lightweight process*.

2. Explain what is meant by each of the following terms:

   - Use-case driven
   - Architecture centric
   - Iterative and incremental

3. Identify the primary modelling elements appearing in a use-case diagram and provide a short explanation for each. Does the actor always represent a human user? If not, then give examples of what the actor might represent.

4. Figure 2.1 describes our lightweight OO process. How do the activities in this diagram relate to the stages in a traditional waterfall lifecycle? What is probably the most significant difference between the two approaches?

5. In figure 2.1 the OO analysis activity is concerned with developing a range of UML diagrams. For what ultimate purpose do we prepare these various diagrams?

6. From which UML diagram do we generate our initial code?

7. Figure 2.1 emphasizes program development as an incremental activity. What is the reasoning behind this and what are its merits?

8. Identify the UML diagrams that are used to reveal how a use-case can be realized. Categorize the diagrams as either static architectural diagrams or dynamic behavioural diagrams.

9. Identify all the elements found in an activity diagram.
10. Show an activity diagram for the logic described by the use-case "Debit account" of figure 2.4 in which a monetary amount is removed from the account if sufficient funds are available.
11. What two kinds of interaction diagrams are available in the UML? How are they similar and how do they differ?
12. Construct the sequence diagram and corresponding collaboration diagram for the use-case "Determine total bank assets" (see figure 2.4).
13. A sequence diagram includes object instances and their lifelines. What other modelling elements are present in a sequence diagram?
14. Carefully distinguish between a collaboration diagram and an object diagram. Both diagrams use a link to connect object instances. Explain the role for a link in such diagrams.
15. What do you understand to be the difference between an object diagram and a class diagram? What are the corresponding elements in a class diagram for an object instance and a link used in an object diagram?
16. What are the architectural relationships that can be shown in a class diagram? Present a clear distinction between them.
17. Explain the terms *multiplicities* and *role names* shown on architectural relationships in class diagrams. Identify four of the more common multiplicities that are used.
18. Complete the following: In a class diagram state is described by ... while behaviour is described by ...

    For each of the following problem definitions:

    • Compose use-cases to capture the required functionality
    • Construct sequence diagrams showing the interactions between objects
    • Construct matching collaboration diagrams
    • Identify the major classes present
    • Construct a class diagram showing any architectural relations
    • Identify the most important operations and attributes for each class

For the moment do not concern yourself with how the operations are actually implemented.
19. A person, identified by a unique social security number and a surname, can own at most one vehicle at any given time. A vehicle is given a maker's name and a registration number. In addition a person must also be able to disown a vehicle and should be able to display details of any vehicle owned.

    Show how a person could own a vehicle, display its details then disown it. Similarly show how the same person could own a replacement vehicle and then display its details.
20. A country has many cities. Each city has a name and a population, while a country has a name and a capital city. It is a requirement that a country should display, on request, its capital city through the operation displayCapital as well as the names of all its cities through the operation displayCities. In addition, it should display the total population of the country through the operation displayTotalPopulation, and the average city population through the operation displayAveragePopulation.

Model this system showing how the capital city, the names of the cities and the average city population of a country could be ascertained.

21. A teaching block consists of several rooms each with a unique room number and specified seating capacity. Rooms may be booked on a particular day for lectures. Each booking must start on the hour and can be of any duration. It must be possible to book a room in the teaching block if it is free and generate a display of the status of each room in the teaching block for a particular day.

    Construct a model showing how a room in the teaching block could be booked.

22. University lecturers may obtain from a library a loan of textbooks. Every lecturer has a name and every book a title. Construct a model of this system showing how a lecturer could ask for a display of books in the library, borrow a book or return a book. Lecturers should also be able to display the books they have out on loan.

23. Diving competitions are scored as follows:

    - Each competitor is given a mark out of 10 for a dive by each of five judges.
    - The score for a dive is calculated by discarding the highest and lowest mark, determining the average, and then multiplying it by a difficulty factor (a number in the range 0.1 to 1.0 in increments of 0.1).

    All competitors complete three dives. Their overall score is the sum of the marks for each dive. The winner is the competitor with the highest score. You are required to construct a system to automate this scoring process.

24. A computer game consists of several players who compete with each other to build a beetle. A complete beetle has two antennae, a head, a neck, a body, six legs and a tail. When a player's beetle is complete that player's name is displayed and he leaves the game. The game continues until each player's beetle is completed.

    The rules of the game are that a beetle:

    - cannot have an antennae unless it has a head
    - cannot have a head unless it has a neck
    - cannot have a neck until it has a body
    - cannot have a leg unless it has a body and
    - cannot have a tail until it has a body

    During the game, each player takes it in turn to be given a random number representing the throw of a die with which to construct his beetle. An integer in the range 1 to 6 represents an antennae, head, neck, body, leg and tail respectively. There should also be a display of the configuration of a player's beetle before and after his turn. Construct a model for the game.

3

Implementing Objects with Java

In the preceding chapters we documented our OOAD with UML diagrams. However, each object identified in an OOAD must eventually be implemented in some programming language such as Java. In this chapter we show how the Java language constructs are used to implement the illustration first introduced in chapter 2. The reader is invited to review the materials in appendices F and G, which summarize most of the Java language features we use throughout this textbook. It is also recommended that specialist textbooks for the Java language be consulted (Eckel 2002, Dietel 2003).

We develop our solution based on the model developed from the previous chapter (repeated here as figure 3.13). We approach the problem by developing a number of versions, each an iterative extension of the preceding one (see figure 2.1). This iterative approach ensures that we have some deliverable at each stage that is demonstrable to all stakeholders involved, including the customer and other developers. Incremental development is applied at each iteration to minimize any risk.

Frequently one activity can progress safe in the knowledge that some requirement has been fully implemented. This is the approach illustrated in this chapter in which we fully develop the Account class before extending the application to consider the implementation of objects with architectural relationships. This way we can grow the Bank class and its relationship with the Account class with little consideration for the latter, confident that it has most of the required behaviours.

3.1   Introduction

Many important decisions are made as part of an OOAD. Clearly all of them should be unambiguously recorded if the design is to be properly implemented with Java. Although the UML is capable of achieving this aim, developers sometimes assume that some fundamental concepts are implicitly understood. This can lead to difficulties for the novice. For this reason, we explain in some detail how objects should be implemented with Java.

3.2   Illustration

Consider the outline for the example from chapter 2:

> *The development of a computer system is required by a community bank. The community bank is a new venture to introduce banking services to a local community*

49

who do not normally use the facilities of the national banks. A system is required whereby customers may open accounts and perform the usual transactions on these accounts (credit the account, debit the account, and obtain the current balance). The bank is also required to provide to government the value for its total assets.

In the previous chapter we characterized the development of OO systems as both itera-tive and incremental. An iteration is a complete development cycle in which we deliver some subset of the overall product such as a single use-case. We then grow toward the final product by adding more use-cases that are developed on each subsequent iteration. Increments introduce small changes as the software is developed.

Such an approach can lead to a significant reduction in the risks involved in software development. Further, we can constantly keep the customer and all other stakeholders informed at each iteration. With the customer always involved we can be sure that we deliver a product that meets their requirements.

Each iteration needs to define its objectives. Without them there is serious danger that our development process degenerates into hacking. Use-cases can play a role in this. Each use-case can be used to specify our aims. Each new iteration will then augment the previous use-cases, developed in terms of the originals.

From the previous chapter we know that this problem will eventually lead to a class diagram with (at least) a one-to-many association relationship between a **Bank** class and an **Account** class. See, for example, figures 2.9(a) to 2.9(c). For the first part of this chapter we simply consider objects of the **Account** class and show how we implement such a class in Java. Later we investigate the realization of architectural relationships and complete our **Bank** class.

Figure 3.1 presents the use-cases that describe the functionality expected of a bank account. Using the clerk as the external stimuli, we create a single bank account object,

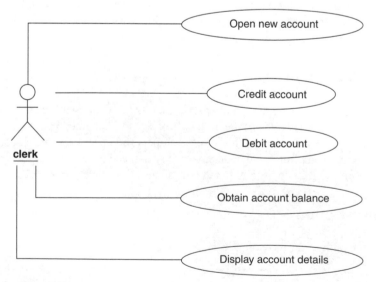

Figure 3.1   *Account use-cases*

perform a variety of transactions on it, and have the account object display its details. Note that there is no implied ordering among these use-cases.

The use-case "Open new account" might be realized in the manner depicted by the sequence diagram in figure 3.2. The clerk is responsible for initiating this request, and through the clerk we will provide the necessary values to establish the account, for example its initial balance.

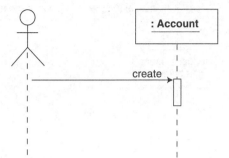

Figure 3.2 *Sequence diagram realization for "Open new account" use-case*

The three separate transactions we consider for a bank account are to credit the account with some additional funds, debit some monetary amount from the account, and obtain the current balance for the account. Figure 3.3 shows a sequence diagram for the particular use-case "Credit account". Specifically, we have shown the monetary amount 200 being deposited into the account.

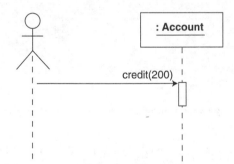

Figure 3.3 *Sequence diagram realization for "Credit account" use-case*

We might develop the activity diagram of figure 3.4 to show the logic behind the "Debit account" use-case. This is obviously a trivial example but it demonstrates how activity diagrams can be deployed at various levels. Elsewhere, we have examples of activity diagrams documenting program code.

Finally, in figure 3.5 we present how we obtain the state information from an Account. In our final programmed implementation for all these use-cases we shall find this message useful for confirming the correctness of our code. In effect, we can use this display message to observe the effect of the other messages. Having credited the account with some amount we can use display both before and after the transaction to check that the balance has changed by the correct amount.

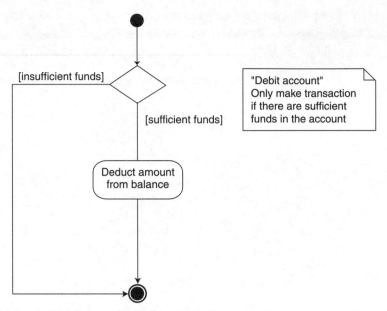

Figure 3.4 *Activity diagram for "Debit account"*

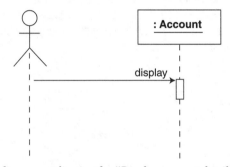

Figure 3.5 *Sequence diagram realization for "Display account details" use-case*

Our class diagram is, of course, trivial having only the single **Account** class. However, from our earlier analysis we can augment the diagram to show the operations that are required to support the various messages that we have identified (see figure 3.6(a)).

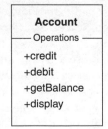

Figure 3.6(a) *Account class diagram*

At this early stage it is sufficient to know that these operations will provide all the required behaviours from **Account** objects and permit us to realize each use-case. Later, we consider further aspects of these operations such as their signatures, e.g. what parameters, if any, must we supply when we send such messages.

One consideration we have not made is the *visibility levels* for the features (both operations and attributes) of a class. The visibility determines which features of one class are accessible from another class and those that are not. In UML the developer can specify one of four levels of visibility:

Visibility	Symbol (prefix)	Description
public	+	available to all other classes
protected	#	available to subclasses (see chapter 5)
private	-	available only to the class itself
package	%	available to all classes in the same package

We see from figure 3.6(a) that all the operations have been given *public access*. This is common practice. When selecting the visibility of features of a class we generally aim to conceal the implementation details and only publicize those features that support the abstraction represented by the class. Since debiting and obtaining the balance are examples of what characterize a bank account, then these operations are given public visibility. Private visibility is discussed below while protected and package visibilities are introduced in later chapters.

The UML class in figure 3.6(a) is implemented by the outline Java *class declaration* (the ellipsis . . . are used to indicate some missing detail to be completed later):

```
public class Account{

    // ----- Operations ----------
    public ... credit( ... ) { ... }
    public ... debit( ... ) { ... }
    public ... getBalance( ... ) { ... }
    public ... display( ... ) { ... }

} // class: Account
```

The operations for a class are realized as Java *methods*. They too are decorated with a *visibility qualifier*. Here the Java keyword **public** provides the required correspondence.

Each *operation signature* is then completed by adding any formal parameters and specifying the class of any return value. The operations credit and debit are each given a single parameter that represents the amount involved in the transaction. Neither operation returns a value to the sender of the corresponding message. We can think of these as *transformer operations* since they modify the state of our **Account** objects. The operation getBalance is an *accessor operation* used to interrogate an **Account** object for its current balance. Such an operation yields a numeric value for the monetary sum. Finally, the operation display simply performs the actions that reveal the content of an **Account** object's state through a set of simple print statements. No parameters are required and no value is returned. A revised UML diagram is given in figure 3.6(b).

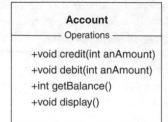

Figure 3.6(b)   *Augmented* Account *class*

The Java class declaration is then:

public class Account{

    // ----- Operations ----------
    public void   credit(**int** anAmount) { ... }
    public void   debit(**int** anAmount) { ... }
    public int   getBalance() { ... }
    public void   display() { ... }

} // class: Account

The internal representation for the class can now be developed. Consideration of the system specification should help us with this task. We can also look to the realization of the use-cases to assist in revealing the necessary state information. For an Account object to support the operations credit, debit and getBalance suggests that an attribute is required to represent the balance. The operations credit and debit will change this attribute while the operation getBalance is an accessor to obtain its value. Not unreasonably we will also associate an account number with each Account object. The String type has been chosen so that we can have account numbers such as ABC123. Figure 3.6(c) is the final class diagram.

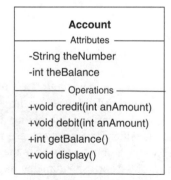

Figure 3.6(c)   Account *class with attributes*

The resulting Java class declaration is shown below. Observe how the attributes have been given private visibility. This is in keeping with the practice of *information hiding* in which the internal representation for a class is not revealed to other classes. Here, we have chosen to list the operations before the attributes. Java does not prescribe any ordering of the features of a class. Our choice is based on the notion that we are primarily concerned with the services offered by the class rather than any representational concerns.

```
public class Account{

    // ----- Operations ----------
    public void     credit(int anAmount) { ... }
    public void     debit(int anAmount) { ... }
    public int      getBalance() { ... }
    public void     display() { ... }

    // ----- Attributes ----------
    private String   theNumber;
    private int      theBalance;

} // class: Account
```

To complete the Java code we must supply the method definition for each operation we have identified. For this class the methods are very simple. For example, the method for the operation debit simply ensures that there are sufficient funds in the account before making the transaction (see figure 3.4). The Account class is almost complete and now reads as:

```
import textio.*;

public class Account {

    // ----- Operations ----------
    public void     credit(int anAmount) {
        theBalance += anAmount;
    } // method: credit

    public void     debit(int anAmount) {
        if(theBalance >= anAmount)
            theBalance -= anAmount;
    } // method: debit

    public int      getBalance() {
        return theBalance;
    } // method: getBalance

    public void     display() {
        ConsoleIO.out.println("Account");
        ConsoleIO.out.println("\t" + "Number: " + theNumber);
        ConsoleIO.out.println("\t" + "Balance: " + theBalance);
    } // method: display

    // ----- Attributes ----------
    private String   theNumber;
    private int      theBalance;

} // class: Account
```

Observe the **import** statement at the start of this Java file. Other than comments and general whitespace (space, tab and newline characters), **import** statements, where present, must appear before all others (excepting a **package** statement, see chapter 4). Here we import the declaration for the classes from the **textio** package, thereby giving access to the classes by which we can undertake simple input and output. The class **ConsoleIO** that supports simple input/output from and to the system console is described in appendix C.

3.2.1   Mandatory profile

A *mandatory profile* specifies the minimum services expected of a class. The Java declaration for class **Account** should include a *constructor method* for proper initialization of objects of this type. We would normally expect a class declaration to include a *parameterized constructor* for the proper initialization of the class attributes.

```
public class Account{
    // parameterized constructor
    public   Account(String aNumber, int anInitialBalance) { ... }
    // ...
} // class: Account
```

We must, of course, provide two actual parameters to this constructor when creating objects of this class. It is sometimes useful to be able to create such an object with some predefined attribute values and subsequently update one or more of its state values. Creating objects which are not properly initialized usually leads to many difficult programming problems. To ensure that an object is safely initialized we may also include a *default constructor* as part of the mandatory profile that is called if no initializing actual arguments are supplied by the user. The latter could be incorporated into the **Account** class above with:

```
public class Account{
    // parameterized and default constructors
    public   Account(String aNumber, int anInitialBalance) { ... }
    public   Account() { ... }                              // default constructor

    // ...
} // class: Account
```

Here, if no parameters are given when an instance of the class **Account** is created, then that object is initialized in the manner described by the method body for the default constructor. Hence we might have:

```
Account ac1 = new Account("ABC123", 1200);   // Parameterized constructor
Account ac2 = new Account();                  // Default constructor
```

The default constructor for this class requires an appropriate behaviour. Here we might consider initializing the number of the **Account** object to the empty string and the balance to zero. Whatever the default behaviour, this ensures that all **Account** objects are properly initialized when they are created. This final additional code for the **Account** class is shown below. Observe how the behaviour for the default constructor is defined in terms of the parameterized constructor. This is common practice when using Java and ensures consistency when developing *overloaded constructors*.

```
public class Account {
   // ----- Operations ----------
   // parameterized constructor
   public   Account(String aNumber, int anInitialBalance) {
      theNumber = aNumber;
      theBalance = anInitialBalance;
   }
   // default constructor
   public Account() {
      this("", 0);
   }
   // ...
} // class: Account
```

The method body for the default constructor is **this("**, 0). Here, the default constructor calls the parameterized constructor with the empty string and zero, as its actual
parameters. Thus the default constructor's behaviour is defined in terms of the parameterized constructor.

Figure 3.7 shows a sequence diagram in which a new Account object is created. In
the note symbol we also show the corresponding Java statement, using a parameterized
constructor.

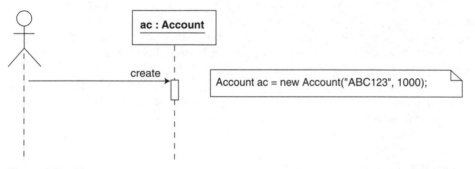

Figure 3.7   *Object creation*

3.3   Building an application

In order that a complete system can be executed there needs to be an object that is able
to respond to a message from the execution environment. For historical reasons with
Java this must be an object that supports a **public static** method named main.

Although it is possible to locate all of the code required for an application in the
main method it is not really very practical or even desirable (see chapter 7). An alternative is to create in the main method a single Application object and send it a message
such as run. Now it is the run method of the Application class that represents the top
level of our software system and not a rather meaningless main method. This results in

the Java code:

```
public class Main{
    // ----- Operations ----------
    public static void main(String[] args) {
        Application app = new Application();
        app.run();
    } // method: main
} // class: Main
```

held in the file Main.java. Figure 3.8 shows a sequence diagram for this, with notes giving the Java code corresponding to the diagram elements.

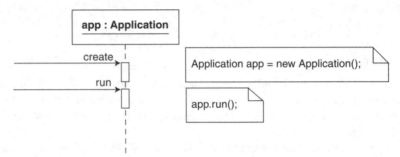

Figure 3.8 *Program start-up*

The method main in Java is a **static** method with an *array* of String parameters and a **void** return type. The array represents the series of Strings given when the program is invoked by the user. We have no use for this. Here, we have made the method a member of a class we call Main. This way, we need only write this class once and employ it unchanged in many further programs.

Typically the run method of the Application object creates the other objects required and sends each the appropriate messages to give the desired system behaviour. The messages we wish to exercise here are those identified by the use-cases. Hence we need to demonstrate that we are able to realize all the use-cases. Figure 3.9 shows part of what this run method will perform. This results in the Java code in the file Application.java and shown as program listing 3.1.

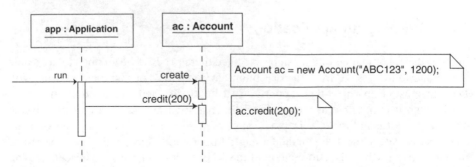

Figure 3.9 *Application object*

Program 3.1 The Application class (model Prog3_1.uml)

```
public class Application{
  public void   run(){
    //
    // Use-case:   Open new account
    //
    Account ac = new Account("ABC123", 1200);
    //
    // Use-case:   Credit account
    //
    ac.credit(200);    // balance now 1400
    //
    // Use-case:   Display account details
    //
    ac.display();
    //
    // Use-case:   Others
    //
    ac.debit(900);    // balance now 500
    ac.debit(700);    // balance unchanged
    ac.display();
  } // method: run
} // class: Application
```

At this stage in the development lifecycle we have moved from analysis through to the implementation in Java of the Account class. We are able to demonstrate, with some confidence, that this class operates as required. Our confidence stems from executable code that respects the requirements captured in the use-cases. The output from this program is:

```
Account
  Number: ABC123
  Balance: 1400

Account
  Number: ABC123
  Balance: 500
```

and shows how the balance has been changed.

3.4 Implementing architectural relationships

In the preceding sections the implementation of a single object was discussed in some detail. We now turn our attention to implementing an object that has an association or aggregation relationship with other objects. As in chapter 2, we first communicate the basic concepts involved by making extensive use of UML analysis diagrams. Having done so we complete our discussions by considering their implementation with detailed Java code.

Further, we see this next phase as a new iteration building on our previous work. If the Account class has all the required features for this new phase, then we have no new work to apply to it and we can make the additional requirements our primary focus. At worst we may have to augment the Account class with some minor additional features. But this is a trivial task since the additional logic can be given to the Account class, independent of any other developments we are engaged in.

3.4.1   Introduction

In a typical object-oriented system most objects do not exist as independent entities. In order that the overall system functionality is achieved, objects enter into architectural relationships, i.e. association or aggregation. This is a consequence of the fact that a well-designed object should be highly cohesive, i.e. it should undertake a small number of relatively simple, closely focused tasks. To be useful in a specific application it needs to be connected (coupled) with others to form a network of co-operating objects.

Recall from chapter 2 that association is a loose form of coupling between objects that are not conceptually related. Essentially the associates have visibility of each other so that they can engage in message passing. Aggregation is a tighter form of coupling in which a whole/part relationship is modelled. Composite aggregation is used when the parts are unique to a single whole and shared aggregation when they are not. Our task is to explore the implementation of these relationships with Java.

3.4.2   The bank problem revisited

The problem specification from section 3.2 introduced the notion of a Bank object to which our Account objects are related. In effect the creation of an Account object occurs when we open an Account with the Bank. Further, when we make a transaction on an account we do so through the bank. The discussion reveals that there is a relationship formed between the Bank object and its many Account objects. Since the Account objects only exist as part of the Bank, then we have an aggregate relationship. Specifically we have composite aggregation since the Account objects have no existence outwith the Bank.

This discussion also reveals that all the use-cases we consider will involve interacting directly with the Bank object. The Account objects are effectively parts of the Bank object with which they are opened. A consequence of this is that the Accounts are private to the Bank and messages destined for an Account object must first be sent to the Bank and then propagated to it.

Consideration of the problem then identifies the following use-cases:

- Create a bank object
- Open a new account
- Perform a transaction on an account
- Obtain the total assets of the bank

The "Create a bank object" use-case is similar to that for creating a single Account object as presented in figure 3.2. The use-case "Open a new account" is more elaborate than that shown in chapter 2. Now we must achieve this in conjunction with the newly created Bank object. Figure 3.10(a) presents a sequence diagram that captures its realization. The message openAccount is sent to the Bank object which, in turn, creates a new Account object.

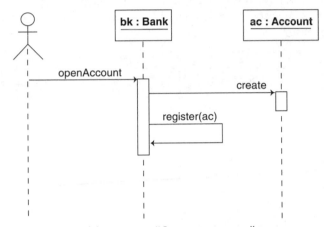

Figure 3.10(a)   *Realization of the use-case "Open new account"*

The accompanying object diagram in figure 3.10(b) further identifies that the newly created Account object has been aggregated into the Bank object. The diagram reveals the relationship that forms between the Bank object and the Account object. Hence figure 3.10(a) must be developed further to show how the architecture is established.

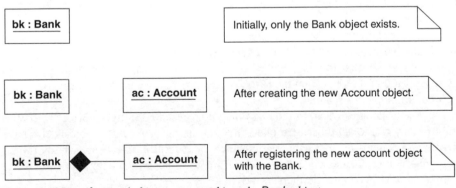

Figure 3.10(b)   *Account object aggregated into the Bank object*

In figure 3.10(b) the object diagram reveals the effect of the two steps in the corresponding sequence diagram. First, only the Bank object exists. It is then instructed to open a new account and, presumably, is given the account details. The Bank object then creates a new Account object suitably initialized. At that point we have the arrangement as shown in the second part of the figure in which two objects now exist but are unrelated. Finally, the new Account object is registered with the Bank object. This has the effect as shown in the third part of the figure with an aggregation relationship formed between the two objects.

The use-cases "Credit an account", "Debit an account" and "Obtain the balance of an account" describe three distinct transactions. A particular example is a credit transaction on a particular account. The sequence diagram of figure 3.11 demonstrates how this might be realized.

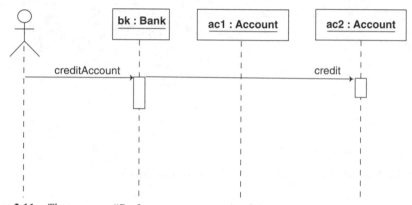

Figure 3.11   *The use-case "Perform a transaction (credit) on an account"*

Here, the Bank object is first sent the message creditAccount. The Bank first needs to identify the correct Account object among its many aggregate members (not shown in the figure). Having done so the Bank can then send the credit message to the particular Account. In this illustration we have deliberately omitted the parameters that would be associated with the messages. A similar arrangement applies to the other two related use-cases.

Our final use-case is to determine the total assets of the bank. This is defined as the sum of the balances for each account opened with the bank. The sequence diagram in figure 3.12 shows that this is achieved by sending the message getBalance to each aggregate Account object from the Bank object, and that the Bank object is responsible for summing the amounts.

These considerations lead to the class diagram presented in figure 3.13. We have a one-to-many relationship between one Bank and its many Account objects. We have arrived at a composite aggregate relation since the Account objects are wholly owned by the Bank object. In essence, the accounts are attributes of the bank, but sufficiently complex to view as objects in their own right.

The class diagram reveals the operations we associate with the Bank class. This set is a direct result of our consideration of the services required by the application. They, of course, have been identified by the various use-case models we have constructed of the system.

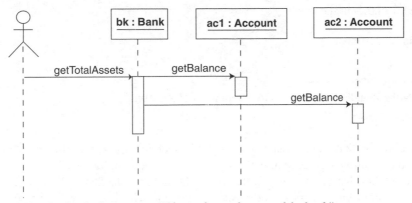

Figure 3.12 *Realizing the use-case "Obtain the total assets of the bank"*

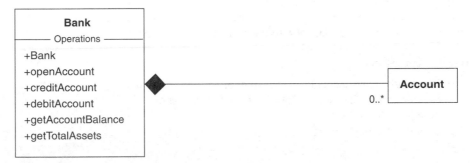

Figure 3.13 *Bank problem class diagram*

3.5   Establishing the architecture

We now investigate the implementation of a one-to-many architecture. As we shall see shortly, this involves using *collection objects*. The reader is invited to review the materials in appendix E which details aspects of the Java Collections Framework.

From the discussions of chapter 2 and that above, it is expected that objects will form architectural relationships. It is worth noting that although the attributes of an object are application independent, architectural relationships depend on the particular configuration of objects. As a consequence they may change from one specific application to another. For example, a Person might be connected to a Company in one application and to a SportsClub in another.

In order that an object has an aggregation relationship with one or more objects it must act as an assembly of its aggregate components. For example, figure 3.13 is a class diagram describing these relationships. As there are many Account objects in the aggregation, a suitable collection is required to hold them. The collection the ArrayList has been chosen based on the assumption that each additional Account object is simply appended to the existing list of Accounts.

The declaration statement:

public class Bank{

    // ...

    // ----- Relations ----------
    private java.util.ArrayList   theAccounts;   // of Account

} // class: Bank

in the class **Bank** declares a suitable **ArrayList** collection, **theAccounts**. The name for this relation attribute is taken directly from the class diagram, figure 3.13.

The relationship between each reference held by the collection and the object it is linked to is modelled as one of composite aggregation. This is a design choice justified by the fact that each **Account** in a **Bank** is considered to be just like an attribute and is therefore not shareable.

In the **Bank** class we have the operation **openAccount** that establishes a new **Account**. Importantly, as well as creating this as a new **Account** for the **Bank**, it also incorporates the **Account** into the collection for this **Bank**.

public class Bank{

    public void   openAccount(String aNumber, **int** anInitialBalance) {
      Account acc = **new** Account(aNumber, anInitialBalance);
      theAccounts.add(acc);
    }   // method: openAccount
    // ...

    // ----- Relations ----------
    private java.util.ArrayList   theAccounts;   // of Account

}   // class: Bank

3.5.1   The mandatory profile revisited

Earlier we discussed the concept of a mandatory profile for a Java class. Essentially it prescribes the **public** operations that a class should have. For example, the **Account** class should have a default constructor and a parameterized constructor as in:

public class Account{

    // ----- Operations ----------
    public Account(String aNumber, **int** anInitialBalance) { ... }
    public Account() { ... }
    // ...

}   // class: Account

Appendix E discusses the need for the equals operation to support containment in a collection. Therefore the **Account** class should have **equals** as a **public** operation. If the Account objects were to be contained by a TreeSet collection, then the operation compareTo is also required.

```
public class Account implements java.lang.Comparable{

    // ----- Operations ----------
    public Account(String aNumber, int anInitialBalance) { ... }
    public Account() { ... }
    public boolean    equals(Object obj) { ... }
    public int    compareTo(Object obj) { ... }
    // ...

}      // class: Account
```

Where a class offers a definition for the operation equals, then it is redefining the version inherited from the implied superclass Object (see chapters 1, 5 and 6). The operation compareTo is a definition from that specified by the interface Comparable (again, see chapters 1, 5 and 6). This is captured by the **implements** clause as shown in the listing above.

Finally, when using a HashSet collection, this container class requires that the objects being contained support the operation hashCode. This method is used to provide a randomizing value for the underlying mechanism employed by HashSet to store and retrieve objects.

```
public class Account implements java.lang.Comparable{

    // ----- Operations ----------
    public Account(String aNumber, int anInitialBalance){ ... }
    public Account(){ ... }

    public boolean    equals(Object obj){ ... }
    public int    compareTo(Object obj){ ... }
    public int    hashCode(){ ... }
    // ...

}      // class: Account
```

By ensuring that all three operations are made part of the mandatory profile for a class, then we have the freedom to change the collection class we use in the implementation.

3.6   The example application

We are now in a position to construct our application. The preceding analysis revealed through a number of use-cases the functionality required by our system. As a consequence, the following Application class is relatively straightforward. The application creates a Bank object and opens a number of Accounts configured according to the class diagram of figure 3.13. Having done so, the Application performs some transactions on one of the accounts, requests the balance for a particular account and obtains the total bank assets. (Further developments of this type of example and a more challenging illustration are discussed in the case study of chapter 4.) This translates to the following Java code:

```
import textio.*;

public class Application {

    // ----- Operations ----------
    public void   run() {
        //
        // Use-case:    Open new bank
        //
        Bank bk = new Bank("Community");
        //
        // Use-case:    Open new accounts
        //
        bk.openAccount("ABC123", 1200);
        bk.openAccount("DEF456", 1000);
        bk.openAccount("GHI789", 2000);
        //
        // Use-case:   Perform transactions on an account
        //
        bk.creditAccount("ABC123", 200);    // balance now 1400
        bk.debitAccount("ABC123", 900);    // balance now 500
        bk.debitAccount("ABC123", 700);    // balance unchanged
        //
        // Use-case:   Display account details
        //
        int balance = bk.getAccountBalance("ABC123");
        ConsoleIO.out.println("Balance for ABC123: " + balance);
        //
        // Use-case:   Obtain total assets
        //
        int totalAssets = bk.getTotalAssets();
        ConsoleIO.out.println("Total assests: " + totalAssets);

    }   // method: run

}   // class: Application
```

The program delivers two outputs, namely, the balance for account with account number ABC123 and the total assets for the bank. The actual values are:

```
Balance for ABC123: 500
Total assets: 3500
```

The Bank class now appears as in program listing 3.2.

Program 3.2   The Bank class (model Prog3_2.uml)

```java
import java.util.*;

public class Bank {

  // ----- Operations ----------
  public   Bank(String aName) {
    theName = aName;
    theAccounts = new ArrayList();
  }   // method: Bank

  public void   openAccount(String aNumber, int anInitialBalance) {
    Account acc = new Account(aNumber, anInitialBalance);
    theAccounts.add(acc);
  }   // method: openAccount

  public void   creditAccount(String aNumber, int anAmount) {
    Iterator iter = theAccounts.iterator();
    while(iter.hasNext() == true) {
      Account acc = (Account)iter.next();
      if(aNumber.equals(acc.getNumber())) {
        acc.credit(anAmount);
        break;
      }
    }
  }   // method: creditAccount

  public void   debitAccount(String aNumber, int anAmount) {
    Iterator iter = theAccounts.iterator();
    while(iter.hasNext() == true) {
      Account acc = (Account)iter.next();
      if(aNumber.equals(acc.getNumber())) {
        acc.debit(anAmount);
        break;
      }
    }
  } // method: debitAccount

  public int   getAccountBalance(String aNumber) {
    Iterator iter = theAccounts.iterator();
    while(iter.hasNext() == true) {
      Account acc = (Account)iter.next();
      if(aNumber.equals(acc.getNumber())) {
        return acc.getBalance();
      }
    }
    return -1;   // denotes no account identified with number
  }   // method: getAccountBalance
```

```
public int    getTotalAssets() {
   int totalAssets = 0;
   Iterator iter = theAccounts.iterator();
   while(iter.hasNext() == true) {
      Account acc = (Account)iter.next();
      totalAssets += acc.getBalance();
   }
   return totalAssets;
} // method: getTotalAssets

// ----- Attributes ----------
private String    theName;
// ----- Relations ----------
private java.util.ArrayList    theAccounts;   // of Account
} // class: Bank
```

3.7 Summary

1. A Java class is a composite structure in which we can define class attributes (properties or instance variables) and class operations (methods). A Java class typically specifies the public services (methods) and the private representation (attributes).
2. The language supports parameterized methods for each class operation. A method body comprises one or more statements. The sentences are assembled into the usual control logic of sequence, selection and iteration. Statements can create local instances and send messages to objects.
3. A collection object is a container for other objects of some arbitrary class. Java offers several different kinds of standard collections. The nature of the application will usually determine the appropriate container to use.
4. The objects to be contained by a collection will generally have to publicize a mandatory profile including the operations compareTo, equals and hashCode.

3.8 Exercises

For the following exercises and those in subsequent chapters we advise that you make use of the ROME object modelling environment supplied with the text. Although its use is not strictly necessary it does make the development of the Java code considerably easier and more enjoyable. Appendix B contains the information you require to use it. You should consult it before proceeding.

1. Complete the following Java code for the class Car by giving implementations for the method bodies. Draw a class diagram in the manner of figure 3.6(c) for this class.

```
public class Car {

   // ----- Operations ----------
   public Car(String aMake, String aModel, int aCapacity) { ... }
```

```
    public String   getMake() { ... }
    public String   getModel() { ... }
    public int   getCapacity() { ... }

    // ----- Attributes ----------
    private String   theMake;
    private String   theModel;
    private int   theCapacity;

} // class: Car
```

Now develop an application that creates an instance of the Car class and displays its details.

2. Complete the following Java code for the class Student by giving implementations for the method bodies. Draw a class diagram in the manner of figure 3.6(c) for this class.

```
public class Student {

    // ----- Operations ----------
    public Student(String aName, String anAddress, String aMatriculationNumber) { ... }
    public String   getName() { ... }
    public String   getAddress() { ... }
    public String   getMatriculationNumber() { ... }

    // ----- Attributes ----------
    private String   theName;
    private String   theAddress;
    private String   theMatriculationNumber;

} // class: Student
```

Now develop an application that creates an instance of the Student class and displays its details.

3. Complete the following Java code for the class House by giving implementations for the method bodies. Draw a class diagram in the manner of figure 3.6(c) for this class.

```
public class House {

    // ----- Operations ----------
    public House (String anAddress, int aNumberOfRooms) { ... }
    public String   getAddress() { ... }
    public int   getNumberOfRooms() { ... }
    public void   extend (int aNumberOfRooms) { ... }   // Add new rooms.

    // ----- Attributes ----------
    private String   theAddress;
    private int   theNumberOfRooms;

} // class: House
```

Now develop an application that creates an instance of the House class, adds some rooms to it, and then displays its details.

When a House instance is created would it thereafter change its address? If not, under what circumstances could we consider moving the attribute theAddress into the public interface?

4. Complete the following Java coding for the class **Point** representing a point in the Cartesian co-ordinate system by giving implementations for the method bodies. In a two-dimensional co-ordinate system a point may be represented by its X and Y co-ordinate values.

```
public class Point {
    // ----- Operations ----------
    public    Point (int anX , int aY) { ... }
    public int    getX() { ... }              // obtain X co-ordinate
    public int    getY() { ... }              // obtain Y co-ordinate
    public void    moveBy (int anX, int aY) { ... }    // displace by amounts

    // ----- Attributes ----------
    private int    theX;
    private int    theY;

} // class: Point
```

Develop a Java **Application** in which a single **Point** object is created, and then displaced by some amount and its new position displayed. Draw a class diagram in the manner of figure 3.6(c) for this class.

5. Using the **Point** class, complete the following Java code for the class **Line** by giving implementations for the method bodies. A line in a two-dimensional co-ordinate space may be represented by its start and end points.

```
public class Line {

    // ----- Operations ----------
    public Line (Point aStart, Point anEnd) { ... }
    public boolean    isHorizontal() { ... }
    public boolean    isVertical() { ... }
    public void    moveBy (int anX, int aY) { ... }    // displace by amounts
    public void    display() { ... }              // print details

    // ----- Attributes ----------
    private Point    theStart;
    private Point    theEnd;

} // class: Line
```

Develop some application code expressed in Java in which a **Line** is established through two **Point** values, is translated through some amount then its new position printed.

Develop a class diagram for this **Line** class showing that it is a composite aggregate of two **Point** objects.

6. Using the **Point** class, a rectangle may be represented by two **Point** values representing, respectively, the lower left vertex and the upper right vertex. Complete the

following Java code for the Rectangle class by giving implementations for the method bodies.

```java
public class Rectangle {

    // ----- Operations ----------
    public Rectangle (Point aLowerLeft, Point anUpperRight) {...}
    public int    getArea() { ... }
    public int    getPerimter() { ... }
    public int    getHeight() { ... }
    public int    getWidth() { ... }
    public void   moveBy (int anX, int aY) { ... }
    public boolean   isPointInRectangle (Point aPoint) { ... }

    // ----- Attributes ----------
    private Point   theLowerLeft;
    private Point   theUpperRight;

} // class: Rectangle
```

Create a Rectangle in some application code expressed in Java, and then test the behaviours of its operations. In a UML object diagram show an instance of a Rectangle object and any other related objects.

7. Prepare the Java code for a Store class. The Store represents some storage device that holds a single **int** value. The Store constructor has a single **int** parameter used to initialize the representation. A single **int** parameter for the write operation is used to update the value of the representation. The read operation returns a copy of the representation value. The operations up and down respectively increment and decrement by one the value of a Store object.

8. For questions 19, 20, 21 and 22 in the exercises of chapter 2, prepare suitable Java files then construct suitable application code to test the system. Make sure that you make full use of the class, collaboration and sequence diagrams.

9. Consider the class Employee with an operation getDateOfBirth that returns the date of birth of an Employee object. Visit the Java website (http://java.sun.com) for a description of the class java.util.GregorianCalendar.

```java
public class Employee {

    // ----- Operations ----------
    public GregorianCalendar getDateOfBirth() {return theDateOfBirth;};

    // ----- Attributes ----------
    private GregorianCalendar theDateOfBirth;

} // class: Employee
```

Then in some application code we have:

```java
Employee            e1 = new Employee ( ... );
GregorianCalendar   dateOfBirth = e1.getDateOfBirth();
```

What danger lies in this code? What principle has been broken?

10. We have stated that every instance of a class has it own set of private attributes. Make a case for every instance sharing the same common attribute. How is this realized in Java?

11. Develop the sample **Bank-Account** application by creating additional accounts. You may use the **ROME** model supplied with this chapter. Extend the **Bank** class with an operation to print the details of only those accounts with a balance over a certain limit given as a parameter to the method.

    In the **Bank** class listing in program 3.2 the methods **creditAccount**, **debitAccount** and **getAccountBalance** all have to find the **Account** object with the required account number. Introduce the private method:

    ```
    //class Bank
    private Account    lookUp(String anAccountNumber) { ... }
    ```

    that will find the required **Account** or return **null** if it does not exist. Now recode the original methods to use this facility.

    Using the examples presented in this and the preceding chapter, develop solutions for the following problems. In all cases adopt the lightweight process introduced in chapter 2. Use the **ROME** modelling tool to develop your solutions.

12. A sports club maintains a register of its members. The member details include their name and address and the name of their preferred sport. A system is required to enrol new members, and to present details of the full membership and a printout of those members interested in tennis.

13. An organization maintains a record for each of its employees. The employee details include their (unique) employment number, their name, age, sex and salary. A system is required to hire new employees and to produce the following reports:
    a) the complete staff list
    b) a list of those staff for a given sex
    c) a list of staff over a certain age
    d) the total wage bill for the company

14. A student has an extensive music collection on CDs. Single artist CDs have a title and the name of the artist. Each track on such a CD has a title. Compilation CDs have a title, and their tracks have both the title and the artist. A system is required to maintain details of the CD collection and to produce a report listing all the CDs in the student collection.

15. A newsagent maintains a list of customers including their name and address. For each customer the newsagent has a list of the names of up to three newspapers to be delivered to that customer's home. Develop a system to list each newspaper and the quantities required.

4

Case Study: A Library Application

This case study uses concepts discussed in the previous chapters to illustrate an OOAD implemented with Java. The specification of the system is unrealistically simple but interested readers can modify it in a variety of ways as indicated in the exercises at the end of the chapter. Throughout the case study we emphasize how the development revolves around a number of iterations, driven by the use-cases. In turn, each use-case mirrors a corresponding set of test-cases that ensure that the required functionality is fulfilled.

All class diagrams and the accompanying Java code were developed using the ROME modelling tool.

4.1   Specification

It is often difficult (or even impossible) to obtain a set of unambiguous, fixed requirements for a software system. Although there are a variety of reasons for this, we assume a sufficient familiarity with the operation of a library to understand the following description:

> *The library has a name, holds a number of books each of which has a title, author and unique catalogue number. There are registered borrowers each with a unique name. A borrower may borrow a book and return it. However, each book transaction must be recorded by a librarian. She is also expected to register a new borrower, be able to display in increasing catalogue number order those books available for loan and those already out on loan.*

We are required to develop an application to support the librarian.

4.2   Iteration 1

In this case study we use the development process described in section 2.1.1. Recall that to implement a particular set of requirements there is an iteration consisting of five main activities. They are as follows:

- Establish a use-case diagram
- Analysis
  - Develop object, collaboration, sequence and activity diagrams

- Design
  - Construct/revise a class diagram
- Implementation
  - Incrementally implement, document and test the Java code for each class
- Reconcile the model diagrams

4.2.1   Establish use-cases

This activity is concerned with discovering and recording what the system is required
to do from a user's perspective. Fortunately the description of the librarian's duties gives
us the information we need. Just a little thought leads us to decide that the librarian
must be able to:

- Register a new borrower
- Add a new book to the stock
- Produce a list of books available for loan
- Produce a list of books out on loan
- Record that a book has been borrowed and
- Record that a book has been returned

In real life we would almost certainly check our conclusions with the librarian but for
this iteration we can safely construct the use-case diagram of figure 4.1.

At some later date we may be presented with additional or modified requirements but
they are implemented in subsequent iterations. Notice that the use-cases are grouped

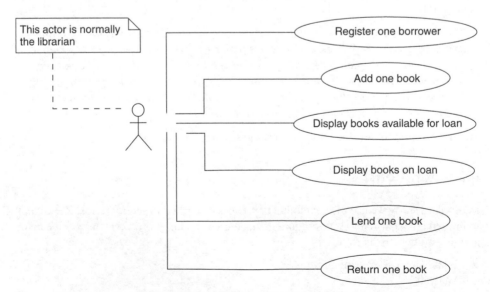

Figure 4.1   *A use-case diagram for the Library application: iteration 1*

together. The first two are concerned with configuration, the next two are displays and the last two record borrower transactions. It is always worthwhile trying to make UML diagrams (and the resulting code) as clear as possible. The note also helps clarify the nature of the actor in each use-case.

While undertaking this activity we can also specify some of the tests that will be implemented. Recall from section 2.1.1 that there should be a least one test-case for each use-case. Therefore we have the following test-cases with a short description of the expected outcome from each:

- test-case 1: Register one borrower

  A borrower is successfully registered with the library.

- test-case 2: Add one book

  A book is successfully added to the library stock and is available for loan.

- test-case 3: Display books available for loan

  A display of each book available for loan when there are zero or more books in the library.

- test-case 4: Display books out on loan

  A display of each book out on loan when there are zero or more books in the library.

- test-case 5: Lend one book

  A book available for loan is recorded as out on loan to a borrower. The book is not available for loan.

- test-case 6: Return one book
  A book on loan to a borrower is recorded as not out on loan to that borrower. The book is available for loan.

In later activities we might add more detail to the test-cases but for the moment it is enough to record that they exist and there is an outline of what we expect from them. Notice that to avoid the risk of any confusion we have numbered the test-cases. A more complex numbering scheme could be used but as testing is not our main concern we do not pursue this further.

4.2.2 Analysis

This activity is concerned with the analysis of the library application, i.e. the "system". As such it is an important part of its development. If it is of poor quality then it is very difficult, or even impossible, to deliver high quality code. The main reason we undertake the analysis is to identify problem domain objects and their classes, the relationships they enter into and the messages they send to each other.

Fortunately it is relatively easy to identify four major objects. Everyday experience indicates the librarian, library, borrower and book. Identification of the classes Librarian, Library, Borrower and Book follows naturally. This is not unusual and is a major strength of object orientation. Real-world experience and intuition can often be used to identify the most important objects and classes. With just a little practice it becomes surprisingly straightforward.

After considering the use-case diagram and the textual description for the library, it becomes clear that the librarian and borrower objects are not part of the system. The librarian is the actor that uses the system and is therefore of no concern to us. Similarly although we expect to record details of each borrower when registered, there is no question of the human borrower interacting with or being part of our system (but see the next section). In real life the interactions would be with the librarian but again that does not concern us.

The object diagram of figure 4.2 helps document this decision.

Figure 4.2   *An object diagram*

Notice that in our view each Book is part of the Library therefore an aggregation relationship is appropriate.

If we now turn our attention to identifying some of the operations we expect the Library to have then the use-case diagram is particularly helpful. There is an operation supported by the Library that corresponds to each. For example, the sequence diagram of figure 4.3 shows the two display operations.

If an actor (the librarian) sends the message displayBooksAvailableForLoan to the Library then it requests each Book available for loan to display itself. On the other hand

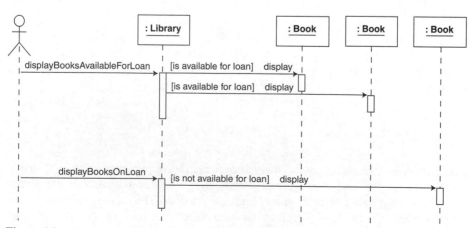

Figure 4.3   *A sequence diagram for the display operations*

if the actor sends the message displayBooksOnLoan to the Library then it requests each Book not available for loan to display itself. Notice that we have implied that the Book class must support a display operation or its equivalent.

An alternative to a sequence diagram is to use an activity diagram as shown for the displayBooksAvailableForLoan operation in figure 4.4.

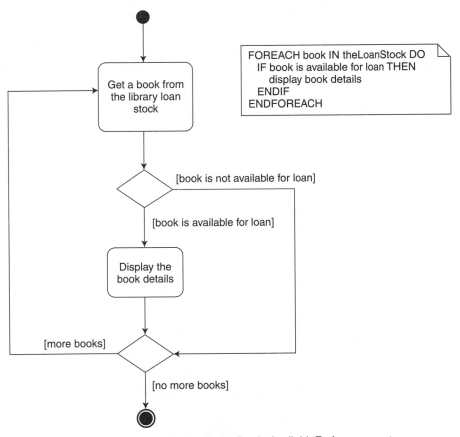

Figure 4.4 *An activity diagram for the displayBooksAvailableForLoan operation*

We might even use both just to make sure that we have carefully documented our intentions. Notice that we have used pseudocode (see appendix H) in the accompanying note to help clarify what the diagram is intended to convey.

A second sequence diagram as shown in figure 4.5 helps document our intentions with the remaining operations.

It shows that the operations registerOneBorrower and addOneBook are the sole responsibility of the Library. However, lendOneBook and returnOneBook have an interaction with a Book object. Although we have no details of how this will be accomplished, we can anticipate the operations attachBorrower and detachBorrower that can be sent to a Book. The operation lendOneBook in the Library uses the former and returnOneBook the latter. A collaboration diagram as shown in figure 4.6 helps makes this point clear.

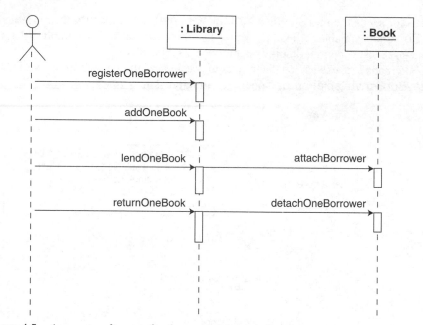

Figure 4.5   *A sequence diagram for the remaining operations*

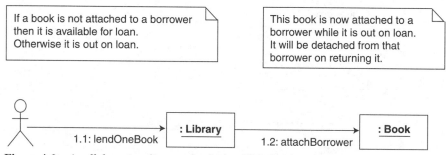

> If a book is not attached to a borrower then it is available for loan. Otherwise it is out on loan.

> This book is now attached to a borrower while it is out on loan. It will be detached from that borrower on returning it.

Figure 4.6   *A collaboration diagram for the lendOneBook operation*

Notice that it is too soon to add details of any parameters on these diagrams. If required, we can do this later when more information is available.

4.2.3   Design

From the object diagram in figure 4.2 we have the initial class diagram shown in figure 4.7.

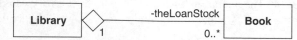

Figure 4.7   *An initial class diagram*

Clearly it must be developed further. For example, two-way traversal of an association or aggregation is the default in the UML. It is likely that there is no need for a Book to send messages to the Library. After all it is just a Book. Therefore the aggregation relation between a Library and a Book should only be traversed from the Library to the Book. (An arrow decoration on the line that connects them captures this decision, as shown in figure 4.8.)

Although it may appear to be a relatively minor detail, the decision to use shared aggregation does have a profound impact. For example, it affects the system's storage requirements and run-time efficiency. Normally composite aggregation is reserved for object-like attributes (see sections 2.5.3 and 2.5.4).

If we consider the nature of the collections used to support the multiplicity of zero or many for Books then recall from appendix E that the default collection is the HashSet. However, we are required to display Books in the order of their unique catalogue numbers. Therefore a TreeSet is an appropriate choice (see appendix E). The role name theLoanStock from the class diagram gives it a suitable identifier.

The library specification indicates a Book is an object with attributes for a catalogue number, author and title. From the preceding activity we also need to be able to decide if a Book is available for borrowing or not. We also need to associate a Book with a registered borrower if it is out on loan to that borrower. Finally, the Library also needs to have a record of each registered borrower.

This leads us to introduce an object of the class BorrowerRecord. Its purpose is to hold the unique name of each borrower and a collection of those Books on loan to that borrower. Notice that this object did not come from the problem domain but is a design object introduced to help implement the use-cases. No doubt other designs are possible but for this iteration it appears to be a good decision. In future iterations it could hold more information about a borrower. For example, personal data such as address and telephone number or information about outstanding fines. The class diagram of figure 4.8 shows its relationships with the Library and Book classes.

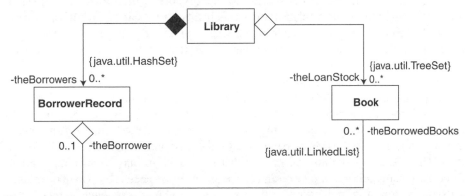

Figure 4.8 *A modified class diagram*

As we anticipate that each BorrowerRecord is for the sole use of the Library then a composite aggregation relationship navigated from the Library to Borrower Record is appropriate. For the moment the default collection of the HashSet is satisfactory as there are no duplicate borrowers and the order in which they are held is not important. The role name of theBorrowers is a suitable identifier for it.

The relationship between the Book and BorrowerRecord classes is interesting. Now we see that each Book on loan is held in a collection belonging to a BorrowerRecord. In keeping with the fact that when a Book is returned, we expect deletions at arbitrary points in this collection, we have chosen a LinkedList (see appendix E). The role name theBorrowedBooks is a suitable identifier for the collection. Given that each Book could be attached to the Library and a BorrowerRecord, a shared aggregation relationship is appropriate. This decision is in keeping with the discussions of section 2.5.4.

Note that the Book has an architectural attribute theBorrower for the BorrowerRecord that has borrowed it. We anticipate that if theBorrower has a value of **null** then by implication a Book is not on loan. Otherwise it is on loan.

If we consider the impact of introducing the BorrowerRecord class on the implementation of the use-cases, then we have:

Register one borrower	Establish a suitably initialized BorrowerRecord object
Add one book	Set theBorrower in the Book to **null**
Display books available for loan	Each Book displayed has theBorrower equal to **null**
Display books out on loan	Each Book displayed has theBorrower not equal to **null**
Lend one book	Attach a BorrowerRecord to a Book and vice versa
Return one book	Detach a BorrowerRecord from a Book and vice versa

Armed with the insights gained we can start to establish the operations and attributes of each class.

If we consider the Book class then we identify the normal attributes theTitle, theAuthor and theCatalogueNumber. They are a String, a String and an **int** respectively, though a String for the latter might have been considered. The class must also have the single-valued architectural attribute theBorrower as shown in the class diagram of figure 4.8.

The mandatory profile operations concerned with object construction, comparison and containment in a collection are easily dealt with. For example, we need a parameterized constructor with a String, a String and an **int** as formal parameters to the Book constructor. We expect the method to initialize theTitle, theAuthor and theCatalogueNumber with its parameter values. It should also set theBorrower to **null**. There is also a default constructor that sets each attribute to an initial default value. As regards object comparison and containment, all we need is the trio of operations equals, compareTo and hashCode (based on the unique catalogue number) discussed in appendix E.

Next we have the operations attachBorrower and detachBorrower to associate and disassociate a Book with a BorrowerRecord. The former needs a BorrowerRecord parameter but the latter does not as it just detaches itself from the BorrowerRecord it is currently attached to.

Finally we have the display operation identified in the previous activity. We anticipate that its method should display the Book attribute values on the system console.

The BorrowerRecord class is relatively straightforward with one normal String
attribute theName and a LinkedList architectural attribute theBorrowedBooks (see
figure 4.8). Its mandatory profile operations consist of a parameterized constructor and
the trio of operations equals, compareTo and hashCode (based on the unique name).
Although the last three are not strictly required we anticipate that they might be
required later and there is no harm including them here. In fact it is good practice
to do so.

The parameterized constructor has a String as a formal parameter to initialize the
attribute theName. It should also initialize theBorrowedBooks to reference a
LinkedList. Note that there is no default constructor therefore it is not possible to create a BorrowerRecord unless a name is supplied as an actual parameter. This is intentional as we consider it undesirable to have default BorrowerRecords in the context of
this application.

Finally we have the operations attachBook and detachBook to associate and disso-
ciate a BorrowerRecord and Book. Both operations need a Book parameter. Recall
that a BorrowerRecord can be associated with many Books.

The Library class has the normal attribute theName that is a String as well as the
architectural attributes theLoanStock and theBorrowers. As there is only one Library
in the problem domain it will never be compared or contained in a collection. It has only
a parameterized constructor for the mandatory profile of operations. It initializes the
attribute theName as well as the architectural attributes theLoanStock and
theBorrowers. The last two reference a TreeSet and HashSet respectively.

The remaining operations of the Library class correspond to the use-cases. The operations displayBooksAvailableForLoan and displayBooksOnLoan require no
parameters. However, addOneBook needs a Book parameter. It is the Book that is to
be added to the Library. Similarly registerOneBorrower needs the name of the borrower to be registered. Therefore it has a String parameter. However, lendOneBook
needs the catalogue number and the borrower's name. Therefore it has an **int** and String
parameter. As noted earlier returnOneBook requires only a Book as a parameter. It is
the Book to be returned.

The class diagrams of figures 4.9(a) and 4.9(b) summarize our design so far.

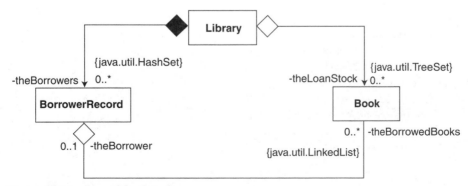

Figure 4.9(a)   *An outline class diagram*

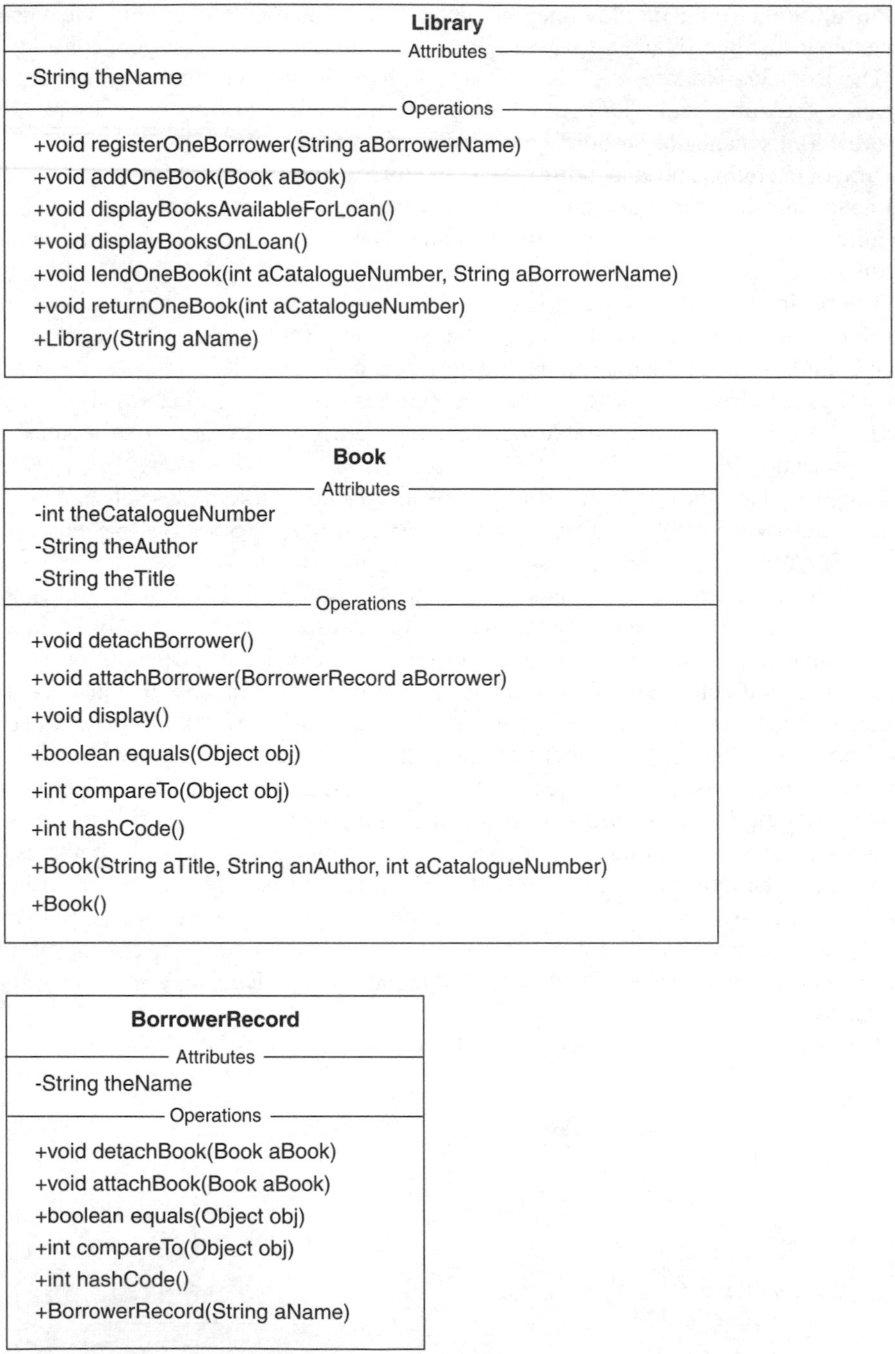

Library

——————————— Attributes ———————————

-String theName

——————————— Operations ———————————

+void registerOneBorrower(String aBorrowerName)

+void addOneBook(Book aBook)

+void displayBooksAvailableForLoan()

+void displayBooksOnLoan()

+void lendOneBook(int aCatalogueNumber, String aBorrowerName)

+void returnOneBook(int aCatalogueNumber)

+Library(String aName)

Book

——————————— Attributes ———————————

-int theCatalogueNumber

-String theAuthor

-String theTitle

——————————— Operations ———————————

+void detachBorrower()

+void attachBorrower(BorrowerRecord aBorrower)

+void display()

+boolean equals(Object obj)

+int compareTo(Object obj)

+int hashCode()

+Book(String aTitle, String anAuthor, int aCatalogueNumber)

+Book()

BorrowerRecord

——————— Attributes ———————

-String theName

——————— Operations ———————

+void detachBook(Book aBook)

+void attachBook(Book aBook)

+boolean equals(Object obj)

+int compareTo(Object obj)

+int hashCode()

+BorrowerRecord(String aName)

Figure 4.9(b)   *Class diagram details*

4.2.4   Implementation

Now we can begin producing some Java code from the class diagram of figure 4.9(a).
We could follow the scheme presented in chapter 3 and first develop then test the
Book class. The Library class could then be introduced and some of its functionality
tested including the addition and display of Books. Finally we might introduce the
BorrowerRecord class to test registering borrowers and borrowing and returning
Books.

   However, as this is a relatively straightforward application it seems more sensible to
push ahead and develop each class in one increment. In chapter 6, where we revisit this
case study, we do indeed take several increments at this corresponding stage.

4.2.4.1   Architectural code

Although there are no method bodies at present we can still produce a significant pro-
portion of the final Java code. The reason is that the mapping between a class diagram
and Java is well defined (see appendix D). For example, in a class diagram:

- an attribute maps to a **private** field declaration
- a public operation maps to a **public** method
- a class maps to a **public** class declaration
- an association or aggregation relationship maps to a **private** field declaration
- the use of a collection maps to an **import** statement naming the class of the
  collection

Therefore we have the following outline Java code for the Book class:

```
public class Book implements Comparable {

    // ----- Operations ----------
    public    Book(String aTitle, String anAuthor, int aCatalogueNumber) { ... }
    public    Book() { ... }
    //
    public boolean    equals(Object obj) { ... }
    public int    compareTo(Object obj) { ... }
    public int    hashCode() { ... }
    //
    public void    detachBorrower() { ... }
    public void    attachBorrower(BorrowerRecord aBorrower) { ... }
    public void    display() { ... }

    // ----- Attributes ----------
    private final int    theCatalogueNumber;
    private String    theAuthor;
    private String    theTitle;
```

```
// ----- Relations ----------
    private BorrowerRecord    theBorrower;
} // class: Book
```

Note that in the class header we have:

```
... implements Comparable {
```

which guarantees that the Book class has a compareTo method used by a collection object such as a TreeSet to order its elements (see appendix E).

The BorrowerRecord and Library classes are treated in exactly the same manner giving the following outline code:

```
import java.util.LinkedList;
public class BorrowerRecord implements Comparable {

    // ----- Operations ----------
    public    BorrowerRecord(String aName) { ... }
    //
    public boolean    equals(Object obj) { ... }
    public int    compareTo(Object obj) { ... }
    public int    hashCode(){ ... }
    //
    public void    detachBook(Book aBook) { ... }
    public void    attachBook(Book aBook) { ... }

    // ----- Attributes ----------
    private final String    theName;

    // ----- Relations ----------
    private java.util.LinkedList    theBorrowedBooks;
} // class: BorrowerRecord
```

and:

```
import java.util.HashSet;
import java.util.TreeSet;

public class Library {

    // ----- Operations ----------
    public    Library(String aName) { ... }
    //
    public void    registerOneBorrower(String aBorrowerName) { ... }
    public void    addOneBook(Book aBook) { ... }
    public void    displayBooksAvailableForLoan() { ... }
    public void    displayBooksOnLoan() { ... }
    public void    lendOneBook(int aCatalogueNumber, String aBorrowerName) { ... }
    public void    returnOneBook(int aCatalogueNumber) { ... }
```

```
// ----- Attributes ----------
private String    theName;

// ----- Relations ----------
private java.util.TreeSet    theLoanStock;
private java.util.HashSet    theBorrowers;

} // class: Library
```

It is worth making the point that this code is not in any sense temporary. Although we must supply method bodies and may add more operations or attributes we do not anticipate changing it. In fact it represents the skeleton of our final model. All we have to do now is to add some flesh to make it executable!

4.2.4.2   Method code

The next step is to implement each method in the three classes. The Java coding for the Book class is relatively straightforward. It is as follows:

```
import textio.*;
public class Book implements Comparable {

    // ----- Operations ----------
    public    Book(String aTitle, String anAuthor, int aCatalogueNumber) {
        // Initialize normal attributes
        theTitle = aTitle;
        theAuthor = anAuthor;
        theCatalogueNumber = aCatalogueNumber;
        //
        // Initialize architectural attributes
        theBorrower = null;
    }   // method: Book

    public    Book() {
        // Initialize attributes using the parameterized constructor
        this("", "", 0);
    }   // method: Book

    public boolean    equals(Object obj) {
        // The method for compareTo expects an Object therefore a cast is not required.
        // Use compareTo as the basis for the equality test.
        return this.compareTo(obj) == 0;
    }   // method: equals

    public int    compareTo(Object obj) {
        // As obj is an Object it must be cast to a Book so that the message
        // getCatalogueNumber can be sent.
```

```java
        Book book = (Book) obj;
        int bookCatalogueNumber = book.getCatalogueNumber();
        //
        // Use the relational operator defined for an int to return the appropriate value.
        int result;
        if(theCatalogueNumber < bookCatalogueNumber)
            result = -1;
        else if(theCatalogueNumber == bookCatalogueNumber)
            result = 0;
        else
            result = 1;
        //
        return result;
    }   // method: compareTo

    public int   hashCode() {
        // Create an Integer and initialize it with theCatalogueNumber – an int.
        Integer integerCatalogueNumber = new Integer(theCatalogueNumber);
        //
        // Hash on this object.
        return integerCatalogueNumber.hashCode();
    }   // method: hashCode

    public void   detachBorrower() {
        theBorrower = null;
    }   // method: detachBorrower

    public void   attachBorrower(BorrowerRecord aBorrower) {
        theBorrower = aBorrower;
    }   // method: attachBorrower

    public void   display() {
        // Note that \t is the tab character. Therefore \t\t tabs the cursor twice to the right.
        // Similarly \n is the new line character. Therefore \n can be used
        // instead of ConsoleIO.out.println().
        //
        ConsoleIO.out.println();
        ConsoleIO.out.println("\t\t" + "Title: " + theTitle);
        ConsoleIO.out.println("\t\t" + "Author: " + theAuthor);
        ConsoleIO.out.println("\t\t" + "Catalogue Number: " + theCatalogueNumber);
        ConsoleIO.out.println();
    }   // method: display

    public int getCatalogueNumber() {
        return theCatalogueNumber;
    }   // method: getCatalogueNumber
```

```
// ----- Attributes ----------
private final int   theCatalogueNumber;
private String   theAuthor;
private String   theTitle;

// ----- Relations ----------
private BorrowerRecord    theBorrower;

}   // class: Book
```

We have introduced the operation getCatalogueNumber for use by the compareTo method. Notice that the comparison and hash code operations use the same attribute (theCatalogueNumber). This gives an internal consistency to these methods. For example, two Books that are equal have the same hash code. To avoid problems that may arise from the inner workings of a collection we try to use an attribute whose value is unique, **final** and set by a constructor.

The Java code for the BorrowerRecord class is slightly more complex as it takes overall control of detaching and attaching a Book. It does this by sending the message detachBorrower or attachBorrower to the Book in question. This is important as we don't want to be in a position where a Book is attached or detached but the corresponding BorrowerRecord is not updated accordingly. It would invalidate our model with disastrous results.

As we compare BorrowerRecords by their names we introduce the operation getName to allow a client to determine the name attribute of a BorrowerRecord. Note that a borrower's name is unique, **final** and set by a constructor:

```
import java.util.LinkedList;
//
public class BorrowerRecord implements Comparable {

  // ----- Operations ----------
  public   BorrowerRecord(String aName) {
    // Initialize normal attributes
    theName = aName;
    //
    // Initialize architectural attributes
    theBorrowedBooks = new LinkedList();
  }   // method: BorrowerRecord

  public boolean   equals(Object obj) {
    return   this.compareTo(obj) == 0;
  }   // method: equals

  public int   compareTo(Object obj) {
    BorrowerRecord   borrowerRecord = (BorrowerRecord) obj;
    String name = borrowerRecord.getName();
    return   theName.compareTo(name);
  }   // method: CompareTo
```

```java
public int    hashCode() {
    return   theName.hashCode();
}    // method: hashCode

public void   detachBook(Book aBook) {
    aBook.detachBorrower();
    //
    // Ensure that the relationship remains consistent
    theBorrowedBooks.remove(aBook);
}    // method: detachBook

public void   attachBook(Book aBook) {
    aBook.attachBorrower(this);
    //
    // Ensure that the relationship remains consistent
    theBorrowedBooks.add(aBook);
}    // method: attachBook

public String getName() {
    return theName;
}    // method: getName

// ----- Attributes ----------
private final String    theName;

// ----- Relations ----------
private java.util.LinkedList    theBorrowedBooks;

}    // class: BorrowerRecord
```

Notice the way in which the detach and attach methods are implemented in the BorrowerRecord class. This is an attempt to make sure that when a Book is borrowed or returned that the BorrowerRecord and the Book are updated at the same time.

If we now turn our attention to the Library class then it is more of a challenge to implement. However, a useful assumption we can make is that perfect data is supplied by the client. Therefore if the Library records that a Book has been returned then we can assume that the Book is actually in the Library and that it was actually borrowed. Similarly when the Library records that a borrower has borrowed a Book then we can assume that the borrower is registered with the Library and that the Book is not out on loan already. This means that for this iteration there is no need to have error checks in our code. Similarly we can ignore any performance issues.

Although we will almost certainly put error checks in place we can do so as part of a series of managed iterations or increments. Also we can modify the model to execute faster or use less memory in a subsequent iteration if it proves to be necessary. By adopting this approach we can reduce the number and complexity of the problems we have to face at any given point in the development. We grow towards the implementation required and therefore reduce the risk of failure.

The first problem that we encounter is that the Library references Books and BorrowerRecords but the client supplies a catalogue number or a name to identify them. Therefore we introduce the operations getBook and getBorrowerRecord to bridge this gap. Notice that they are used internally by the Library and given **private** visibility. In general, it is good practice to limit direct access to an operation or attribute as much as possible.

Our second problem is that we need a method for the operations displayBooksOnLoan and displayBooksAvailableForLoan. After a little thought it becomes clear that they can make use of a **private** operation displayBooks that takes a **boolean** parameter. If the actual parameter is **true** then the method displays the Books on loan otherwise it displays those not on loan. This makes sense as it is easier to maintain our code. The alternative would be to have two similar methods: one in displayBooksOnLoan and the other in displayBooksAvailableForLoan. Apart from unnecessary duplication there is a danger that we might change one but not the other in later iterations.

Finally we must import the Iterator class to enable us to iterate over the various collections (see appendix E). The resulting code for the Library class is:

```
import java.util.HashSet;
import java.util.TreeSet;
import java.util.Iterator;

public class Library {

    // ----- Operations ----------
    public    Library(String aName) {
        // Initialize normal attributes
        theName = aName;
        //
        // Initialize architectural attributes
        theLoanStock = new TreeSet();
        theBorrowers = new HashSet();
    }   // method: Library


    public void    registerOneBorrower(String aBorrowerName){
        // Assume it has not been registered before.
        BorrowerRecord record = new BorrowerRecord(aBorrowerName);
        theBorrowers.add(record);
    }   // method: registerOneBorrower


    public void    addOneBook(Book aBook) {
        // Assume it has not been added before.
        theLoanStock.add(aBook);
    }   // method: addOneBook


    public void    lendOneBook(int aCatalogueNumber, String aBorrowerName) {
        // Find the correct Book. Assume it is present and is not out on loan.
        Book book = this.getBook(aCatalogueNumber);
```

```java
    //
    // Find the correct BorrowerRecord. Assume it is present.
    BorrowerRecord record = this.getBorrowerRecord(aBorrowerName);
    //
    // Attach them to each other
    record.attachBook(book);
  }    // method: lendOneBook


  public void    returnOneBook(int aCatalogueNumber) {
    // Find the correct Book. Assume it is present and was out on loan.
    Book book = this.getBook(aCatalogueNumber);
    //
    // Find the correct BorrowerRecord. Assume it is present.
    BorrowerRecord record = book.getBorrower();
    //
    // Detach them from each other.
    record.detachBook(book);
  }    // method: returnOneBook


  public void    displayBooksOnLoan() {
    //
    this.displayBooks(true);
  }    // method: displayBooksOnLoan


  public void    displayBooksAvailableForLoan() {
    //
    this.displayBooks(false);
  }    // method: displayBooksAvailableForLoan


  private Book    getBook(int aCatalogueNumber) {
    Book foundBook = null;
    //
    Iterator    iter = theLoanStock.iterator();
    while(iter.hasNext() == true) {
      Book book = (Book) iter.next();
      int catalogueNumber = book.getCatalogueNumber();
      if(catalogueNumber == aCatalogueNumber) {
        foundBook = book;
        break;
      }
    }
    //
    return foundBook;
  }    // method: getBook
```

```java
private BorrowerRecord   getBorrowerRecord(String aBorrowerName) {
   BorrowerRecord foundRecord = null;
      //
      Iterator    iter = theBorrowers.iterator();
      while(iter.hasNext() == true) {
         BorrowerRecord record = (BorrowerRecord) iter.next();
         String name = record.getName();
         if(name.equals(aBorrowerName) == true) {
            foundRecord = record;
            break;
         }
      }
      //
      return foundRecord;
}   // method: getBorrowerRecord

private void   displayBooks(boolean onLoan) {
   ConsoleIO.out.println("\n" + "Library: " + theName + "\n");
   //
   boolean found = false;
   Iterator iter = theLoanStock.iterator();
   //
   if(onLoan == true ) {
      ConsoleIO.out.println("\t" + "Books out on loan");
      while(iter.hasNext() == true) {
         Book book = (Book) iter.next();
         if(book.getBorrower() != null) {
            found = true;
            book.display();
         }
      }
   }
   else {
      ConsoleIO.out.println("\t" + "Books available for loan");
      while(iter.hasNext() == true) {
          Book book = (Book) iter.next();
          if(book.getBorrower() == null) {
             found = true;
             book.display();
          }
      }
   }.
   //
   if(found == false)
      ConsoleIO.out.println("\n\t\t" + "None");
}   // method: displayBooks
```

```
// ----- Attributes ----------
private String    theName;

// ----- Relations ----------
private java.util.TreeSet    theLoanStock;
private java.util.HashSet    theBorrowers;

}   // class: Library
```

In the returnOneBook method and the displayBooks method there is a need to request a reference to the BorrowerRecord attached to a Book. For example, in the returnOneBook method we have:

```
BorrowerRecord    record = book.getBorrower();
```

Clearly we must support getBorrower in the Book class as:

```
// class Book
public BorrowerRecord    getBorrower() {
   return theBorrower;
} // method: getBorrower
```

Notice that to iterate over a collection we have code with the following form:

```
Iterator iter = theCollectionIdentifier.iterator();
while(iter.hasNext() == true) {
   ClassName objectIdentifier = (ClassName) iter.next();
   // Send messages through the objectIdentifier
}
```

This is important as it brings consistency of approach to our code. Essentially, we solve the same problem in the same way each time. At this stage all of the Java classes developed should compile with no errors or warnings reported.

4.2.4.3   Testing

As discussed in section 3.3 we use an Application object as the effective point of entry into the program. It responds to a message run that originates from the execution environment. The run method creates the major objects, establishes the object architecture and has the control logic necessary to stimulate the model.

Recall that we identified a test-case for each use-case in an earlier activity. Most of the code for our model is in place and now is the time to execute each test-case in the Application run method. As this is the first iteration the tests are as straightforward as possible. Therefore we create a Library object and send the minimum messages to execute each test-case with no human user involvement. We do not send messages that could give rise to an error.

It seems reasonable to start with test-case 3 in combination with test-case 2. This means that we can demonstrate that we can display the Books available for loan when there are zero or more Books in the Library. This leads to the following code for the Application class that represents the first increment:

```
public class Application {

    // ----- Operations ----------
    public Application () {
        // No initialization required
    }   // method: Application

    public void run() {
        //
        // Create the Library
        Library library = new Library("Napier");
        //
        // Test-case 3: Display books available for loan
        library.displayBooksAvailableForLoan();
        //
        // Test-case 2: Add one Book
        Book b1 = new Book("Java", "Ken", 1);
        library.addOneBook(b1);
        //
        // Test-case 3: Display books available for loan
        library.displayBooksAvailableForLoan();
    } // method: run

}   // class: Application
```

Now we can prepare a Main class:

```
public class Main {

    // ----- Operations ----------
    public static void main(String [] args) {
        Application app = new Application();
        app.run();
    }   // method: main
}   // class: Main
```

then compile all of the Java files (Book.java, BorrowerRecord.java, Library.java, Application.java and Main.java). The execution of Main gives a typical output produced as follows:

```
Library: Napier
      Books available for loan
            None
```

Library: Napier

    Books available for loan

        Title: Java
        Author: Ken
        Catalogue Number: 1

This is our first increment and it is successful. Had it been unsuccessful it would have been necessary to investigate the detailed coding of the methods for the appropriate operations.

If we now consider test-case 1 then we encounter a major problem. It is that nowhere in the system specification are we required to display details of the borrowers registered with the Library. Therefore we introduce a displayBorrowers operation to the Library coded as follows:

```
// class Library
public void   displayBorrowers() {
    ConsoleIO.out.println("\n" + "Library: " + theName + "\n");
    ConsoleIO.out.println("\t" + "Details of the borrowers");
    //
    if(theBorrowers.isEmpty() == false) {
        Iterator    iter = theBorrowers.iterator();
        while(iter.hasNext() ) {
            BorrowerRecord    record = (BorrowerRecord) iter.next();
            String name = record.getName();
            ConsoleIO.out.println("\n\t\t" + name);
        }
    }
    else
        ConsoleIO.out.println("\n\t\t" + "None");
} // method: displayBorrowers
```

Notice that we intend discussing the lack of a display of borrower details with the librarian and anticipate that it may well become a use-case in its own right in a later iteration. However, for the moment, it is sufficient to use the name of a borrower for the display.

Now we can modify the Application run method to:

```
// class Application
public void run() {
    //
    // As for increment 1
    // ...
    // Test-case 1: Register one borrower
    library.displayBorrowers();
    library.registerOneBorrower("Peter");
    library.displayBorrowers();
}   // method: run
```

and compile and execute as before. This is our second increment and it produces the successful outcome shown as follows:

```
// As for increment 1
// ...
Library: Napier
          Details of the borrowers
                    None

Library: Napier
          Details of the borrowers
                    Peter
```

Having added one more Book, test-case 4 is now combined with test-case 5 for our third increment. We check that a Book that was available for loan is no longer available and is now on loan. Therefore we modify the Application run methods as:

```
// class Application
public void run() {
    //
    // As for increment 2
    // ...
    Book b2 = new Book("Basic", "Sally", 2);
    library.addOneBook(b2);
    //
    // Test-case 4: Display books on Loan
    library.displayBooksOnLoan();
    library.displayBooksAvailableForLoan();
    //
    // Test-case 5: Lend a book
    library.lendOneBook(2, "Peter");
    //
    library.displayBooksOnLoan();
    library.displayBooksAvailableForLoan();
}    // method: run
```

and then execute it to produce the following output:

```
// As for the increment 2
// ...
Library: Napier
          Books out on loan
                    None

Library: Napier
          Books available for loan
```

        Title: Java
        Author: Ken
        Catalogue Number: 1

        Title: Basic
        Author: Sally
        Catalogue Number: 2

Library: Napier

    Books out on loan
        Title: Basic
        Author: Sally
        Catalogue Number: 2

Library: Napier

    Books available for loan
        Title: Java
        Author: Ken
        Catalogue Number: 1

As it is successful we only have test-case 6 to implement and we are finished with this iteration. We combine it with test-cases 4 and 3 for the fourth increment. It checks that a **Book** on loan is returned and is then available for further loan. The modified **Application run** method is:

```
// class Application
public void run() {
    //
    // As for increment 3
    // ...
    // Test-case 6: Return a book
    library.returnOneBook(2);
    //
    library.displayBooksOnLoan();
    library.displayBooksAvailableForLoan();
}   // method: run
```

and the output when it is compiled and executed is as follows:

```
// ... As for the increment 3
// ...
Library:  Napier

        Books out on loan
            None

    Library:  Napier
```

Books available for loan
    Title: Java
    Author: Ken
    Catalogue Number: 1

    Title: Basic
    Author: Sally
    Catalogue Number: 2

The reader is invited to review McGregor 2001 for a more detailed account of testing OO systems.

4.2.5   Reconcile model diagrams

Although all of the use-cases have been successfully implemented it does not mark the end of the first iteration. We must ensure that our diagrams are brought up to date with decisions made during each activity. No changes to the use-case diagram of figure 4.1 are required as it covered all of the requirements for this iteration. However, we should make a mental note that we have introduced the operation **displayBorrowers** to the **Library** and that it might become a use-case in the next iteration. On reflection, the only diagrams that would benefit from an update are the sequence diagram of figure 4.5 and the class diagram of figure 4.9. The others are accurate enough for our purposes.

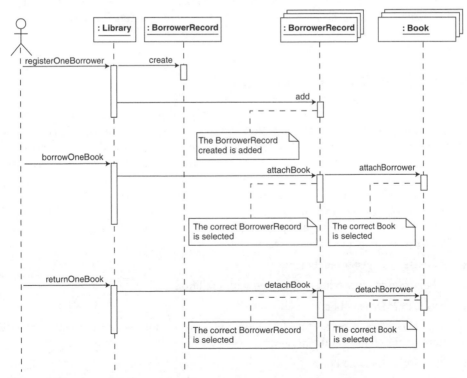

Figure 4.10   *An updated sequence diagram*

The sequence diagram of figure 4.5 now introduces a **BorrowerRecord** object when a new borrower is registered. Its purpose is to illustrate the attach and detach operations on borrowing and returning a **Book** as shown in figure 4.10. Observe how we have shown the implementation multi-objects that represent the collections of **BorrowerRecords** and **Books** held by the **Library**. Normally, we would only show the model objects such as **Book** and **BorrowerRecord** objects held by the **Library**. Here, we wish to explain and emphasize the underlying details of each method.

If we turn our attention the class diagram of figure 4.9 then all we need do is to update the **Library**, **BorrowerRecord** and **Book** classes to include the operations we introduced. Figure 4.11 illustrates.

This marks the end of the first iteration. Full listings of the resulting code are given with the accompanying software (model **Lib4_1.uml**). See also appendices A and B.

4.3   Iteration 2

After demonstrating the execution of iteration 1 and some discussion with the librarian it turns out that a display of the status of all borrowers is required. Essentially she wants a display of the borrowers to be ordered by each borrower's name. Accompanying each should be the catalogue numbers of each book out on loan in catalogue number order. We also learn that a text-based user interface controlled by a simple menu is acceptable at this stage. However, the librarian must have a system that can accommodate erroneous data input. Ideally it should give helpful error messages.

4.3.1   Establish use-cases

We do not have much work to do here as there is only one new use-case to deal with and we have anticipated it in the previous iteration. It is the use-case "Display borrowers". However, we must have a test-case to match it. Iteration 1 resulted in six test-cases therefore we might be tempted to make this one test 7. This is not advisable as it is important to group use-cases and test-cases sensibly otherwise our work can easily become difficult to understand. Therefore we have "Display borrowers" as use-case 2 and test-case 2. Although we are forced to renumber the use and test-cases it is worth the effort. Figure 4.12 shows the updated use-case diagram for iteration 2.

The corresponding test-cases are now:

- test-case 1: Register one borrower

  A borrower is successfully registered with the library.

- test-case 2: Display borrowers

  A display of each borrower when there are zero or more borrowers with zero or more books on loan in the library.

- test-case 3: Add one book

  A book is successfully added to the library and is available for loan.

Library

—————— Operations ——————

+void registerOneBorrower(String aBorrowerName)

+void addOneBook(Book aBook)

+void displayBooksAvailableForLoan()

+void displayBooksOnLoan()

+void lendOneBook(int aCatalogueNumber, String aBorrowerName)

+void returnOneBook(int aCatalogueNumber)

+void displayBorrowers()

-void displayBooks(boolean onLoan)

-BorrowerRecord getBorrowerRecord(String aBorrowerName)

-Book getBook(int aCatalogueNumber)

+Library(String aName)

Book

—————— Operations ——————

-int compareTo(Object obj)

+boolean equals(Object obj)

+int hashCode()

+int getCatalogueNumber()

+BorrowerRecord getBorrower()

+void attachBorrower(BorrowerRecord aBorrower)

+void detachBorrower()

+void display()

+Book(String aTitle, String anAuthor, int aCatalogueNumber)

+Book()

BorrowerRecord

—————— Operations ——————

+int compareTo(Object obj)

+boolean equals(Object obj)

+int hashCode()

+String getName()

+void attachBook(Book aBook)

+void detachBook(Book aBook)

+BorrowerRecord(String aName)

Figure 4.11   *Updated classes*

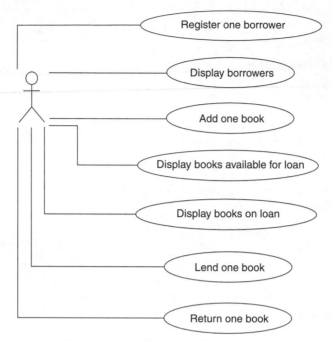

Figure 4.12   *A use-case diagram for the Library application: iteration 2*

- test-case 4: Display books available for loan

  A display of each book available for loan when there are zero or more books in the library.

- test-case 5: Display books out on loan

  A display of each book out on loan when there are zero or more books in the library.

- test-case 6: Lend one book

  A book available for loan is recorded as out on loan to a borrower. The book is not available for loan.

- test-case 7: Return one book

  A book on loan to a borrower is dissociated from that borrower. The book is now available for loan.

The requirements that the system must "accommodate erroneous data input" and "give as helpful error messages as possible" are typical. However, they are difficult to deal with as they are non-specific and might just be "wishful thinking" on the part of the librarian. However, after some discussion with her we arrive at what is meant by them. Essentially the system must inform the human user of a problem that arises by displaying a short textual message on the screen. Specifically the system should take no other action. The human user is expected to try again having understood why the problem has arisen.

Again after discussion with the librarian we arrive at the following test-cases (but no new use-cases) that reflect commonly occurring problems:

- test-case 8: Attempt to add a book with a catalogue number that already exists

  Display a message indicating that the book with the given catalogue number is already in the library and that it cannot be added.

- test-case 9: Attempt to register a borrower with a name that already exists

  Display a message indicating that the borrower with the given name is already registered in the library and it cannot be registered.

- test-case 10: Attempt to borrow a book with a catalogue number that does not exist

  Display a message indicating that the book with the given catalogue number is not in the library and that it cannot be borrowed.

- test-case 11: Attempt to borrow a book for a borrower that is not registered

  Display a message indicating that the borrower with the given name is not registered in the library and that it cannot be borrowed.

- test-case 12: Attempt to borrow a book that is already out on loan

  Display a message indicating that the book with the given catalogue number is out on loan and that it cannot be borrowed.

- test-case 13: Attempt to return a book with a catalogue number that does not exist

  Display a message indicating that the book with the given catalogue number is not in the library and that it cannot be returned.

- test-case 14: Attempt to return a book that is not out on loan

  Display a message indicating that the book with the given catalogue number is not out on loan and that it cannot be returned.

Clearly many more such test-cases could be developed but these are sufficient for this iteration. Others can be added in subsequent iterations as necessary.

4.3.2 Analysis

As we have added a relatively straightforward use-case that requires no new objects or messages then there is no extra work to do in this activity. This is a good sign as it means that our architecture is stable. In general for a well-designed system we tend to have less and less work to get more and more functionality.

However, the run method of the Application class needs more thought as its effects are controlled by using a simple text-based menu. Essentially the Application presents a menu of options to the user and a suitable selection is made. On the basis of this selection one or more messages are sent to the Library.

As the run method of the Application can become excessively large it is wise to separate the action of obtaining a selection from the user and that of stimulating the Library.

Therefore we introduce a **private** method getSelection for use by the run method of the Application. It displays a menu and returns the choice made.

It is important to understand that the Application class is not part of the model. Its only purpose is to create and then control the model by sending messages to it. This point is developed later when we replace the text-based user interface with a graphical user interface (see chapter 7).

We must be sure that this iteration can do at least as much as the previous one and that each use-case can be successfully executed. The difficulty is that the input is from a human user and it is therefore outside our control. One solution is to introduce a **private** method testUseCases similar to the run method of the previous iteration. It executes the "hard-wired" versions of test-cases 1 to 7. As no errors are expected to arise from these test-cases we extend the approach to introduce a **private** method testErrorConditions. It executes "hard-wired" versions of test-cases 8 to 14. If both

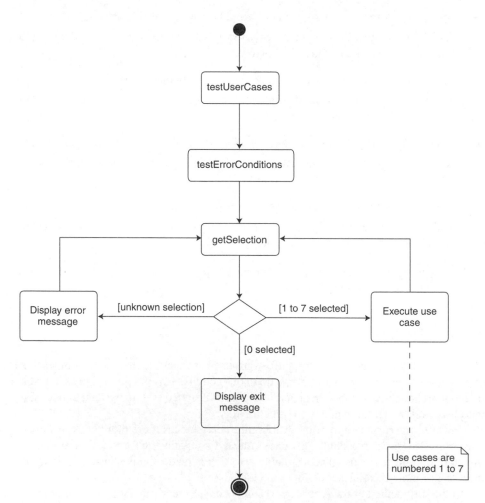

Figure 4.13   *An activity diagram for the run method of the Application*

private methods execute successfully then we can be reasonably confident that the human user should not have any unforeseen difficulties when using the menu. These two **private** operations can be removed or reserved for use by the system developers when the software is delivered to the librarian.

Finally, if we consider the run method of the Application then the design of the control logic becomes important. Essentially we need to ensure that the correct action follows from the choice made by the user. Clearly a selection statement is at the heart of this method. Rather than get too detailed at this stage the activity diagram shown in figure 4.13 helps document this decision.

4.3.3   Design

All we have to do in this activity is to update the class diagram from iteration 1. As we require an alphabetical list of borrower names in the operation displayBorrowers, one solution is to change the HashSet used to hold them in iteration 1 to be a TreeSet.

We must also add the **private** methods getSelection, testUseCases and testErrorConditions to the Application class. Figure 4.14 illustrates.

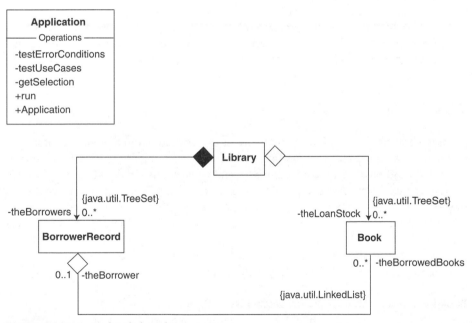

Figure 4.14   *An updated class diagram*

4.3.4   Implementation

As with iteration 1 we are confident that we can code each class in a single increment. Therefore there is no need to consider a more cautious approach. With more complex examples this would not be the case.

4.3.4.1   Architectural code

It should come as no surprise that there is very little to do in this activity. All that
is necessary is to declare **theBorrowers** as a reference to a TreeSet in the Library class
rather than a HashSet. Its constructors should initialize it to refer to a TreeSet. This
reflects the new requirement that the Library displays each borrower in name order.
An outline of the Java code using italicized characters to indicate the two changes
required is:

```
// ...
public class Library {

    public Library(String aName) {
        // ...
    theBorrowers = new TreeSet();
    }   // method: Library

    // ...
    private    TreeSet theBorrowers;
}   // class: Library
```

An Iterator is used to visit each element in a collection therefore no other changes to
our code are necessary. Whichever collection we choose delivers a suitable Iterator. This
is a significant benefit as it makes such code virtually maintenance-free.

4.3.4.2   Method code

For this iteration we intend revisiting many of the methods from the previous iteration
with a view to detecting and reporting error conditions. This is significant amount of
work that must be carried out methodically and consistently. To do otherwise is to run
the risk of incomplete updates and confusing code. Our aim is to increase the quality of
the code from the previous iteration. Therefore we need to develop a strategy.

For example, we must be able to determine if a Book is in the Library or if a borrower
is registered. We can decide to do this with the **private** Library operations getBook and
getBorrower. Recalling that they return **null** if no Book or BorrowerRecord is found,
typical outline examples from Library methods are:

```
Book foundBook = this.getBook(aCatalogueNumber);
//
if(foundBook != null) {
    // Messages that assume that the Book with aCatalogueNumber is in the Library
    // ...
} else {
    // Messages that assume that the Book with aCatalogueNumber is not in the Library
    // ...
}
```

and

```
BorrowerRecord foundRecord = this.getBorrowerRecord(aName);
//
if(foundRecord != null) {
    // Messages that assume that the borrower with aName is registered with the Library
    // ...
} else {
    // Messages that assume that the borrower with aName is not registered with the Library
    // ...
}
```

This approach has the merit of using just one operation (getBook or getBorrowerRecord) to make a decision. Therefore we need only change a single method if the code requires any changes. The impact of this approach on maintenance can be significant.

When an Object is added to a collection a **boolean** value of **true** is returned if it is successful and **false** otherwise (see appendix E). We can make use of this return value to detect the fact that something unforeseen has gone seriously wrong. For example, we may have inadvertently tried to add a duplicate element to a HashSet. Therefore the following code is typical.

```
boolean result = theLoanStock.add(aBook);
if(result == false) {
    // Report an error
    // ...
}
```

To bring consistency to our code we use the same structure of Java code for similar problems. For example, in the Library class when we detect and report errors we have:

```
// class Library
public void   addOneBook(Book aBook) {
    int aCatalogueNumber = aBook.getCatalogueNumber();
    Book foundBook = this.getBook(aCatalogueNumber);
    //
    // Check that the Book is not the Library
    if(foundBook == null) {
        boolean result = theLoanStock.add(aBook);
        // Check that the Book has been added to the collection
        if(result == false) {
            ConsoleIO.out.println("\n\t" + "Book not added - unknown error");
        }
    } else {
        ConsoleIO.out.println("\n\t" + "Book already present - cannot add");
    }
} // method: addOneBook
```

and

```java
public void    registerOneBorrower(String aBorrowerName) {
   BorrowerRecord foundRecord = this.getBorrowerRecord(aBorrowerName);
   //
   // Check that the Borrower is not already registered
   if(foundRecord == null) {
      BorrowerRecord borrowerRecord = new BorrowerRecord(aBorrowerName);
      boolean result = theBorrowers.add(borrowerRecord);
      // Check that the Borrower has been added to the collection
      if(result == false) {
         ConsoleIO.out.println("\n\t" + "Borrower not registered - unknown error");
      }
   } else {
      ConsoleIO.out.println("\n\t" + "Borrower already present - cannot register");
   }
} // method: registerOneBorrower
```

and

```java
public void    lendOneBook(int aCatalogueNumber, String aBorrowerName) {
   Book foundBook = this.getBook(aCatalogueNumber);
   //
   // Check that the Book is in the Library
   if(foundBook != null) {
      // Check that the Book is borrowable
      if(foundBook.getBorrower() == null) {
         // Check that the borrower is registered
         BorrowerRecord foundRecord = this.getBorrowerRecord(aBorrowerName);
         if(foundRecord != null) {
            // Always attach the Book to the BorrowerRecord
            foundRecord.attachBook(foundBook);
         } else {
            ConsoleIO.out.println("\n\t" + "Borrower not registered - cannot lend");
         }
      } else {
         ConsoleIO.out.println("\n\t" + "Book out on loan - cannot lend");
      }
   } else {
      ConsoleIO.out.println("\n\t" + "Book not present - cannot lend");
   }
} // method:  lendOneBook
```

Notice that the various tests map directly to the test-case for the use-case that each operation represents. Even the format of the output is the same to give the same "look and feel" to each method.

All of these approaches are consistently used throughout this activity. Full listings of the code for the Library class are given with the accompanying software (model Lib4_2.uml).

4.3.4.3   Testing

Finally we should consider the Application class. First of all we need a method for getSelection. It is quite straightforward (if a little tedious):

```
// class Application
private String   getSelection() {
   // Display menu to the human user
   ConsoleIO.out.println();
   ConsoleIO.out.println( "0: Quit");
   ConsoleIO.out.println( "1: Register one borrower");
   ConsoleIO.out.println( "2: Display borrowers");
   ConsoleIO.out.println( "3: Add one book");
   ConsoleIO.out.println( "4: Display books available for loan");
   ConsoleIO.out.println( "5: Display books on loan");
   ConsoleIO.out.println( "6: Lend one book");
   ConsoleIO.out.println( "7: Return one book");
   //
   // Get and return the choice made
   ConsoleIO.out.print( "\n\t" + "Enter the choice >>> " );
   return ConsoleIO.in.readString();
} // method: getSelection
```

Notice that a Quit option is offered and that the others map directly to the use-cases identified. Also the selection is a String rather than an **int** to permit any character (not just a digit) to be input. This makes our software less prone to run-time failure.

Given the work done in the previous iteration, the coding for testUseCases and testErrorConditions is quite straightforward:

```
// class Application
private void   testUseCases() {
   // Create the Library
   Library library = new Library("Napier");
   // Test-case 4: Display books available for loan
   library.displayBooksAvailableForLoan();
   // Test-case 3: Add one Book
   Book b1 = new Book("Java", "Ken", 1);
   library.addOneBook( b1 );
   // Test-case 4: Display books available for loan
   library.displayBooksAvailableForLoan();
   // Test-case 2: Display borrowers
   library.displayBorrowers();
   // Test-case 1: Register one borrower
   library.registerOneBorrower("Peter");
   // Test-case 2: Display borrowers
   library.displayBorrowers();
```

```
      //
      Book b2 = new Book("Basic", "Sally", 2);
      library.addOneBook(b2);
      // Test-case 5: Display books on Loan
      library.displayBooksOnLoan();
      library.displayBooksAvailableForLoan()
      // Test-case 6: Lend a book
      library.lendOneBook(2, "Peter");
      //
      library.displayBooksOnLoan();
      library.displayBooksAvailableForLoan();
      // Test-case 7: Return a book
      library.returnOneBook(2);
      //
       library.displayBooksOnLoan();
      library.displayBooksAvailableForLoan();
  } // method: testUseCases

  private void    testErrorConditions() {
      // Create the library.
      Library library = new Library("Napier");
      // Add 3 books to the library.
      library.addOneBook(new Book("C++", "Jim", 1));
      library.addOneBook(new Book("Java", "Peter", 2));
      library.addOneBook(new Book("Basic", "Jane", 3));
      // Register 2 borrowers.
      library.registerOneBorrower("John");
      library.registerOneBorrower("Ken");
      // Inspect the results.
      library.displayBooksAvailableForLoan();
      // Test-case 8: Attempt to add a book with same catalogue number
      library.addOneBook(new Book("XML", "Mike", 2));
      // Test-case 9: Attempt to register borrower with same name
      library.registerOneBorrower("Ken");
      // Test-case 10: Attempt to borrow non-existent book.
      library.lendOneBook(10, "John");
      // Test-case 11: Attempt to borrow a book for non-existent borrower
      library.lendOneBook(1, "James");
      // Test-case 12: Attempt to borrow book already on loan.
      library.lendOneBook(1, "John");
      library.lendOneBook(1, "Ken");
      // Test-case 13: Attempt to return non-existent book
      library.returnOneBook(10);
      // Test-case 14: Attempt to return a book not on loan.
      library.returnOneBook(2);
  } // method: testErrorConditions
```

Although the run method is relatively complex we have the activity and sequence diagram of figure 4.13 to guide us. Great care has been taken to ensure that the formatting and general approach taken are as consistent as possible. This minimizes the risk of making mistakes.

```java
// class Application
public void run() {
  this.testUseCases();
  //
  this.testErrorConditions();
  //
  // Create and initialize a Library
  Library library = new Library("Books-R-Us");
  //
  String choice = "";
  do {
    // Get the human user's choice.
    choice = this.getSelection();
    ConsoleIO.out.println();
    //
    // Action the human user's choice.
    if(choice.equals("0") ) {
      ConsoleIO.out.println("\n\t\t" + "SYSTEM CLOSING" + "\n" ) ;
      ConsoleIO.out.println();
    }
    else if(choice.equals("1") ) {
      // Get the borrower details from the human user
      ConsoleIO.out.print( "\t" + "Enter the borrower name  >>>  " );
      String borrowerName = ConsoleIO.in.readString();
      // Register the borrower with the Library
      library.registerOneBorrower( borrowerName );
    }
    else if(choice.equals("2") ) {
      library.displayBorrowers();
    }
    else if(choice.equals("3") ) {
      // Get the Book details from the human user
      ConsoleIO.out.print("\t" + "Enter the title >>> " );
      String title = ConsoleIO.in.readString();
      //
      ConsoleIO.out.print("\t" +"Enter the author >>> " );
      String author = ConsoleIO.in.readString();
      //
      ConsoleIO.out.print("\t" + "Enter the catalogue number >>> " );
      int catalogueNumber = ConsoleIO.in.readInt();
      // Add the Book to the Library
```

```
        library.addOneBook( new Book(title, author, catalogueNumber) );
      }
      else if(choice.equals("4")) {
        library.displayBooksAvailableForLoan();
      }
      else if(choice.equals("5")) {
        library.displayBooksOnLoan();
      }
      else if(choice.equals("6")) {
        // Get the Book details from the human user
        ConsoleIO.out.print( "\t" + "Enter the catalogue number >>> " );
        int catalogueNumber = ConsoleIO.in.readInt();
        //
        // Get the borrower details from the human use
        ConsoleIO.out.print( "\t" + "Enter the borrower name >>> " );
        String borrowerName = ConsoleIO.in.readString();
        // Loan the Book to the borrower
        library.lendOneBook( catalogueNumber, borrowerName );
      }
      else if(choice.equals("7")) {
        // Get the Book details from the human user
        ConsoleIO.out.print("\t" + "Enter the catalogue number >>> " );
        int catalogueNumber = ConsoleIO.in.readInt();
        // Return it to the Library
        library.returnOneBook( catalogueNumber );
      }
      else {
        ConsoleIO.out.println("\n\t\t" + "Unknown selection - try again" + "\n");
      }
      //
    } while(choice.equals("0") == false);
} // method: run
```

On successful compilation the run method of the Application is executed with the output from testUseCases and testErrorConditions as expected. Samples from each are:

```
// testUseCases
Library: Napier
      Details of the borrowers
          None
Library: Napier
      Details of the borrowers
          Peter
// ...

// testErrorConditions
```

Book already present – cannot add
Borrower already present – cannot register
Book not present – cannot lend
// ...

The human user can make use of the menu as normal (user input is shown as bold type):

0: Quit
1: Register one borrower
2: Display borrowers
3: Add one book
4: Display books available for loan
5: Display books on loan
6: Lend one book
7: Return one book

   Enter the choice >>> **2**

Library: Books-R-Us

     Details of the borrowers
      None

0: Quit
1: Register one borrower
2: Display borrowers
3: Add one book
4: Display books available for loan
5: Display books on loan
6: Lend one book
7: Return one book
      Enter the choice >>> **0**
        SYSTEM CLOSING

After a thorough use of the menu without any problems we consider this iteration to be completed. As the **private** operations testUseCases and testErrorConditions have served their purpose we comment them out of the Application run method and deliver this iteration to the librarian for evaluation. The accompanying software has full listings of the resulting code (model Lib4_2.uml).

4.3.5   Reconcile model diagrams

As expected there is nothing to do in this activity as we have not changed the model or the Application class that stimulates it. However, it is important to understand that we have made a positive decision that this is the case. In no sense is it a waste of time as it explicitly documents the fact that no changes to the model diagrams are required. The alternative is to run the risk of analysis and design decisions that do not map to Java code.

4.4   Iteration 3

Having demonstrated iteration 2 to the librarian it turns out that she wants to be able to initialize the system from a text file. We anticipate that this new requirement is not especially difficult to implement. However, she indicated that at some point soon there should be a version of the software with a graphical user interface (GUI). There may even be a requirement for one that operates over the World Wide Web. Worryingly any updates should apply to the text-based, GUI and World Wide Web versions. Finally she asked for more feedback from the system when changes are made to the information held by the system.

4.4.1   Establish use-cases

If we start with the requirement for initialization from a file then we can introduce a use-case "Load books from file" as shown in figure 4.15.

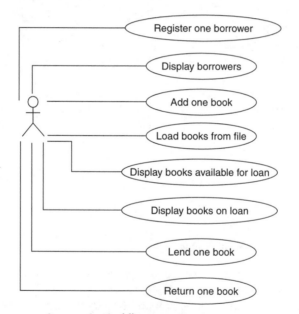

Figure 4.15   *A use-case diagram for the* Library *application: iteration 3*

There should be a corresponding test-case such as:

- test-case 4: Load books from file

  Using initialization data from a named file, zero or more books are added to the library and are available for loan.

Notice that although the use-cases from iteration 2 do not change, many of them must be renumbered. Although this is unfortunate, it is unavoidable as we insist that use-cases and the test-cases that follow are sensibly grouped together. The reason is that we strive to help others who will maintain our designs and code.

There are no other new use-cases that follow from the new requirements as they involve the internal workings of the system.

4.4.2 Analysis

The analysis of the new use-case "Load Books from file" is relatively straightforward. All we need do is to support an operation loadBooksFromFile that initializes each new Book with data input from a text file before adding it to the Library. If for any reason the file cannot be opened then an error should be reported to the user. The activity diagram and pseudocode of figure 4.16 help illustrate.

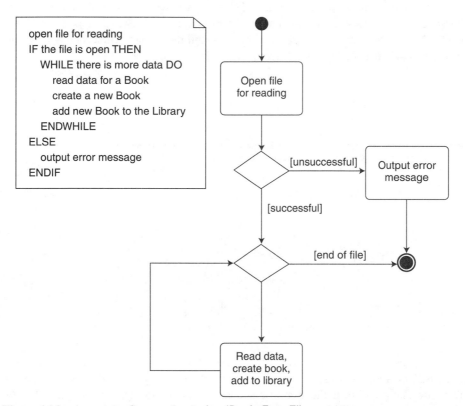

```
open file for reading
IF the file is open THEN
    WHILE there is more data DO
        read data for a Book
        create a new Book
        add new Book to the Library
    ENDWHILE
ELSE
    output error message
ENDIF
```

Figure 4.16 *An activity diagram for the loadBooksFromFile operation*

It is more difficult to make an analysis of the provision for different kinds of user interfaces. If we consider the model developed in the previous iteration then a design flaw becomes apparent. It is that the Application, Library, BorrowerRecord and Book classes all have a significant responsibility for input and/or output of data. Essentially they are all concerned with some element of the user interface. Quite simply we cannot contemplate having this as a feature of our new model if we are to achieve the flexibility to have different user interfaces. The problem is that a change of user interface would necessitate significant changes to these classes.

Our intention is that those classes from the model, i.e. Library, BorrowerRecord and Book, should have no responsibility for input or output. Collectively we refer to them as the *domain model*. In the text-based versions developed so far the Application run method allows the human user to interact with the domain model. Therefore we can ensure that it has some responsibility for input and output.

Further consideration of the model from iteration 2 leads us to conclude that over time the Application run method will become ever more complex. It seems sensible that while undertaking this major revision we should try to simplify it as much as possible and avoid its complexity overwhelming us.

A useful approach is to have an object whose responsibility is to interact with the domain model. This means that the Application run method has no direct involvement with the domain model but uses this object instead. For this reason we introduce an Action class an instance of which presents a straightforward set of operations corresponding to each use-case. Much of the complexity associated with domain model interactions are hidden in its methods. The sequence diagram of figure 4.17 illustrates the approach.

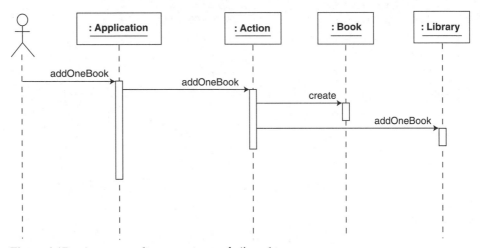

Figure 4.17   *A sequence diagram using an* Action *object*

In response to a menu presented by the Application, the user makes a request to add a Book. This request is passed on to an Action object that gets the details of the Book from the user (input from the user interface) and then creates a suitably initialized Book object. It is this Book object that is passed to the Library (part of the domain model) to be added.

Notice that we have separated two concerns. The first is what the system needs to do and the second is how it should do it. The Application class is largely concerned with the first and the Action class the second. This is a very important point and is the hallmark of a good design.

Finally we should ensure that whatever design we eventually develop for the user interface it must be able to interrogate the domain model for the outcome of any state changes. For example, when a Book is returned to the Library then the Action object should report completion of this task by displaying a message that originates from the

Library. This should enable us to give feedback to the human user as recorded in the requirements of this iteration.

4.4.3 Design

From the analysis it is clear that the domain model classes the Library, BorrowerRecord and Book must all be encapsulated by the Action class. This guarantees that the domain model can only be accessed by it. Therefore we establish a composite aggregation relationship between the Action and Library classes as shown in figure 4.18.

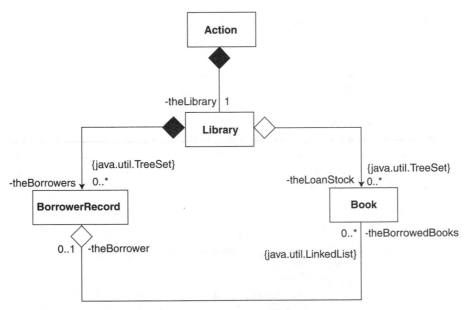

Figure 4.18 *A modified class diagram*

Notice that the architecture of the domain model is unchanged from the previous iteration. All we have done is to introduce the Action class as a "front-end" (façade) to it.

In the previous discussions we implied that the domain model and its façade can be treated as a single entity. Taken together they are a subsystem of the overall system with the Application as the other. We can make this clear by using a UML package symbol as shown in figure 4.19.

The UML package symbol is similar to the class symbol except that it has a tab and it contains classes. In this case it is named librarysubsystem and contains the Library, BorrowerRecord, Book and Action classes.

The + symbol associated with the Action class indicates that it is available to the package's clients, i.e. it is public. In contrast the % symbol associated with the other classes indicates that they are not available to clients but have visibility of each other. This is normally referred to as *package access* and is discussed in more detail in the

The Application creates a single Action object.

The Action object creates a single Library and Book objects as required.

The Library creates BorrowerRecord objects as required.

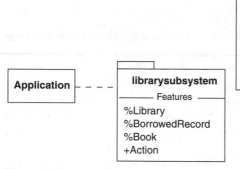

Application	- - - -	**librarysubsystem**

——— Features ———

%Library
%BorrowedRecord
%Book
+Action

Figure 4.19 *A modified class diagram using a package*

section 4.4.4.1. For the moment it is enough to realize that our use of public and package access modifiers reflects our decision that the Application should only access the domain model through the Action object.

Ignoring the removal of input/output responsibilities from the domain model we do not anticipate any major changes are necessary. However, we must consider the Action class in more detail. As indicated earlier, it should have a method for each use-case and any others that the Application may require when interacting with the domain model. This leads us to the class diagram of figure 4.20.

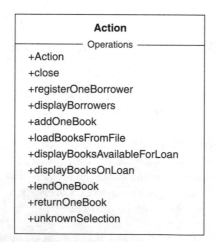

Action

——— Operations ———

+Action

+close

+registerOneBorrower

+displayBorrowers

+addOneBook

+loadBooksFromFile

+displayBooksAvailableForLoan

+displayBooksOnLoan

+lendOneBook

+returnOneBook

+unknownSelection

Figure 4.20 *The Action class in more detail*

Notice that we intend that the operation loadBooksFromFile in the Action class makes use of addOneBook in the Library. There is no loadBooksFromFile operation in the Library. This is a good example of the Action class helping us keep the underlying domain model as simple as possible.

4.4.4 Implementation

Having established the design for iteration 3, it is now time to implement it. Before doing so it is wise to decide on a general approach to the problems we will encounter. Essentially as in section 4.3.4.2, we intend solving the same kind of problems in the same way even though several alternatives exist. For example, the domain model classes have no input/output responsibilities but we must be able to get textual information about them. We adopt one way of doing this.

Before adopting a strategy to deal with this problem we should discuss the operation toString. There are two important points that we should understand about it. The first is that it is defined in the class Object to return basic information about the recipient as a String. However, it can be redefined in any other Java class to give more detailed information such as the attribute values of the recipient. The second point is that the String concatenation operator + can take an Object as its second parameter. All it does is to send the message toString to it resulting in the second String parameter it actually requires. If toString has been redefined in that object's class then that method will execute. If not then the method defined in Object will execute. This is an example of *polymorphic substitution* discussed earlier in section 1.3.2 and later in the next chapter.

Bearing this in mind our strategy is:

- For each class in the domain model replace the display operations with toString. The information in the String it returns should relate to its normal attributes and if appropriate general information about its associates or aggregate components, e.g. the number of Books in the Library but not their details.
- If there are any aggregate components or associates then an Iterator to the collection that holds their references should be returned by a suitable method. Using this Iterator a client can visit each element and send it the message toString (implicitly or explicitly) as required. In addition the Iterator should only permit read access to the collection so as to limit the danger of unwarranted changes to it. Although this is not "bulletproof" it is safer than making the collection available to a client for reading and writing.

We must also ensure that the user is given more feedback. One way to achieve this is for the Library to have a String attribute theMessage and an accessor operation getStatus to return its value to a client (in this iteration it is the Action object). Our intention is that any operation that changes the state of the Library should update theMessage with a suitable value. The Library can then be interrogated as required.

4.4.4.1 Architectural code

The introduction of the librarysubsystem package causes no major implementation difficulties as the package construct is supported by Java. All that is necessary is to name the package that a class belongs to as in:

```
package librarysubsystem;
```

By convention the package name has only lower-case letters. It must be the first state-ment preceding a class declaration. For example, we have:

```
package librarysubsystem;
// ...
public class Action {
    // ...
} // class Action
```

and

```
package librarysubsystem;
// ...
class Library {
    // ...
} // class Library
```

Notice that the default access modifier for a class gives it package access. Therefore the Library, BorrowerRecord and Book classes have no explicit access modifier. As with any other Java package it must be imported by another class. No other architectural changes are required.

4.4.4.2   Method code

If we start with the changes necessary to the domain model classes then in keeping with the approach discussed previously we replace each display method with toString. As the Book class has no aggregate components it is quite straightforward.

```
// class Book
public String    toString() {
    return theCatalogueNumber + ": " + theTitle + " by " + theAuthor;
} // method: toString
```

In the BorrowerRecord class its toString method is:

```
// class BorrowerRecord
public String    toString() {
    int borrowedBooksCount = theBorrowedBooks.size();
    //
    return theName + " :" + borrowedBooksCount + " book(s)";
} // method: toString
```

Notice how the number of Books it is associated with is reported by using the size method of collection that holds references to them. The Library method is treated sim-ilarly as in:

```
// class Library
public String    toString() {
    int bookCount = theLoanStock.size();
```

```
   int borrowerCount = theBorrowers.size();
   //
   return theName + ": " + bookCount + " book(s) :" + borrowerCount + " borrower(s)";
} // method: toString
```

The BorrowerRecord and Library classes should also return an Iterator to their aggregate components. They do this by converting the appropriate collections to be read only with the **static** method unmodifiableCollection from the Collections class then return an Iterator to it. The methods are as follows:

```
// class BorrowerRecord
public Iterator   getBorrowedBooksIterator() {
   Collection readOnlyCollection = Collections.unmodifiableCollection(theBorrowedBooks);
   return readOnlyCollection.iterator();
} // method: getBorrowedBooksIterator
```

```
// class Library
public Iterator   getBorrowersIterator() {
   Collection readOnlyCollection = Collections.unmodifiableCollection(theBorrowers);
   return readOnlyCollection.iterator();
} // method: getBorrowersIterator
```

```
// class Library
public Iterator   getBooksIterator() {
   Collection readOnlyCollection = Collections.unmodifiableCollection(theLoanStock);
   return readOnlyCollection.iterator();
 } // method: getBooksIterator
```

Note that it is the references in the unmodifiable collection that are read-only, not the objects they reference. This means that as a consequence of Java's reference semantics they could be modified through another shared reference despite our best intentions!

To finish the domain model changes we introduce into the Library class the **private** String attribute theMessage and the accessor operation getStatus that returns its value. Each state-changing operation updates theMessage as appropriate. For example, we have:

```
public class Library {
   // ...
   public void   addOneBook(Book aBook) {
      int aCatalogueNumber = aBook.getCatalogueNumber();
      Book foundBook = this.getBook(aCatalogueNumber);
      //
      if(foundBook == null) {
         boolean result = theLoanStock.add(aBook);
         if(result == true) {
            theMessage = "Book added";
         }
```

```
        else {
            theMessage = "Book not added";
        }
    }
    else {
        theMessage = "Book already present - cannot add";
    }
} // method: addOneBook
    public String getStatus() {
        return theMessage;
    } // method: getStatus
    private String   theMessage;
    // ...
} // class Library
```

The method **getStatus** is used by the **Action** class as follows:

```
// class Action
public void   addOneBook() {
    // Get book details from the user.
    // ...
    // Add the book to the library.
    theLibrary.addOneBook(new Book(title, author, catalogueNumber));
    //
    // Display the outcome
    ConsoleIO.out.println( "\n\t" + theLibrary.getStatus() );
}
```

If we now turn our attention to the **Action** class then we must implement each of its methods. As it has visibility of the **Library**, **Book** and **BorrowerRecord** classes it can create an instance of them as required. For example, its constructor creates the **Library** for which it acts as a façade. Similarly the method **addOneBook** creates the **Book** that is to be added. No other class outside the **librarysubsystem** package (including the **Application** class) can do this.

The **Action** class methods that output details of the domain model use **toString** and an **Iterator**. A typical method is:

```
// class Action
public void   displayBooksAvailableForLoan() {
    // Display information about the Library
    ConsoleIO.out.println("\n" + theLibrary);
    //
    // Display information about the Books available for loan
    ConsoleIO.out.println("\n\t" + "Books available for loan");
    boolean bookFound = false;
    Iterator iter = theLibrary.getBooksIterator();
    //
    while(iter.hasNext() == true) {
```

```
      Book book = (Book) iter.next();
      if(book.getBorrower() == null) {
         ConsoleIO.out.println("\n\t\t" + book);
         bookFound = true;
      }
   }
   if(bookFound == false)
      ConsoleIO.out.println("\n\t\t" + "None");
} // method: displayBooksAvailableForLoan
```

The completed model (Lib4_3.uml) is given with the supplied software.

4.4.4.3   Testing

The Application class now is much simpler than in the previous iteration. The reason of course is that much of its complexity is now in the Action class. All the run method has to do is to present the human user with a menu, get a response and then pass it on to an Action object.

As in iteration 2 we could make use of the **private** operations such as testUseCases and testErrorConditions to help automate the burden of testing. However, we leave this as an exercise to the reader and assume that the following outline code could be used for manual testing.

```
import librarysubsystem.*;
import textio.*;

public class Application {

   public    Application() {
            // No initialization required
   } // method: Application

public void    run() {
   // Create a unified interface (facade) for the Library
   Action action = new Action("Books-R-Us");
   //
   // Get and process the user's choice
   String choice = "";
   do {
      // Get the user's selection
      choice = this.getSelection();
      ConsoleIO.out.println();
      // Action the user's choice
      if(choice.equals("0") ) {
         action.close();
      } else if(choice.equals("1")) { // Register one borrower.
         action.registerOneBorrower();
      }
```

```
      // Other choices as per the use-cases
      // ...

      else {
         action.unknownSelection();
      }
   } while(choice.equals("0") == false);

} // method: run

private String    getSelection() {
   // Display menu to the user.
   ConsoleIO.out.println();
   ConsoleIO.out.println("0: Quit");
   ConsoleIO.out.println("1: Register one borrower");
   //
   // Other choices as per the use-cases
   // ...
} // method: getSelection

} // class: Application
```

4.4.5   Reconcile model diagrams

For this activity there is very little to do as only one increment was necessary to implement the new use-case. However, it is always worthwhile making the decision explicit. Although the capability of iteration 3 is the very similar to iteration 2 it represents a major improvement on the earlier designs. Undoubtedly we will benefit from our hard work in later iterations.

Finally, the following listing (from the model Lib4_3.uml) gives the Action class in its entirety.

```
package librarysubsystem;

import java.io.*;
import textio.*;
import java.util.*;

public class Action {

   public Action(String aLibraryName) {
      theLibrary = new Library(aLibraryName);
   } // method: Action
   public void    close() {
      ConsoleIO.out.println("\n\t" + "SYSTEM CLOSING" + "\n");
   } // method: close

   public void    registerOneBorrower() {
      // Get the borrower details from the user.
```

```
    ConsoleIO.out.print("\t" + "Enter the borrower name >>> ");
    String borrowerName = ConsoleIO.in.readLine();
    //
    // Register the borrower with the library
    theLibrary.registerOneBorrower(borrowerName);
    //
    // Display the outcome
    ConsoleIO.out.println("\n\t" + theLibrary.getStatus());
} // method: registerOneBorrower

public void    displayBorrowers() {
    // Display information about the Library
    ConsoleIO.out.println("\n" + theLibrary);
    //
    // Display information about the Borrowers
    boolean borrowerFound = false;
    Iterator iter1 = theLibrary.getBorrowersIterator();
    //
    while(iter1.hasNext() == true) {
        //
        // Display information about the Borrower
        BorrowerRecord borrowerRecord = (BorrowerRecord) iter1.next();
        ConsoleIO.out.println("\n\t" + borrowerRecord);
        borrowerFound = true;
        //
        // Display information about the Books on loan to the Borrower
        ConsoleIO.out.println("\n\t\t" + "Books on loan");
        boolean bookFound = false;
        Iterator iter2 = borrowerRecord.getBorrowedBooksIterator();
        //
        while(iter2.hasNext() == true) {
            Book book =(Book) iter2.next();
            ConsoleIO.out.println("\n\t\t\t" + book);
            bookFound = true;
        }
        if(bookFound == false)
            ConsoleIO.out.println("\n\t\t\t" + "None");
    }
    if(borrowerFound == false)
            ConsoleIO.out.println("\n\t" + "No registered borrowers");
} // method: displayBorrowers

public void    addOneBook() {
    // Get book details from the user.
    ConsoleIO.out.print("\t" + "Enter the title >>> ");
    String title = ConsoleIO.in.readLine();
    ConsoleIO.out.print("\t" + "Enter the author >>> ");
```

```java
    String author = ConsoleIO.in.readLine();
    ConsoleIO.out.print("\t" + "Enter the catalogue number >>> ");
    int catalogueNumber = ConsoleIO.in.readInt();
    //
    // Add the book to the library.
    theLibrary.addOneBook(new Book(title, author, catalogueNumber));
    //
    // Display the outcome
    ConsoleIO.out.println("\n\t" + theLibrary.getStatus() );
} // method: addOneBook

public void   loadBooksFromFile() {
    // Get the file name from the user.
    ConsoleIO.out.print("\t" + "Enter the file name >>> ");
    String fileName = ConsoleIO.in.readString();
    //
    // Initialize the Library from a file
    try {
        // Open the file
        FileTextReader inputFile = new FileTextReader(fileName);
        while(inputFile.isEOF() == false) {
            // Get the data from the file
            String title = inputFile.readLine();
            String author = inputFile.readLine();
            int catalogueNumber = inputFile.readInt();
            //
            // Create and add a Book to the Library
            theLibrary.addOneBook(new Book(title, author, catalogueNumber));
            //
            // Display the outcome
            ConsoleIO.out.println( "\n\t" + theLibrary.getStatus() );
        }
    }
    catch(IOException exception) {
        ConsoleIO.out.println("\n\t" + "Cannot open " + fileName);
    }
} // method: loadBooksFromFile

public void   displayBooksAvailableForLoan() {
    // Display information about the Library
    ConsoleIO.out.println("\n" + theLibrary);
    //
    // Display information about the Books available for loan
    ConsoleIO.out.println("\n\t" + "Books available for loan");
    boolean bookFound = false;
    Iterator iter = theLibrary.getBooksIterator();
    //
```

```
    while(iter.hasNext() == true) {
       Book book = (Book) iter.next();
       if(book.getBorrower() == null) {
          ConsoleIO.out.println("\n\t\t" + book);
          bookFound = true;
       }
    }
    if(bookFound == false)
       ConsoleIO.out.println("\n\t\t" + "None");
} // method: displayBooksAvailableForLoan

public void    displayBooksOnLoan() {
   // Display information about the Library
   ConsoleIO.out.println("\n" + theLibrary );
   //
   // Display information about the Books on loan
   ConsoleIO.out.println("\n\t" + "Books on loan");
   boolean bookFound = false;
   Iterator iter = theLibrary.getBooksIterator();
   //
   while(iter.hasNext() == true) {
      Book book = (Book) iter.next();
      if(book.getBorrower() != null) {
         ConsoleIO.out.println("\n\t\t" + book);
         bookFound = true;
      }
   }
   if(bookFound == false)
      ConsoleIO.out.println("\n\t\t" + "None");
} // method: displayBooksOnLoan

public void    lendOneBook() {
   // Get details from the human user
   ConsoleIO.out.print("\t" + "Enter the catalogue number >>> ");
   int catalogueNumber = ConsoleIO.in.readInt();
   ConsoleIO.out.print("\t" + "Enter the borrower name >>> ");
   String borrowerName = ConsoleIO.in.readLine();
   //
   // Lend the book to the borrower
   theLibrary.lendOneBook(catalogueNumber, borrowerName);
   //
   // Display the outcome
   ConsoleIO.out.println("\n\t" + theLibrary.getStatus());
} // method: lendOneBook

public void    returnOneBook() {
   // Get details from the human user
```

```
ConsoleIO.out.print("\t" + "Enter the catalogue number >>> ");
int catalogueNumber = ConsoleIO.in.readInt();
//
// Return it to the library
theLibrary.returnOneBook(catalogueNumber);
//
// Display the outcome
ConsoleIO.out.println("\n\t" + theLibrary.getStatus());
} // method: returnOneBook

public void   unknownSelection() {
ConsoleIO.out.println("\n\t" + "UNKNOWN SELECTION" + "\n");
} // method: unknownSelection

private Library   theLibrary;

} // class: Action
```

A typical output from the Application is now (user input is shown as bold type):

```
0: Quit
1: Register one borrower
2: Display borrowers
3: Add one book
4: Load books from file
5: Display books available for loan
6: Display books on loan
7: Lend one book
8: Return one book

    Enter the choice >>> 5

Books-R-Us :0 book(s) :0 borrower(s)

    Books available for loan
        None
// ...
```

Full listings for all of the classes involved are given in the software supplied (model Lib4_3.uml).

4.5   Summary

1. The development process revolves around a number of iterations that deliver the final application. The iterations are driven by the use-cases identified at each cycle.
2. The application code is realized by successive increments. Each adds a small piece of functionality whose introduction can be tested thereby reducing the risk posed by new code.

3. The class diagram derived from other UML diagrams developed during the analysis activity acts as the architectural framework on which the application development hangs. The use-case iterations and code increments are conducted within that guiding architecture.

4. Each use-case is accompanied by a corresponding test-case. Further test-cases can be introduced to check any special situations that need to be considered.

5. The combined use of Iterators and the trio of operations equals, compareTo and hashCode makes the code more resilient to change. One collection can substitute for another without code revisions.

6. The domain model should have no responsibility for any input and output. This we achieve by overloading the toString method and by providing non-modifiable Iterators.

4.6   Exercises

1. Revisit the object diagram in figure 4.2 to show three distinct Book objects (rather than a multi-object) and their relationship with the Library object. Show exploded versions for the Library and Book object instances to suggest possible attribute values.

2. Prepare two collaboration diagrams for the operations displayBooksAvailableForLoan and displayBooksOnLoan presented as sequence diagrams in figure 4.3. What observation can be made about the similarities and differences between these two types of interaction diagrams? Under what circumstances would you use one rather than the other?

3. Figure 4.4 is an activity diagram for the operation displayBooksAvailableForLoan. Prepare a corresponding activity for the operation displayBooksOnLoan. Factor out the similarities between the two and propose how we might simplify the logic.

4. The class diagram for the first iteration of the case study is given in figure 4.8. The BorrowerRecord class and the Book class have a one-to-many aggregation relationship. Make a case for replacing this with a one-to-many association relationship.

5. With reference to the class diagram of figure 4.8, reflect on the multiplicity of 0..1 for a BorrowerRecord in the BorrowerRecord–Book shared aggregation relationship. Why is it not just 0..*?

6. Architectural relations in a class diagram are ultimately realized as Java code. Offer a scheme for translating one-to-one and one-to-many relations into Java.

7. Section 4.2.4.2 presents the method code for the Book class. Give an explanation for the inclusion of the methods equals, compareTo and hashCode. Why are they included in the class? Why should they be implemented consistently? What other classes in this case study should define these methods?

8. Section 4.2.4.2 also presents the method for the displayBooks operation in the Library. Although the logic used is clear, it is unwieldy in that it repeats code that iterates over a collection of Books displaying each as appropriate. Revisit this method and eliminate its duplicated code.

9. Again in section 4.2.4.2, the Library adds a Book object created by a client with a method:

```
// class Library
public void   addOneBook(Book aBook) {
   // ...
}// method: addOneBook
```

Replace it with one that creates a Book object before adding it. It should have the following form:

```
// class Library
public void   addOneBook(String aTitle, String anAuthor, int aCatalogueNumber) {
   // ...
}// method: addOneBook
```

What are the factors that affect using one rather than the other?

10. The Application class shown at the beginning of section 4.2.4.3 created a Library object then immediately displayed the books available for loan. Given that no Books had been added to the Library, what was the rationale for doing this?

11. In the same code fragment discussed in question 7, the method displayBooksAvailableForLoan is being used for a purpose other than its functional requirement identified in the use-case analyses. What is that additional purpose?

12. In section 4.3.4.1 we replaced the architectural relationship that was realized with a HashSet with a TreeSet. This is the only change made to the code. How is it possible to change the container yet not make any further revisions to the code?

13. Section 4.3.4.3 presents a method getSelection that returns a String. Amend it to return an int. What are the implications of this decision?

For these remaining exercises you should modify the final model (Lib4_3.uml) developed in the case study. It is probably wise to add one increment at a time and not to try everything at once. As with the exercises of chapter 3, it is recommended that you use the ROME modelling environment supplied.

14. Amend the Book class so that it has an International Standard Book Number (ISBN) as one of its attributes. Ensure all appropriate methods in the Book class are also updated.

15. How might we revise the specification so that a borrower is restricted to a maximum of five Books on loan? Implement this change.

16. Modify the system to include, at program start-up, a login procedure for the librarian. At the program start the librarian is invited to enter a system password. Allow for the librarian to have three attempts at providing the correct password before the program closes.

17. Amend the addOneBook method of the Library so that a unique catalogue number is given to each Book on addition to the Library. It is expected that these catalogue numbers should be in ascending numeric sequence starting at 1.

18. Amend the BorrowerRecord class so that each borrower has a unique registration number. Again it should be automatically allocated by the Library in the method registerOneBorrower.

19. Amend the **BorrowerRecord** class so that it has more attributes, e.g. an address.

20. It is not long before software developers start to recognize similarities between the systems they build. Although many of the details may be different their overall design and implementation is often very similar (we discover later in chapter 8 that common design patterns have been documented). In our case study we have a **Library** with many **Books**. However, we might have used a **Doctor** with many **Patients** or a **University** with many **Students**. The design and the issues that arise from it would have been much the same.

    Using this case study as a guide consider the following:

    - A hire company has several cars. Each car has a unique registration number, model name and year of registration. Cars may be hired out to a customer registered in the company. Each customer has a name and a unique number associated with him or her. We are asked to support the company owner by logging which cars are out on hire to which customer.
    - A video shop has a large number of videos for hire to customers. Each video has a unique title and each customer a unique registration number. As before we are asked to log which videos are out on loan to which customers.
    - A hospital has many doctors and patients. Each doctor and patient has a name and unique number associated with them. Doctors look after many patients but a patient has exactly one doctor. We are asked to develop a patient monitoring system by recording which patient is associated with a particular doctor.
    - A university has many students each with a unique matriculation number as well as a name and the course of study. We are asked to be able to log each student and obtain a display of all students as well as those on a particular course.

    Now develop a suitable design and implementation based on one (or all) of them. You can add as much detail as you think is appropriate.

21. You are required to develop software to support the administration of a hotel. The major features of the hotel are:

    - there are three floors numbered 1, 2 and 3 each of which has up to five rooms
    - not all floors have the same combination of rooms
    - each room has a room number, e.g. 201 for room 1 on floor 2
    - each room has a maximum occupancy, i.e. the number of people that can use it.

    Staff should be able to obtain details of each room on each floor in a variety of ways. As a minimum, the user must be able to request for a report for:

    - all the rooms on all floors
    - all the rooms on a particular floor or
    - a particular room on a particular floor

    and to decommission a room:

    - remove a particular room from a given floor.

22. A software house employs programmers each of whom has an expertise in a par-
     ticular programming language, e.g. Cobol, C++ or Java. All programmers are paid
     a basic monthly salary of around 1000 pounds per month. However, it may vary
     from programmer to programmer.

     As they are in demand a 10% enhancement of the basic salary is paid to each pro-
     grammer that specializes in Java. However, programmers may change their special-
     ist language and their salary enhancement should change accordingly. For example,
     after suitable training, a Cobol programmer could become a Java programmer.

     When a new programmer joins the staff a more experienced programmer is
     assigned to him as a mentor. Both must specialize in the same programming
     language. The basic idea behind this practice is that the new recruit (the mentee)
     will benefit from the experience of the mentor. As this is extra work for the mentor
     he is awarded 5% of current salary enhancement for every mentee under his super-
     vision. When a programmer no longer needs a mentor the mentor's salary is
     changed accordingly.

     For administrative purposes each employee has a name and a unique payroll number.
     You are required to develop software that supports the administration of the com-
     pany. As a minimum it should produce a detailed report on each programmer and a
     total monthly salary bill for the company.

     The report should show for each programmer:

     • his payroll number and name
     • the specialist programming language and current monthly salary

together with:

     • details of any programmers that he is mentoring or is mentored by.

5

Specialization

Chapter 2 introduced the architectural relationships of association and aggregation. In this chapter we develop those discussions further with the introduction of the *specialization* relationship that may also exist between classes. It is widely used in an OOAD and brings to our designs a powerful feature unique to object orientation.

5.1 Specialization

Consider a software house that employs several programmers. Each is given a unique payroll number. His name and monthly salary are recorded. The programming language used is also recorded but it is expected to change from time to time. It should be possible to get a display showing the details of each programmer and a monthly salary bill for the software house.

Using the knowledge gained from the preceding chapters we might arrive at the class diagram shown in figure 5.1.

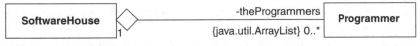

Figure 5.1   *An initial class diagram*

We might also develop the following outline Java code for the Programmer class.

```java
public class Programmer {
    // ----- Operations ----------
    public    Programmer(int aPayrollNumber, int aMonthlySalary, String aName,
                        String aLanguage) { ... }
    public int    getMonthlySalary() { ... }
    public void    setLanguage(String aLanguage) { ... }
    public void    display() { ... }

    // ----- Attributes ----------
    private int    thePayrollNumber;
    private int    theMonthlySalary;
    private String    theName;
    private String    theLanguage;
}    // class: Programmer
```

Although there is nothing intrinsically wrong with the design we can improve it sig-nificantly. For example, it seems likely that all of the software house employees, not just programmers, are given a payroll number. Similarly we expect their name and salary to be recorded. Finally we expect it to be necessary to be able to access the salary of each employee and to display its details. On reflection, the only attribute special to a Programmer is theLanguage and the only operation is setLanguage.

We can think of a Programmer as a special kind of Employee and amend the class diagram as shown in figure 5.2.

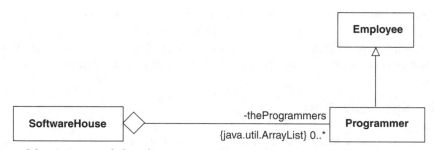

Figure 5.2   *An improved class diagram using the specialization relationship*

The arrow symbol shows that the Employee class is related to the Programmer class by specialization. The Employee class is usually referred to as the *parent* and the Programmer class as the *child* or *descendant*. It is implicit in this relationship that the parent must be able to respond to the same set of messages as the child. In other words a Programmer is in effect an Employee. However the reverse is not true as the child may introduce operations not present in the parent. We cannot assume that an Employee is a Programmer.

The Java reserved word **extends** specifies that a class is a specialization (extension) of another. Its use leads to the following outline code for the Employee and Programmer classes:

```
public class Employee {
    // ----- Operations ----------
    public    Employee(int aPayrollNumber, int aMonthlySalary, String aName) { ... }
    public int    getMonthlySalary() { ... }
    public void    display() { ... }

    // ----- Attributes ----------
    private int    thePayrollNumber;
    private int    theMonthlySalary;
    private String    theName;

} // class: Employee:
```

and

```
public class Programmer extends Employee {

    // ----- Operations ----------
    public    Programmer(int aPayrollNumber, int aMonthlySalary, String aName,
                          String aLanguage) { ... }
    public void    setLanguage(String aLanguage) { ... }

    // ----- Attributes ----------
    private String    theLanguage;

} // class: Programmer
```

The Programmer class now only has those features special to it. In contrast the Employee class has those relevant to employees in general. This makes the Programmer class much easier to develop. In this programming context the parent is often referred to as the *superclass* and the child as the *subclass*.

It is important to realize that the Employee class could be reused in a different application. For example, we might declare a class Administrator as:

```
public class Administrator extends Employee {

    // ----- Operations ----------
    public    Administrator(int aPayrollNumber, int aMonthlySalary, String aName,
                          String aDepartment) { ... }
    public void    display() { ... }

    // ----- Attributes ----------
    private String    theDepartment;

} // class: Administrator
```

which uses the Employee class unchanged.

5.2   Inherited methods

With Java all of the features (methods and attributes) declared in a superclass are inherited by a subclass. This means that the Programmer class need only declare those methods and attributes special to it. In this case it is the method setLanguage and the attribute theLanguage. Clearly in more complex examples this would represent a major saving in effort. In this example it means that we need only introduce the attribute theLanguage and the method setLanguage to the Programmer class.

Typically we have:

```
Employee employee = new Employee(123, 2000, "John");
int salary = employee.getMonthlySalary();
```

giving the variable salary a value of 2000 and:

```
Programmer programmer = new Programmer(234, 3000, "Ken", "C++");
salary = programmer.getMonthlySalary();
```

giving it a value of 3000. In both cases the method for getMonthlySalary executed is
defined in the Employee class.

```
// class Employee
public int    getMonthlySalary() {
   return theMonthlySalary;
} // method: getMonthlySalary
```

However, we might also have:

```
programmer.setLanguage("Java");
```

which sets the attribute theLanguage to a value of "Java". The method for
setLanguage executed is defined in the Programmer class as:

```
// class Programmer
public void    setLanguage(String aLanguage) {
   theLanguage = aLanguage;
}    // method: setLanguage
```

Note that:

```
employee.setLanguage("Java");
```

is illegal. An Employee cannot respond to the message setLanguage as it is not
declared in the Employee class.

   Clearly the construction of a subclass object must involve the use of a constructor
declared in its class. However, a subclass constructor method may also make use of its
superclass constructor. For example, we have:

```
// class Employee
   public    Employee(int aPayrollNumber, int aMonthlySalary, String aName) {
      thePayrollNumber = aPayrollNumber;
      theMonthlySalary = aMonthlySalary;
      theName = aName;
} // method: Employee
```

and

```
// class Programmer
   public    Programmer(int aPayrollNumber, int aMonthlySalary, String aName,
                        String aLanguage) {
      super(aPayrollNumber, aMonthlySalary, aName);
      theLanguage = aLanguage;
}    // method: Programmer
```

Notice the use of the reserved word super in the Programmer constructor method
that refers to the immediate superclass. This is how its superclass constructor with three
parameters (an int, an int and a String) is executed. It is important to understand that even
though two constructors are used in the construction of a Programmer there is only one
object created. It is a Programmer not a Programmer and an Employee.

5.3   Redefined methods

Some words of caution are necessary at this point. The Java access modifiers **public** and **private** are also inherited by a subclass. This means that a method body (it is implicitly **private**) or any **private** attributes in a superclass cannot be accessed in a subclass method. This may appear strange as they are inherited.

The consequence is that the method body for display in the Employee class cannot be modified by the Programmer class to reflect the fact that each Programmer has a language as an attribute. Clearly we need two display methods: one for an Employee and one for a Programmer. In other words the display operation is *variant over specialization*.

However, with Java, a method inherited by a subclass can be *redefined* to have a different behaviour. An obvious strategy for the display method required in the Programmer class is to make use of the display method in the Employee class but to augment it with additional logic. Therefore we have:

```
// class Employee
public void    display() {
      ConsoleIO.out.println("payroll number:" + "\t" + thePayrollNumber);
      ConsoleIO.out.println("monthly salary:" + "\t" + theMonthlySalary);
      ConsoleIO.out.println("first name:" + "\t" + theName);
}    // method: display
```

and

```
// class Programmer
   public void    display() {
      super.display();
      ConsoleIO.out.println("language:" + "\t" + theLanguage);
}    // method: display
```

Notice again the use of the reserved word **super** in the Programmer class. It ensures that the display method defined in its immediate superclass is executed before the extra attribute is displayed. In fact the Programmer class could not display the attributes declared in Employee. The reason is that they are declared as **private** in the Employee class. Even though they are inherited by the Programmer class they are not accessible in its methods. As a consequence it is unwise to introduce an attribute with the same name into two classes in the same hierarchy.

This apparently odd behaviour is an example of good software engineering practice. One reason is that the logic required to display the Employee attributes is located in just one method. Therefore any changes are much easier to implement consistently. Another is that access to Employee attributes is restricted to the Employee class preventing the possibility of unexpected changes by methods in another class.

Using the declarations for **employee** and **programmer** in section 5.2, execution of
the statements:

employee.display();
programmer.display();

produces the output:

payroll number: 123
monthly salary: 2000
first name: John

payroll number: 234
monthly salary: 3000
first name: Ken
language: C++

The first uses **display** defined in the **Employee** class and the second **display** defined in
the **Programmer** class.

5.4   Polymorphism

A defining characteristic of object-oriented systems is the *polymorphic effect*. A message sent to an object of some class is received as normal. However the same message
may also be received by an object of a descendant class. Use of the polymorphic effect
results in systems that are apparently simple but have complex execution behaviours.

To illustrate it we introduce project leaders to the software house. A project leader is
a programmer responsible for a team of programmers. He is given a 10% monthly
salary bonus for each member in the team. The display of a project leader should
include a display of each programmer assigned. To make the problem more interesting
we also specify that if a programmer uses Java then a 20% bonus is awarded.

We might propose the class diagram of figure 5.3.

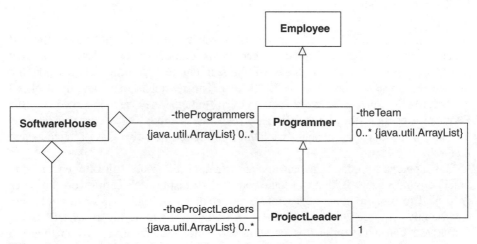

Figure 5.3   *An initial class diagram with specialization*

The class diagram documents an inheritance hierarchy with the Employee class specialized into a Programmer that is in turn specialized into a ProjectLeader. The SoftwareHouse maintains two collection objects. The first (referenced by theProgrammers) holds Programmer references and the second (referenced by theProjectLeaders) holds ProjectLeader references. In addition each ProjectLeader has a collection (referenced by theTeam) with a reference to each Programmer in his team. Our intention is that the SoftwareHouse sends messages to its ProjectLeaders and Programmers. For example, the method for displayStaff is:

```
// class SoftwareHouse
public void    displayStaff() {
    ConsoleIO.out.println();
    ConsoleIO.out.println("Staff list for " + theName);
    ConsoleIO.out.println();
    //
    Iterator iter1 = theProgrammers.iterator();
    while(iter1.hasNext()) {
        Programmer programmer = (Programmer) iter1.next();
        programmer.display();
        ConsoleIO.out.println();
    }
    //
    Iterator iter2 = theProjectLeaders.iterator();
    while(iter2.hasNext()) {
        ProjectLeader projectLeader = (ProjectLeader) iter2.next();
        projectLeader.display();
        ConsoleIO.out.println();
    }
} // method: displayStaff
```

Although using two collections is not actually wrong, it leads to code duplication which is almost certainly unnecessary. We can improve the design significantly by using the polymorphic effect. However, before doing so, we must be clear about two points. The first is that a descendant class must have at least the same set of methods as its parent. Therefore an object of a descendant class can be used in place of an object of a parent class. In other words a child can be *substituted* for its parent. In this example it follows that we can have a heterogeneous collection of Programmer reference and ProjectLeader reference values. This is because the class ProjectLeader is a descendant of the class Programmer and so a ProjectLeader object can substitute for a Programmer object.

The second point is that with some programming languages the binding of a message to a corresponding method may occur before the software starts its execution. Alternatively the binding may occur when the software is actually running. The former is usually referred to as *static* or *early binding*, and the latter as *dynamic*, *late* or *virtual binding*. Late binding is always assumed in Java.

If descendant class substitution is combined with late binding then the polymorphic effect is the result. This means that a message sent from a SoftwareHouse object through a reference to a Programmer object, can now be received by a Programmer object or a ProjectLeader object. This simplifies the class diagram as shown in figure 5.4.

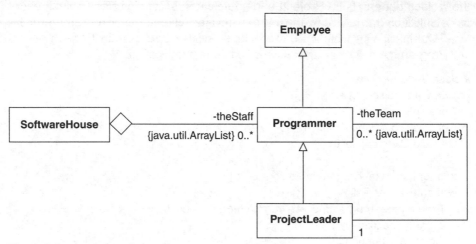

Figure 5.4   *An improved class diagram using polymorphism*

There is no longer any need to have an explicit relationship between the SoftwareHouse and its ProjectLeaders. As a result the coding in the SoftwareHouse methods is simplified. For example, we now have:

```
// class SoftwareHouse
public void    displayStaff() {
   ConsoleIO.out.println();
   ConsoleIO.out.println("Staff list for " + theName);
   ConsoleIO.out.println();
   //
   Iterator iter = theStaff.iterator();
   while(iter.hasNext()) {
      Programmer programmer = (Programmer) iter.next();
      programmer.display();
      ConsoleIO.out.println();
   }
}   // method: displayStaff
```

where the message display sent through a Programmer reference may be received by a Programmer or a ProjectLeader object. Similarly we have:

```
// class SoftwareHouse
public void    addProgrammer(Programmer aProgrammer) {
   // ...
} // method addProgrammer
```

where the formal parameter is declared to be a **Programmer** reference but the actual parameter may reference a **Programmer** or ProjectLeader.

As there are some new requirements we should revisit the implementation of the **Employee** and **Programmer** classes. There are no major changes necessary in the **Employee** class as its display method just displays the attributes set by its constructor. However, we should recognize that the monthly salary is now subject to change in descendants therefore it is more accurately named as theBasicMonthlySalary. Its value is returned by getMonthlySalary.

```
// class Employee
public int   getMonthlySalary() {
   return theBasicMonthlySalary;
} // method: getMonthlySalary
```

The **Programmer** class is more interesting as it has a redefined method for getMonthlySalary that checks the language attribute and awards a bonus as appropriate. This leads to the method:

```
// class Programmer
public int   getMonthlySalary() {
    int bonus = 0;
    int monthlySalary = super.getMonthlySalary();
    //
    if(theLanguage.equals("Java") == true) {
       bonus = (int) (monthlySalary * 0.2);
    }
    //
    return monthlySalary + bonus;
} // method: getMonthlySalary
```

Note the use of the reserved word **super** to specify the method getMonthlySalary as defined in the immediate superclass **Employee**.

If we consider the ProjectLeader class then it is clear that it must also redefine the method getMonthlySalary inherited from **Programmer**. It checks for the number of **Programmer**s in the team and awards a bonus. It is coded as follows:

```
// class ProjectLeader
public int   getMonthlySalary() {
    int monthlySalary = super.getMonthlySalary();
    int bonus = 0;
    int numberOfProgrammers = theTeam.size();
    //
    bonus = (int) (numberOfProgrammers * 0.1 * monthlySalary) ;
    //
    return monthlySalary + bonus;
} // method: getMonthlySalary
```

Again the reserved word **super** is used. Now it specifies the method getMonthlySalary defined in its immediate superclass Programmer in which a check of the language used is made. It awards a bonus as appropriate. Therefore the bonus awarded to a ProjectLeader depends on the language used and the number of Programmers in the team. Notice the chaining of method calls down through a class hierarchy. It is typical of the complex run-time behaviours of object-oriented systems.

The remaining methods are relatively straightforward and full listings are given in program 5.1. It is included in the software supplied.

Program 5.1   Using the polymorphic effect (model Prog5_1.uml)

```
import textio.*;

public class Employee {
  // ----- Operations ----------
  public   Employee(int aPayrollNumber, int aMonthlySalary, String aName) {
    thePayrollNumber = aPayrollNumber;
    theBasicMonthlySalary = aMonthlySalary;
    theName = aName;
  } // method: Employee

  public int   getMonthlySalary() {
    return theBasicMonthlySalary;
  } // method: getMonthlySalary

  public void   display() {
    ConsoleIO.out.println("payroll number:" + "\t" + thePayrollNumber);
    ConsoleIO.out.println("monthly salary:" + "\t" + this.getMonthlySalary() );
    ConsoleIO.out.println("first name:" + "\t" + theName);
  } // method: display

  // ----- Attributes ----------
  private final int   thePayrollNumber;
  private int   theBasicMonthlySalary;
  private String   theName;
} // class: Employee

import textio.*;

public class Programmer extends Employee {
  // ----- Operations ----------
  public   Programmer(int aPayrollNumber, int aMonthlySalary, String aName,
                      String aLanguage)
  {
    super(aPayrollNumber, aMonthlySalary, aName);
    theLanguage = aLanguage;
  } // method: Programmer

  public void   display() {
    super.display();
    //
```

```
      ConsoleIO.out.println("language:" + "\t" + theLanguage);
   } // method: display

   public int   getMonthlySalary() {
      int bonus = 0;
      int monthlySalary = super.getMonthlySalary();
      //
      if(theLanguage.equals("Java") == true) {
         bonus = (int) (monthlySalary * 0.2);
      }
      //
      return monthlySalary + bonus;
   }  // method: getMonthlySalary

   // ----- Attributes ----------
   private String   theLanguage;
}   // class: Programmer


import java.util.*;
import textio.*;

public class ProjectLeader extends Programmer {
   // ----- Operations ----------
   public   ProjectLeader(int aPayrollNumber, int aMonthlySalary, String aName,
                          String aLanguage)
   {
      super(aPayrollNumber, aMonthlySalary, aName, aLanguage);
      theTeam = new ArrayList();
   }   // method: ProjectLeader

   public void   display() {
      super.display();
      //
      ConsoleIO.out.println("\n" + "Team members are:" + "\n");
      ConsoleIO.out.println("-------------" + "\n");
      //
      Iterator iter = theTeam.iterator();
      while(iter.hasNext() == true) {
         //
         // As next returns an Object it is cast to a Programmer
         // so that the message display can be sent.
         Programmer programmer = (Programmer) iter.next();
         programmer.display();
         ConsoleIO.out.println();
      }
      ConsoleIO.out.println("-------------" + "\n");
   } // method: display
```

```
  public int   getMonthlySalary() {
    int monthlySalary = super.getMonthlySalary();
    int bonus = 0;
    int numberOfProgrammers = theTeam.size();
    //
    bonus = (int) (numberOfProgrammers * 0.1 * monthlySalary);
    //
    return monthlySalary + bonus;
  } // method: getMonthlySalary

  public void   addProgrammer(Programmer aProgrammer) {
    theTeam.add(aProgrammer);
  } // method: addProgrammer

  // ----- Relations ----------
  private java.util.ArrayList   theTeam;   // of Programmer
} // class: ProjectLeader

import textio.*;
import java.util.*;
public class SoftwareHouse {
  // ----- Operations ----------
  public   SoftwareHouse(String aName) {
    theName = aName;
    theStaff = new ArrayList();
  } // method: SoftwareHouse

  public void   addProgrammer(Programmer aProgrammer) {
    theStaff.add(aProgrammer);
  } // method: addProgrammer

  public void   displayStaff() {
    ConsoleIO.out.println();
    ConsoleIO.out.println("Staff list for " + theName);
    ConsoleIO.out.println();
    //
    Iterator iter = theStaff.iterator();
    while(iter.hasNext()) {
      Programmer programmer = (Programmer) iter.next();
      programmer.display();
      ConsoleIO.out.println();
    }
  } // method: displayStaff

  public int   getMonthlySalaryBill() {
    int totalMonthlyBill = 0;
    Iterator iter = theStaff.iterator();
    while(iter.hasNext()) {
      Programmer programmer = (Programmer) iter.next();
```

```
      totalMonthlyBill += programmer.getMonthlySalary();
    }
    return totalMonthlyBill;
  } // method: getMonthlySalaryBill

  // ----- Attributes ----------
  private    String theName;

  // ----- Relations ----------
  private    java.util.ArrayList    theStaff;    // of Programmer

} // class: SoftwareHouse

import textio.*;

public class Application {
  // ----- Operations ----------
    public void    run() {
      // Create a new organisation.
      SoftwareHouse sh = new SoftwareHouse("Objects-R-Us");
      //
      // Create some new programmers.
      Programmer p1 = new Programmer(123, 2000, "John", "Ada") ;
      Programmer p2 = new Programmer(234, 2500, "Ken", "C++");
      Programmer p3 = new Programmer(456, 3000, "Peter", "Java") ;
      //
      // Create some new project leaders.
      ProjectLeader pl1 = new ProjectLeader(567, 4000, "Jon", "C") ;
      ProjectLeader pl2 = new ProjectLeader(789, 4000, "Jessie", "Java");
      //
      // Assign each programmer to a project leader
      pl1.addProgrammer(p3);
      pl1.addProgrammer(p2);
      //
      pl2.addProgrammer(p1);
      //
      // Hire each programmer and project leader
      sh.addProgrammer(p1);
      sh.addProgrammer(p2);
      sh.addProgrammer(p3);
      //
      sh.addProgrammer(pl1);
      sh.addProgrammer(pl2);
      //
      // Display some details of the staff.
      sh.displayStaff();
      //
      // Obtain and print the wage bill.
```

```
        ConsoleIO.out.println("The monthly salary bill is: " + sh.getMonthlySalaryBill());
        ConsoleIO.out.println();
    } // method: run
} // class: Application
```

The output produced is:

Staff list for Objects_R_Us

payroll number:	123
monthly salary:	2000
first name:	John
language:	Ada

payroll number:	234
monthly salary:	2500
first name:	Ken
language:	C++

payroll number:	456
monthly salary:	3600
first name:	Peter
language:	Java

payroll number:	567
monthly salary:	4800
first name:	Jon
language:	C

Team members are:

----

payroll number:	456
monthly salary:	3600
first name:	Peter
language:	Java

payroll number:	234
monthly salary:	2500
first name:	Ken
language:	C++

----

payroll number:	789
monthly salary:	5280
first name:	Jessie
language:	Java

Team members are:

----

payroll number:	123
monthly salary:	2000
first name:	John
language:	Ada

The monthly salary bill is: 18180

The collaboration diagram in figure 5.5 for the operation displayStaff helps clarify the configuration of objects and the messages they send to each other.

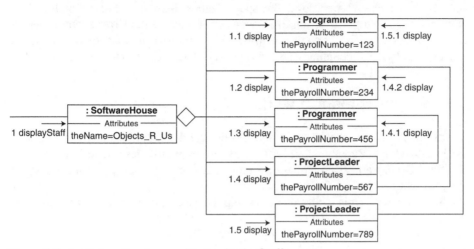

Figure 5.5 *A collaboration diagram for the displayStaff operation*

Notice that it emphasizes the inherent run-time complexity in object-oriented systems. If we reflect on the fact that the code for the display methods executed by a Programmer or ProjectLeader involves dynamic binding to code located in a super-class then the point is reinforced.

Before leaving this section we should be aware that it is possible that a subclass of ProjectLeader such as SystemManager could further redefine an inherited operation such as display and engage in polymorphic behaviour. Clearly this requires dynamic binding of the display message to its method. As this is the default with Java no further action is required on our part. Note that constructors cannot be redefined, therefore the terms polymorphic or redefined do not apply to them.

Sometimes we don't want an operation to be subject to redefinition in a specialized class. Our intention is that all descendants should respond in exactly the same manner. In other words the operation should be *invariant over specialization*. For example, we have setLanguage in the class Programmer that just sets the value of the language used. We don't want a subclass to do anything different. Such an operation is described as *frozen* when using the UML and the corresponding keyword in Java is **final** as in:

```
// class Programmer
public final void    setLanguage(String aLanguage) {
   theLanguage = aLanguage;
}
```

As its name suggests an operation that is **final** cannot be redefined in a subclass. The Java qualifier **final** can also be applied to a class declaration as in:

```
public final class ProjectLeader { ... }
```

In this context it means that the class cannot be extended. It is used when the class is question is at the bottom of a hierarchy that should not be deepened further.

It is our normal practice to redefine an operation only in an immediate descendant and not some class lower down the specialization hierarchy. For example, if the display operation inherited from Employee is not redefined in Programmer then we do not redefine it in Project Leader. Our intention is that there should be no gaps in a chain of redefinitions. It would almost certainly lead to unexpected behaviours.

5.5   Polymorphism at work

Polymorphism is remarkably useful and it is no surprise that it is an important part of the Java language. Recall that all Java classes are implicitly descended from the class Object. Like any other class it has operations that are inherited by its subclasses with methods that may be redefined.

One such operation is toString used during the case study of chapter 4 to return the textual representation of a Book. As it is declared in Object any object of any Java class must be able to respond to the message toString. However, if its method has not been redefined then the method originally defined in Object will execute.

Clearly we can do much better by arranging that toString returns a String with the attribute values of the object in question. For example, using the Employee class from section 5.1 we might have:

```
// class Employee
public String    toString() {
   return  " payroll number: " + thePayrollNumber +
            " monthly salary: " + theBasicMonthlySalary +
            " first name: " + theName;
} // toString
```

Notice how toString redefined in Employee makes no use of the parent method from class Object. To be more accurate this is method replacement rather than method redefinition. Now if we create and initialize an Employee object with:

```
Employee employee = new Employee( 123, 2000, "John");
```

then:

```
employee.toString()
```

produces the String representation of the Employee object. It can then be output with:

```
ConsoleIO.out.println( employee.toString() );
```

giving:

```
payroll number: 123   monthly salary: 2000  first name:  John
```

Alternatively we might have:

```
ConsoleIO.out.println( "Employee details:" + employee );
```

giving:

Employee details: payroll number: 123   monthly salary: 2000 first name: John

The reason for the alternative is that the concatenation operator + accepts an Object reference as its second actual parameter. It sends the message toString to it and then uses the String it returns.

Some judgement should be used when redefining toString as the method may become too complex. If a simple String made up of a few attributes is all that is required then its use is justified. However, if the method has a complex logic then it is usually better to have a named method such as displayStaff in the SoftwareHouse class of program 5.1.

A second example of polymorphism at work in the Java language is the operation:

public boolean equals(Object obj)

declared in the class Object. Its method compares the reference supplied as an actual parameter with a reference to the object executing the method. It delivers **true** if they are the same and **false** otherwise. In other words two objects are equal if they are identically the same. Notice that the formal parameter is declared as an Object. Polymorphic substitution allows any class of object to be supplied as the actual parameter. This is a common strategy used with Java to develop generally reusable (generic) methods. For example, the add method advertised by a collection declares its formal parameter to be an Object and so any class of object to be added to it (see appendix E).

The default behaviour for equals that compares objects by identity is not always what is required. In fact it can sometimes be confusing as the following code illustrates:

```
// Create two distinct objects
Employee e1 = new Employee(123, 2000, "John");
Employee e2 = new Employee(123, 2000, "John");
//
if(e1.equals(e2) )
    ConsoleIO.out.println("Equal");
else
    ConsoleIO.out.println("Not equal");
```

It produces Not equal as its output even though we might believe that the two Employees are clearly equal.

Usually we expect two objects to be considered equal if they have one or more equal attribute values. For example, two distinct Employee objects with the same payroll number are equal in this sense. To make this possible we redefine the method for equals (inherited from Object) and use the polymorphic effect. If we introduce:

```
// class Employee
public final int   getPayrollNumber() {
    return thePayrollNumber;
} // method: getPayrollNumber
```

to the Employee class we might have:

```
// class Employee
public final int   compareTo(Object obj) {
   Employee emp = (Employee)obj;
   return thePayrollNumber - emp.getPayrollNumber();
} // method: compareTo

public final boolean   equals(Object obj) {
   return this.compareTo(obj) == 0;
} // method: equals
```

Now two Employees are equal if they have the same payroll number and the previous code sample produces the output Equal as expected. For more discussion on the method equals see section 5.9 and appendix E.

Notice that both methods are qualified as **final**. This ensures that they cannot be redefined in a subclass and so all Employees, Programmers and ProjectLeaders are compared for equality on the same basis. Also there is no possibility of duplicated code in subclasses.

5.6   Protected features

An object of a descendant class and its parent are very closely related. In fact they are so close that the polymorphic effect allows them to be treated as being the same. Often we find that a descendant requires privileged access to its parent's features. In order to make this possible Java provides the **protected** keyword to modify access to operations and attributes. Descendants have access to those that are **public** or **protected** but associates or aggregate parts can access only those that are **public**. As stated earlier, **private** features are not accessible to clients or a subclass method.

Assume that the Employee class is amended so that the date of birth of an Employee is held as an attribute. As this is sensitive information, it is not available to normal clients and is not displayed. However, we intend that a Programmer subclass method should be able to access it and display the year of birth.

One possibility is to qualify the date of birth attribute in the Employee class as **protected** and give subclasses direct access to it. We take the view that this is not always advisable as it violates the rule that an object should encapsulate its own state. Therefore we qualify the date of birth as **private** and provide a **protected** method to access it. The following code fragments illustrate:

```
// class Employee
public Employee(int aPayrollNumber, int aMonthlySalary, String aName,
                java.util.GregorianCalendar aDateOfBirth) {
   theDateOfBirth = aDateOfBirth;
   // ...
} // method: Employee
```

```
// class Employee
protected java.util.GregorianCalendar   getDateOfBirth() {
  return theDateOfBirth;
} // method: getDateOfBirth
```

```
// class Employee
private java.util.GregorianCalendar   theDateOfBirth;
```

```
// class Programmer
public int   getYearOfBirth() {
  // Use the protected operation getDateOfBirth
  GregorianCalendar dateOfBirth = this.getDateOfBirth();
  return dateOfBirth.get(Calendar.YEAR);
} // method: getYearOfBirth
```

Execution of the statements:

```
Programmer p1 = new Programmer(123, 2000, "John", "Ada",
                    new GregorianCalendar(1980,1,14));
ConsoleIO.out.println("Year: " + p1.getYearOfBirth());
```

results in the output:

Year: 1980

Notice that a GregorianCalendar object is created as part of the call to the Programmer constructor method. An anonymous reference to it is passed as the actual parameter. This is a common Java programming idiom equivalent to:

```
GregorianCalendar dateOfBirth = new GregorianCalendar(1980,1,14);
Programmer p1 = new Programmer(123, 2000, "John", "Ada", dateOfBirth);
```

It avoids having to creating an unnecessary reference to an object.

5.7   The abstract class

It is often useful to be able to define a class that acts as a basis for establishing others. There is no intention to make an instance of it. It is a way of guaranteeing that all descendants share a common set of operations on their **public** interface. This kind of class is referred to as an *abstract class*.

For example, consider the software house from our previous discussions. Assume that there will never be an instance of an Employee as we have Programmers or ProjectLeaders but never just Employees. We intend that all employees of the SoftwareHouse must share common operations such as getPayrollNumber, getMonthlySalary, display and equals. Therefore we decide that the class Employee is stereotypical of an abstract class.

Experience has shown that an abstract class is so useful that the UML has a standard *stereotype*, **abstract** used to adorn a class box (we also render it grey). In Java the keyword **abstract** is used. For example, we might have:

and

public abstract class Employee { ... }

Specialization of an abstract class and qualification of its operations as frozen, polymorphic redefined are unaffected. Anticipating the use of the polymorphic effect, a declaration such as:

Employee employee;

is allowed. The reason is that **employee** is a reference type. All we need do is to ensure that the object it references belongs to a concrete subclass of Employee. Therefore we can have:

Employee employee = **new** Programmer(234, 2500, "Ken", "C++");

but not:

Employee employee = new Employee(123, 2000, "John");

The formal parameters for methods are treated similarly therefore:

public void   addEmployee(Employee anEmployee) { ... }

is permitted and we expect the actual parameter to reference a Programmer or ProjectLeader. We can now modify the previous class diagram to that shown in figure 5.6.

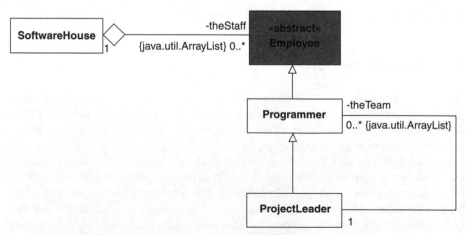

Figure 5.6   *A modified class diagram using an **abstract** class*

Only minor changes are required to the resulting Java code. Essentially all we need do is to replace **Programmer** with **Employee** references in the **SoftwareHouse** class. For example, we have:

```
// class SoftwareHouse
public void   addEmployee(Employee anEmployee) {
   theStaff.add(anEmployee);
} // method: addEmployee
```

and

```
// class SoftwareHouse
public void   displayStaff() {
   ConsoleIO.out.println();
   ConsoleIO.out.println("Staff list for " + theName);
   ConsoleIO.out.println();
   //
   Iterator iter = theStaff.iterator();
   while(iter.hasNext()) {
      Employee employee = (Employee) iter.next();
      employee.display();
      ConsoleIO.out.println();
   }
} // method: displayStaff
```

and

```
// class SoftwareHouse
public int   getMonthlySalaryBill() {
   int totalMonthlyBill = 0;
   Iterator iter = theStaff.iterator();
   while(iter.hasNext()) {
      Employee employee = (Employee) iter.next();
      totalMonthlyBill += employee.getMonthlySalary();
   }
   return totalMonthlyBill;
}   // method: getMonthlySalaryBill
```

Program 5.2 (**Prog5_2.uml**) in the software supplied gives complete listings. The **Application run** method is unchanged and so it has the same output as program 5.1.

Having made this change to the design we can now introduce other kinds of **Employee**s to the software house. For example, we might have administrators. The class diagram of figure 5.7 shows a modified class diagram.

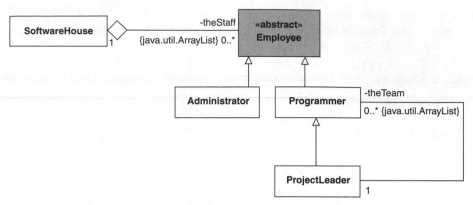

Figure 5.7 *A class diagram with an **abstract** class*

Crucially the SoftwareHouse class requires no changes at all. As far as it is concerned an Administrator is just another Employee. It is simply coded for illustration as:

import textio.*;

public class Administrator **extends** Employee {
   // ----- Operations ----------
   public   Administrator(**int** aPayrollNumber, **int** aMonthlySalary, String aName,
                  String aDepartment) {
     super(aPayrollNumber, aMonthlySalary, aName);
     theDepartment = aDepartment;
   }   // method: Administrator

   public void   display() {
     super.display();
     //
     ConsoleIO.out.println("department:" + "\t" + theDepartment);
   }   // method: display

   // ----- Attributes ----------
   private String theDepartment;
}   // class: Administrator

If the Application run method is modified to be:

// class Application
public void   run() {
   // Create a new organisation.
   SoftwareHouse sh = **new** SoftwareHouse("Objects-R-Us");
   //
   // As for programs 5.1 (Prog5_1.uml) and 5.2 (Prog5_2.uml)
   //
   // Create some new administrators.
   Administrator a1 = **new** Administrator(901, 5000, "Alice", "Accounts") ;

```
    Administrator a2 = new Administrator(012, 4500, "James", "Personnel") ;
    //
    // Hire each administrator
    sh.addEmployee(a1);
    sh.addEmployee(a2);
    //
    // Display some details of the staff.
    sh.displayStaff();
    //
    // Obtain and print the wage bill.
    ConsoleIO.out.println("The monthly salary bill is: " + sh.getMonthlySalaryBill());
    ConsoleIO.out.println();
} // method: run
```

then it produces an output of:

```
// As for program 5.2 (Prog5_2.uml)
```

payroll number:	901
monthly salary:	5000
first name:	Alice
department:	Accounts
payroll number:	902
monthly salary:	4500
first name:	James
department:	Personnel

The monthly salary bill is: 27680

Full listings are given in program 5.3 (Prog5_3.uml) in the software supplied.

Observe how the class Employee has been declared **abstract** ensuring that no instance can be created. A class is also **abstract** when one or more of its methods is qualified as **abstract**, i.e. no method definition is possible. For example, our Employee class might have the operation getSalaryEnhancement that delivers the enhancement to the basic salary. This cannot be defined for this class as it is dependent on the nature of the subclass. Hence class Employee might appear as:

```
pubic abstract class Employee {
    // ----- Operations ----------
    public abstract int    getSalaryEnhancement();
    // ...
} // class: Employee
```

This introduces a protocol (see next section) which concrete subclasses must define.

5.8   The interface class

It is possible to have an **abstract** class with **public** operations none of which have a defined method. They are all deferred to a subclass for their implementation. As it has no method bodies it presents only a specification of its behaviours. At first sight this may seem to be a rather strange class but it turns out be extremely useful. Often it is used as the specification of an interface (or protocol) that a subclass must conform to. For this reason it is normally referred to as an *interface class*.

Java supports the concept of an interface class with the keyword **interface**. Although it is similar to an abstract class with no defined methods, it is important to realize that it is different in one important respect. It is that a class that implements the interface, i.e. one that provides methods for its *deferred operations*, can belong to any class. Specifically such classes need not belong to the same class hierarchy. Although they may implement other methods and have different parents, if they implement those operations advertised by the interface then they can substitute for it. This simple fact makes the Java interface an extremely powerful facility that gives the designer more flexibility than the abstract class allows.

Consider the software house and its employees. We can insist that we must be able to get the payroll number and monthly salary of any employee, display his details and compare him for equality with another employee. Clearly the class to which an employee belongs must have implementations for the **public** operations getPayrollNumber, getMonthlySalary, display and equals. However, there is no requirement that each class is part of the same specialization hierarchy. This is an important point that makes a critical difference to our design. All that matters is that the SoftwareHouse is able to send the messages getPayrollNumber, getMonthlySalary, display and equals to each of its employees. It may be possible to send other messages but to be employable by the SoftwareHouse these four are the minimum required.

We can model this situation with a Java interface as shown in figure 5.8. Note that the interface class name ends in "able". This a common idiom used to imply the capabilities that classes that *implement* the interface must have. The dashed specialization arrows connecting Consultant and Employee to Employable denote that Consultant and Employee (or its descendants) implement Employable. Finally we use the UML stereotype <<**interface**>> to adorn the Employable class box.

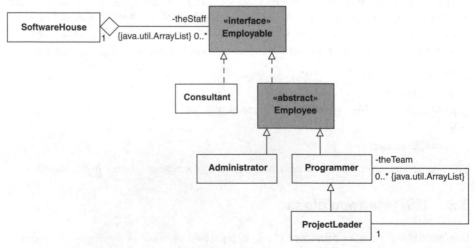

Figure 5.8 *Class diagram with an **interface***

Now the SoftwareHouse has an aggregation relationship with Employable objects that can belong to any class hierarchy. As before the Programmer and ProjectManager classes are descendants of Employee but Consultant forms it own hierarchy and is unrelated to them. This frees us from having to construct an unnatural single hierarchy just to make use of the polymorphic effect.

Java code for the interface class Employable is:

```
public interface Employable {

  // ----- Operations ----------
  public abstract int    getPayrollNumber();
  public abstract int    getMonthlySalary();
  public abstract void   display();
  public abstract boolean    equals(Object obj);
} // interface: Employable
```

Note that use of the keyword **public** in the interface declaration is optional but we prefer to make it explicit as it provides additional documentation for the code. Similarly the use of **abstract** in the operation signatures is optional but it does emphasize that an operation advertised by an interface can have no method body.

The primary change (but see the next section) required to the Employee class is to its header as in:

public abstract class Employee **implements** Employable { ... }

It specifies that:

- it is available to all other objects (**public**)
- no instances of it can be created (**abstract**)
- it is a declaration for the class Employee (**class**) and
- it conforms to the protocol of the Employable interface class (**implements**)

Although the list is rather long it documents the class extremely well and is a major strength of the Java language. The Programmer, ProjectLeader and Administrator classes are unchanged.

The Consultant class that follows is coded rather simply for illustration purposes. Normally it would have many more attributes and operations.

```
import textio.*;

public final class Consultant implements Employable {

  // ----- Operations ----------
  public   Consultant(int aPayrollNumber, String aSpeciality) {
    thePayrollNumber = aPayrollNumber;
    theSpeciality = aSpeciality;
  }    // method: Consultant

  public int    getPayrollNumber() {
    return thePayrollNumber;
  }    // method: getPayrollNumber
```

```
public int   getMonthlySalary() {
   return 5000;
}   // method: getMonthlySalary

public void   display() {
   ConsoleIO.out.println("payroll number:" + "\t" + thePayrollNumber);
   ConsoleIO.out.println("monthly salary:" + "\t" + this.getMonthlySalary() );
   ConsoleIO.out.println("speciality:" + "\t" + theSpeciality);
}   // method: display

public boolean   equals(Object obj) {
   return this.compareTo(obj) == 0;
}   // method: equals

public int   compareTo(Object obj) {
   // ...
}

// ----- Attributes ----------
private int       thePayrollNumber;
private String   theSpeciality;
}   // class: Consultant
```

The SoftwareHouse is interesting in that its Employee references now become
Employable references. For example, we have:

```
// class SoftwareHouse
public void   addEmployee(Employable anEmployee) {
   theStaff.add(anEmployee);
} // method: addEmployee
```

and

```
// class SoftwareHouse
public void   displayStaff() {
   ConsoleIO.out.println();
   ConsoleIO.out.println("Staff list for " + theName);
   ConsoleIO.out.println();
   //
   Iterator iter = theStaff.iterator();
   while(iter.hasNext()) {
      Employable employee = (Employable) iter.next();
      employee.display();
      ConsoleIO.out.println();
   }
} // method:  displayStaff
```

This may appear to be a small change but it has a profound effect. There is no need for the
SoftwareHouse to differentiate between a Consultant, Programmer, Administrator or
ProjectLeader. Each will respond to the operations advertised by the Employable inter-
face according to its class.

A suitable outline Application run method is:

```
// class Application
public void   run() {
    // Create a new organisation.
    SoftwareHouse sh = new SoftwareHouse("Objects-R-Us");
    //
    // As for program 5.3 (Prog5_3.uml)
    //
    // Create a new consultant.
    Consultant c1 = new Consultant(903, "OOAD") ;
    //
    // Hire the Consultant
    sh.addEmployee(c1);
    //
    // Display some details of the staff.
    sh.displayStaff();
    //
    // Obtain and print the wage bill.
    ConsoleIO.out.println("The monthly salary bill is: " + sh.getMonthlySalaryBill());
    ConsoleIO.out.println();
}// method: run
```

producing an output of:

```
// As for program 5.3 (Prog5_3.uml)
```

payroll number: 903
monthly salary: 5000
speciality: OOAD

The monthly salary bill is: 32680

Full listings are given in program 5.4 (Prog5_4.uml) in the software supplied.

5.9   The interface at work

Collection objects are extremely useful but they do have a cost associated with them. It arises from the fact that a collection can hold any class of object. Unfortunately it is unavoidable that as part of their inner workings (methods), collections must carry out manipulations on the objects they hold. For example, they will almost certainly be compared if they need to be ordered.

With some classes such as Integer, Decimal, Boolean and String this does not present any difficulty as the meanings (semantics) of the various comparisons are already defined in Java. However, with objects of user-defined classes, the problem needs careful consideration. The reason is that the meaning of testing for equality or being less than another object may be far from clear. What does it mean for one Employee to be less than another?

The decision cannot be the responsibility of the collection, as it cannot anticipate which class of object it may be required to hold. It follows that it must be the responsibility of the class of the object held by the collection to provide the necessary operations.

Methods in the Java collections use the operation compareTo when making comparisons of objects for ordering purposes. They do this by casting the Object they receive as an actual parameter to be Comparable, where Comparable is an interface class available as part of the Java environment.

```
public interface Comparable {
    // ----- Operations ----------
    public abstract int compareTo(Object obj);
} // interface: Comparable
```

Assuming that the cast is successful, then a method for compareTo is guaranteed to exist. It is expected to return an integer less than zero if the receiving object is less than object referenced by the actual parameter, 0 if it is equal to it and an integer greater than 0 if it is greater than it. The collection's ordering algorithm makes use of this fact.

Crucially a collection is not concerned about the actual class of its objects when ordering them. All that is required is that they can be successfully cast to be Comparable. This means that any object of any class can be ordered by a collection. All it has to do is to implement the Comparable interface by providing a suitable method for compareTo (see appendix E).

In the example programs presented so far in this chapter, the collection used to implement one-to-many associations and aggregations has been the ArrayList. It was carefully chosen because its add method adds its Object reference parameter to the end of the list. It makes no use of compareTo and so it does not cast its parameter to reference an object that implements the Comparable interface. Therefore the Employee class of program 5.1 need not implement the Comparable interface to be added to an ArrayList.

However, there are other collections such as the TreeSet whose elements must implement the Comparable interface. This means that if our example programs are modified so that a TreeSet is used then they will not execute successfully. Therefore it is wise to ensure that any object held by a collection implements the Comparable interface. The Employee class now becomes:

```
public abstract class Employee implements Employable {
    // ----- Operations ----------
    public int   compareTo(Object obj) {
        // Cast the actual parameter to be an Employable
        // so that the message getPayrollNumber can be sent.
        Employable   employable = (Employable) obj;
        //
        // get the payroll number of  the parameter
        int  parameterPayrollNumber = employable.getPayrollNumber();
        //
```

```
        return thePayrollNumber - parameterPayrollNumber;
    }   // method: compareTo
    // ...
}   // class: Employee
```

Since the Consultant class is not part of the Employee hierarchy, then it too must implement a similar compareTo method.

The declaration of collection reference such as:

```
// class SoftwareHouse
private java.util.ArrayList   theProgrammers;
```

should also be changed as it names a concrete class (type) ArrayList. A more flexible approach is to name the Collection interface that specifies the methods an implementation class (type) must implement. Therefore we have:

```
// class SoftwareHouse
private java.util.Collection   theProgrammers;
```

Program 5.4 is interesting in that the SoftwareHouse holds Employable references. Therefore we add compareTo to the Employable interface. As an interface cannot implement a method it **extends** the Comparable interface as in:

```
public interface Employable extends java.lang.Comparable {
    // ----- Operations ----------
    // Further Employable operations
    // ...
}   // interface: Employable
```

The Consultant and Employee classes provide a suitable method for compareTo. Note that they cast the Object reference parameter to Employable not Consultant or Employee. This is because the collection may hold an arbitrary mixture of Consultant, Programmer and ProjectLeader references.

These changes are illustrated in program 5.5 (Prog5_5.uml) included with the software supplied. As expected, its output is the same as program 5.4 except that each employee is displayed in the order of his payroll number.

As a second illustration, we show how an interface can be used to restrict the functionality of an existing class. The Java language currently provides the GregorianCalendar class to support dates. However, this class is really too complex for our purposes. All we require is to be able to create a date (given a day, month and year) and obtain its textual representation with toString.

Of course we cannot disinherit a method on specialization but we can achieve a similar effect by specifying an interface. For example, we have:

```
public interface DateIF {
    // ----- Operations ----------
    public abstract String  toString();
}   // interface: DateIF
```

implemented with:

import java.util.GregorianCalendar;

public class DateImp **implements** DateIF {

    // ----- Operations ----------
    public DateImp(**int** aDay, **int** aMonth, **int** aYear) { ... }
    public String   toString() { ... }

    // ----- Attributes ----------
    private GregorianCalendar   theDate;

} // DateImp

Now the DateImp class supports only the functionality we require with an encapsulated (**private**) GregorianCalendar object, theDate. The actual details of how it achieves it are of no concern to us here. The main point is that we can now declare a reference to DateIF that is initialized to reference a DateImp object with:

DateIF date = **new** DateImp(1, 5, 2003);

and display the date quite simply with:

ConsoleIO.out.println("The date is" + date);

The case study that follows in chapter 6 makes extensive use of this interface and its implemenation.

5.10   Summary

1. A class is a specialization (extension) of a parent class if it can be considered as an example of the parent. Although the descendant (subclass) normally has additional behaviours not present in the parent (superclass) it must respond to the same messages as the parent.
2. A descendant class has privileged access to its parent through a **protected** interface.
3. The polymorphic effect permits a message sent through a reference to an object of a parent class to be received and interpreted by an object of a descendant class.
4. An operation may be initially documented as frozen. It is qualified in Java as **final** in which case it cannot be redefined by a descendant. This qualification cannot be changed in a descendant class.
5. An operation may be initially documented as polymorphic in which case its method may be redefined by a descendant. No qualification is required in Java as this is the default behaviour.
6. An operation may subsequently be documented as redefined in which case its method has been redefined by a descendant. It must have been initially documented as polymorphic (or deferred) and may be subject to further redefinition by other descendants. No qualification is required in Java.
7. An operation may be initially documented as deferred in which case it must be redefined by a descendant if its class is to be concrete. It is polymorphic by implication and may be subject to further redefinition by other descendants. It is qualified

in Java as **abstract** and the class to which it belongs must also be qualified as **abstract**.

8. An **abstract** class can have no instances but acts only to define the protocol of its descendants. It may have some of its methods defined.

9. An interface class can have no instances but acts only to define all or part of the interface of its descendants. None of its methods are defined. A class that implements an interface must define its inherited methods or it is itself **abstract**.

5.11 Exercises

1. Animals that populate the earth can be categorized according to the outline class diagram in figure 5.9.

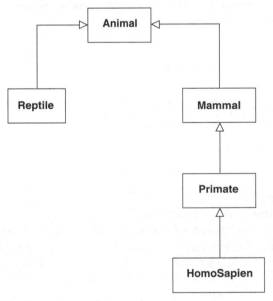

Figure 5.9   *An outline class diagram for animals*

   a) Extend this diagram to encompass other animals, e.g. apes, snakes, birds and dogs.

   b) Consider which classes should be designated as **abstract**

   c) Consider an operation move advertised on the **public** interface of the class Animal representing the movement of an animal. It could be documented as frozen, deferred, polymorphic or redefined. Select and justify the most appropriate documentation.

2. A university has a records system to store and retrieve information about its staff and students. Typically records are added and viewed. Staff have a name, address, employee number and department while students have a name, address, matriculation number and course name.

   a) Construct a class diagram for this record system.

    b) Develop Java code for each class and a suitable Application class to exer-cise them.

    c) Extend the details held on staff and students.

3. A bank offers two kinds of account to its customers that they can make withdrawals from, deposit funds into and enquire as to the current balance. The first is an ordin-ary account and the second a current account. Both have an account number and hold the current balance. However, the current account has an overdraft limit that is normally agreed with the bank manager when the account is created. Withdrawals can be made up to the overdraft limit. There is no overdraft limit available for ordin-ary accounts. Clearly there is no limit on deposits that can be made in either case.

    a) Construct a class diagram for the bank accounts.

    b) Develop Java code to define each class.

    c) Construct the Java coding for an Application object to exercise the system.

4. Academic staff in a university are designated as either a lecturer or a senior lecturer. Both teach students but a lecturer does a small amount of research while a senior lecturer is expected to do significantly more.

    a) Construct a class diagram for academic staff.

    b) Define outline Java code for each class assuming the polymorphic effect is to be used to its full extent. Method bodies should consist of simple insert messages to the screen such as,

```
ConsoleIO.out.println("Doing some teaching");
ConsoleIO.out.println("Doing a little research");
```

    and

```
ConsoleIO.out.println("Doing extra research");
```

    c) What changes would be necessary if senior lecturers carry out administrative tasks but lecturers do no administration?

5. A class diagram consists of several symbols. They may be either a class symbol or a relation symbol. With the former the symbol may describe a concrete or abstract class. With the latter the symbol may describe an association, aggregation or spe-cialization relation. A relation symbol always connects two class symbols and a class symbol may have zero or more relation symbols connected to it.

    a) Construct a class diagram to describe a class diagram.

    b) Develop outline Java code for each class. In particular focus on an operation draw that draws a complete class diagram on a computer screen.

6. Object-oriented software is commonly developed as successive versions. The first version might be quite rudimentary and later versions increasingly sophisticated. Although this approach is very useful it does pose the problem of version control. We must always be sure that a later version can be used in place of an earlier one. In other words a new version must be compatible with earlier ones. Assume that the versions are numbered as 1.1, 1.2, 1.3, . . . then 2.1, 2.2, 2.3 and so on. Version 1 objects should have the same interface as should version 2 objects.

    a) Construct a class diagram and corresponding Java that ensures that the Application object for each version complies with this requirement.

    b) Develop an Application object to implement a simple version controller.

7. If we return to exercise 4 we find that university staff consist of not just lecturers and senior lecturers but readers and professors as well. They all teach students to the same extent but in addition:
   - a lecturer does a small amount of research
   - a senior lecturer does a moderate amount of research and some administration
   - a reader does a lot of research but no administration
   - a professor does a lot of research and some administration.
   a) Construct a class diagram for academic staff.
   b) Define outline Java code for each class.

8. Sometimes an aggregate component is composed of another aggregate component. We refer to it as a composite object. For example, we might have a picture object that is composed of some text, lines, rectangles and another picture. Clearly a picture that is embedded in a picture may have some text, lines, rectangles or another picture. In other words a component of a composite object can be just a single component (a leaf) such as a piece of text, a line or a rectangle or it may have another composite such as another picture.

   When constructing a composite object it is important that we develop a simple way of treating a leaf and a composite in the same way. The class diagram of figure 5.10 solves the problem.

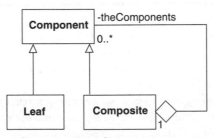

Figure 5.10   *Class diagram for a composite object*

The class **Component** acts as a common interface for a **Leaf** or **Composite** object. Clients refer to **Component**s and by using the polymorphic effect we can implement operations advertised by it in the **Leaf** and **Composite** classes as appropriate. Typical operations allow clients to **add**, **remove** and **display Component**s. The first two allow a suitable architecture to be constructed and the last allows its inspection. A **Leaf** object would normally implement operations fully while a **Composite** object would iterate over its components sending messages to each as appropriate.
   a) Construct an instance diagram showing a **Component** composed of several **leaf**es and **Composite**s.
   b) Develop sequence diagrams to show how a **Component** object might implement the operations **add** and **display**.

As this is a rather complex situation that occurs quite frequently we can capture it as a design pattern (Gamma 1994) which we can name as the *composite design pattern*. To accomplish this we specialize the **Leaf** class into **Text, Line** and **Rectangle** classes and the Composite class into the Picture class. They have responsibility for implementing operations such as display that are unique to them. Operations that are not specific are implemented by the classes Component, Leaf and Composite as appropriate.

c) Develop Java code for the composite design pattern.

d) Show how it might be used with the picture example.

9. In a hospital there are a number of wards each of which may be empty or have on them one or more patients. Each ward has a unique name. The hospital has an administration department that is responsible for recording information about the hospital's wards and the patients that are on each ward.

The doctors in the hospital are organized into teams, each of which has a unique team code. Each team is headed by a consultant doctor who is the only consultant doctor in the team. The rest are all junior doctors at least one of which must be at grade 1. Each doctor is in exactly one team. The administration department keeps a record of these teams and the doctors allocated to each team.

Each patient is under the care of a single team of doctors. The consultant who heads that team is responsible for the patient. A patient may be treated by any number of doctors but they must all be in the team that cares for the patient. A doctor can treat any number of patients.

The system should provide support for the administration department by recording information about:

• wards, including their name

• doctors, including their name, address (and for junior doctors their grade) and the way in which they are organized into teams

• patients, from admission to discharge, including their name, address, date of birth and the ward the patient is on

Patients are identified by the name of the ward they are on together with their name. Similarly doctors are identified by the team code and their name.

The system should be able to do the following:

• display the name of the consultant doctor responsible for a patient

• display the team code of the team caring for a patient

• list the details of all of the doctors who treated a patient

• list the details of each patient, including their personal data and the ward they are on.

On admission to the hospital the patient's personal details are recorded together with the name of the ward she is on and the team code for the team that is caring for her. When a patient is treated by a doctor the system is given the name of the patient, ward and doctor together with the team code for the doctor. When a patient is discharged the system is given the name of the patient and the ward the patient is on. It should then remove all information relating to that patient.

10. You are required to develop software to support the administration of a hotel. The major features of the hotel are:

• there are three floors numbered 1, 2 and 3 each of which has up to five rooms

• most rooms are ordinary bedrooms but some are used for conferences

• conference rooms may have study rooms associated with them

- study rooms are simply modified bedrooms
- not all study rooms are associated with a conference room
- not all floors have the same combination of rooms
- each room has a room number, e.g. 201 for room 1 on floor 2
- each room has a maximum occupancy, i.e. the number of people that can use it
- conference rooms also have a name

Staff should be able to obtain details of each room on each floor in a variety of ways. As a minimum, the user must be able to request for a report for:

- all the rooms on all floors
- all the rooms on a particular floor or
- a particular room on a particular floor

and to decommission a room:

- remove a particular room from a given floor

If the room is a bedroom then its number and maximum occupancy are displayed. However, if it is a conference room then its name and the room number of each study room associated with it must also be given. A study room displays the same information as a bedroom.

11. Revisit exercise 22 from chapter 4. Reflect on your solution given the new material from this chapter. Now develop the problem using specialization.

Case Study: The Library Application Revisited

This case study uses the same problem domain as that of chapter 4. It is intended to highlight more advanced elements of an OOAD. As with the first case study we emphasize how the development revolves round a number of iterations. All diagrams and the accompanying Java code were developed using the ROME modelling tool.

6.1 Specification

In the case study of chapter 4 the library held only books. In this version our library stocks both books and journals. An outline specification for the system is:

> *A library has a name and holds a number of stock items that may be either books or journals. Books and journals both have a title and a unique catalogue number. However, each book has an author and each journal has its date of publication and the name of the editor. The system should be able to display the stock items available for loan and those that are out on loan. At some point in the future the library will hold other stock items such as videos and compact disks.*
>
> *There are registered borrowers each with a unique name. A borrower may borrow and return a book or journal. The system should record each transaction. To record the borrowing of a book or journal the name of the borrower and the catalogue number for the publication is required. To record that a book or journal has been returned only the catalogue number is required.*
>
> *The system should also be able to display details of the stock items out on loan to borrowers.*

We are required to develop an application to support the librarian.

6.2 Iteration 1

This version has two aims. The first is to make sure that the initial object model developed is a good reflection of the problem domain. Clearly if it is not correct then the rest of the development effort is severely jeopardized. The second aim is more concerned with the implementation of the final system. It is to demonstrate that the polymorphic effect described in chapter 5 can be used successfully.

6.2.1   Establish use-cases

As in chapter 4 we start by developing a use-case diagram, supporting documentation and test-cases. However we can take this opportunity to explore use-cases in more detail. For example, we use the term *scenario* to refer to a specific sequence of inter-actions between the librarian (an actor) and the system.

As a scenario is just one path through a particular use-case it is often referred to as its *use-case instance*. More generally:

- a use-case is a collection of scenarios that define specific behaviours of the use-case and
- a use-case diagram is a collection of use-cases that define specific behaviours of the system.

This observation is helpful as it gives us a better understanding of the relationship between a system, its use-cases and their scenarios. Bearing this is mind, if we consider how we might document a use-case then it is clear that we should take into account the various scenarios it supports.

The most important scenario is the one that leads to a successful outcome without any difficulties arising. It represents the *basic flow* of events through the use-case. For example, in the "Add one book" use-case from chapter 4, the basic flow occurs when the librarian creates a new book that is added to the library. The creation of the book and adding it to the library happen successfully at the first attempt.

However, there may be other scenarios that may arise and some may result in failure. For example, the catalogue number of a new book entered by the librarian may not be unique and therefore the book created cannot be added to the library. Others may require additional actions to be taken but still result in success. For example, an invalid catalogue number may be entered by the librarian and then re-entered before the new book or journal can be created. Having done so it can be added to the library. Clearly these scenarios should be identified and documented as they represent an *alternative flow* of events through the use-case.

An important point to realize is that the basic flow is most often just a straightforward sequence of steps that lead to success. This makes it easy to understand and stresses that it is the main success path, i.e. there are no "problems". By way of contrast, alternative flows often take account of more complex situations, e.g. repeated data entry. Therefore they can be more difficult to understand. This means that we must be careful when docu-menting a use-case with more than one scenario. We might use the following template (Larman 2002):

Use-case: Name

> **Basic flow**
>> The scenario that is a typical success path

> **Alternative flows**
>> All other scenarios (successful or otherwise)

When applied to the "Add one book" use-case we have:

Use-case: Add one book

Basic flow
1. User enters the title of the new book
2. User enters the name of the author of the new book
3. User enters the catalogue number of the new book
4. A new book is created
5. The new book is added to the library and is available for loan
6. The user is informed that the addition was successful

Alternative flows
1a, 2a, 3a The system detects an invalid data entry:
1. The system informs the user
2. The user re-enters the data
   Steps 1–2 (alternative flows) are repeated until
   the data entered is valid

Notice that the steps (events) are numbered to indicate the time order in which they occur. Alternative flows are annotated with respect to the branch point in the main flow. For example, 1a indicates a deviation from the success scenario at step 1. The system detects an invalid data entry and informs the actor. The numbered actions that follow specify the subsequent steps that are taken before returning to the basic flow at the branch point. For brevity we group three alternative flows as a comma-separated list, i.e. 1a, 2a, 3a, as they take the same sequence of remedial actions.

On completion of an alternative flow, the main flow continues at the next step unless the scenario indicates otherwise, e.g. cancellation. For example, if a new book or journal cannot be added to the library then we might have:

Alternative flows
// ...
5a The system detects a duplicate catalogue number:
1. The system informs the user
2. The scenario is abandoned

At this point it is worth remembering that non-specialists need to understand use-cases when they help develop system requirements. We expect them to be able to make an informed judgement on a use-case. This means that we try to avoid using the full power of the UML in their construction. For example, we have chosen not to use an activity diagram to document a use-case where it may have been expected. We should also resist the strong temptation to add too much detail at this early stage. Even worse, we must not attempt to decompose use-cases into simpler ones. To do so may well lead to an analysis that is not object-oriented but is based on a top-down (divide-and-conquer) structured approach.

Bearing these points in mind, UML use-cases may have relationships with each other. The most common one is where a use-case shares some part of its functionality with another. It is called the *include relationship*. For example, we can anticipate that the "Add one book" and "Add one journal" use-cases involve the librarian entering basic details, i.e. the title and catalogue number. We capture this common behaviour as shown in figure 6.1 in which the include relationship is depicted as a stereotype on a directed line. For the textual documentation we now have:

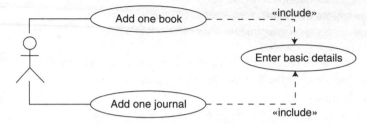

Figure 6.1   *Use-cases with an include stereotype*

Use-case: Enter basic details

Basic flow
1. User enters the title of the new book
2. User enters the catalogue number of the new book

Alternative flows
1a, 2a The system detects an invalid data entry:
1. The system informs the user
2. The user re-enters the data
   Steps 1–2 (alternative flows) are repeated until the data entered is valid

Use-case: Add one book

Basic flow
1. Include Enter basic details
2. User enters the name of the author of the new book
3. A new book is created
4. The new book is added to the library and is available for loan
5. The user is informed that the addition was successful
Alternative flows
   // ...

Use-case: Add one journal

Basic flow
1. Include Enter basic details
2. User enters the name of the editor of the new journal
3. User enters the date of publication of the new journal
4. A new journal is created
5. The new journal is added to the library and is available for loan
6. The user is informed that the addition was successful

Alternative flows
   // ...

Other more advanced use-case features exist in the UML, e.g. the *extend relationship*. However, it is our opinion that for the vast majority of circumstances the complexity they introduce conflicts with the primary aim of a use-case. We don't make use of them even though our use-case diagrams may sometimes be less elegant and the resulting documentation verbose. The important point is that both have an obvious meaning that can be readily understood. Interested readers should consult more advanced texts for a more extensive discussion of use-case relationships (Larman 2002).

We are now in a position to develop a use-case diagram as shown in figure 6.2.

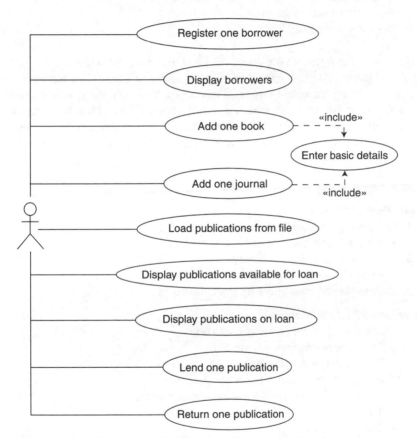

Figure 6.2   *A use-case diagram for the* Library *application: iteration 1*

When compared to the corresponding diagram in the case study of chapter 4, the main difference is that a book or journal is borrowed and returned rather than just a book. Typical supporting documentation is:

Use-case: Add one book

>   **Basic flow**
>   1. Include <u>Enter basic details</u>
>   2. User enters the name of the author of the new book

   3. A new book is created
   4. The new book is added to the library and is available for loan
   5. The user is informed that the addition was successful

Alternative flows

2a The system detects an invalid data entry:
   1. The system informs the user
   2. The user re-enters the data
      Steps 1-2 are repeated until the data entered is valid
4a The system detects a duplicate catalogue number:
   1. The system informs the user
   2. The scenario is abandoned

Hopefully any intelligent person could read and understand it without too much difficulty. Also bear in mind that some readers may choose to ignore the alternative flows.

As in the previous case study we must develop preliminary test-case documentation for each use-case. However, we should take into account the fact that there may be several different scenarios to test. Fortunately this is not too difficult as we have summarized the expected outcome of each test-case in the use-case documentation. For example, we might have:

Test-case: Add one book

   Basic flow outcomes
      1. The user is prompted to enter data
      2. A new book with the data entered is created
      3. The book is added to the library.
      4. The book is available for loan
      5. The user is informed of a successful outcome

   Alternative flows outcomes
      1a The system detects an invalid data entry:
         1. The system informs the user
         2. The user re-enters the data
            Steps 1–2 are repeated until the data entered is valid
      3a The system detects a duplicate catalogue number:
         1. The system informs the user
         2. The scenario is abandoned

Each numbered outcome is effectively the test we wish to demonstrate that we have achieved.

6.2.2 Analysis

As in the chapter 4 analysis, we start by considering the use-case diagrams. It clearly shows that a user must be able to borrow, return and display a book or journal from the library.

We readily identify the **Library** class as before. The main issue is whether a book and a journal are essentially the same kind of entity or whether they are sufficiently distinct to designate them as objects. This decision is crucial as it determines the fundamental structure of the final system.

Clearly journals and books have similarities in that they are both a kind of publication with a title and catalogue number in common. However, they are also quite different in some important respects. For example:

- a book has an author
- a journal has an editor and a date of publication
- users of the library perceive books and journals to be both conceptually and physically different
- the library may treat books and journals differently, e.g. there may be different rules for lending and returning them or storing them in different parts of the library

Our conclusion is that their differences are sufficient to merit their designation as distinct object types. Therefore we can identify the classes **Book** and **Journal**. The object diagram of figure 6.3 documents this decision.

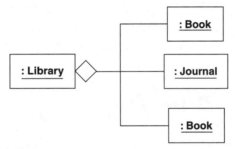

Figure 6.3 *An object diagram*

6.2.3 Design

It is important to recognize that two distinct relationships are developed during this design phase. The first is based on specialization and gives information about the relationships between a class and its ancestors. The second is based on aggregations and gives information about the scope and visibility of classes. Usually both relationships are combined in the same class diagram.

If we consider the first relationship, then as we intend using the polymorphic effect, it is reasonable to decide that **Book** and **Journal** objects are examples of a more general class. From the text of the outline system specification we can name it **Publication**. Our intention is to send messages through a **Publication** reference but we anticipate that each message is actually received by a **Book** or **Journal**. Figure 6.4 describes this specialization hierarchy.

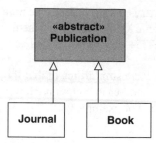

Figure 6.4 *A specialization hierarchy*

Notice that we have determined that the class Publication should be **abstract** and therefore no instance of it can be made. As discussed earlier in chapter 4, we can declare a reference to a Publication as normal but it must reference a concrete object such as a Book. This means that much of our code does not distinguish between Books and Journals. We just treat them uniformly as Publications.

If we consider the second relationship then we can reuse the model developed for version 3 from chapter 4. Recall that the aim of this iteration is to demonstrate that we can make use of the polymorphic effect using a basic model. If necessary we can change the model in a later iteration. Therefore the Library object references zero or more Publications and BorrowerRecords. Each BorrowerRecord references zero or more Publications and each Publication may reference a BorrowerRecord. With this new information the architecture can be updated as shown in figure 6.5.

Figure 6.5 *Initial class diagram (from model Lib4_3.uml)*

At this corresponding point in the chapter 4 case study an object diagram was particularly useful. With the introduction of polymorphism it is even more important. Figure 6.6 illustrates a typical configuration of objects. It shows a Library with three Books and one Journal. There are two registered borrowers. One has borrowed a Book and the other a Book and a Journal. The remaining Book is available for loan. Note that there are no Publication objects present: only Books or Journals.

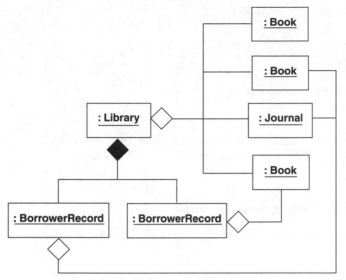

Figure 6.6 *An object diagram*

6.2.4 *Implementation*

This part of the development is relatively straightforward. We can use much of the work done in chapter 4. The essential structure is unchanged. Therefore we have an Application object whose run method communicates with the user through a text-based interface. As before, the run method creates an Action object that is a façade for the Library. However, there is the one crucial difference in that each Book reference parameter used in the chapter 4 solution now becomes a Publication reference. Similarly, methods declare Publication references rather than Book references. This simplifies much of the code as we do not have to distinguish between Books and Journals: they are just Publications.

6.2.4.1 Architectural code

Following on from the preceding section we have the following outline code for the Library:

```
package librarysubsystem;
import java.util.*;
public final class Library {
    // ----- Operations ----------
    public void    addOnePublication(Publication aPublication) {
        int catalogueNumber = aPublication.getCatalogueNumber();
        Publication foundPublication = this.getPublication(catalogueNumber);
        //
```

```
      if(foundPublication == null) {
         // ...
      } else {
         // ...
      }
   } // method: addOnePublication
   // ...
   // ----- Relations ----------
   private java.util.Collection    theBorrowers;    // of BorrowerRecord
   private java.util.Collection    theLoanStock;    // of Publication

} // class: Library
```

Similarly the BorrowerRecord class code is:

```
package librarysubsystem;

import java.util.*;

public final class BorrowerRecord implements Comparable {
   // ----- Operations ----------
   public void    attachPublication(Publication aPublication) {
      theBorrowedPublications.add(aPublication);
      aPublication.attachBorrower(this);
    } // method: attachPublication

   public void    detachPublication(Publication aPublication) {
      theBorrowedPublications.remove(aPublication);
      aPublication.detachBorrower();
    } // method: detachPublication

   public final int    compareTo(Object obj) {
      //...
    } // method: compareTo

   public final boolean    equals(Object obj) {
      //...
    } // method: equals

   public final int hashCode()
    //...
} // method: hashCode
   // ...
   // ----- Relations ----------
   private java.util.Collection    theBorrowedPublications;    // of Publication
} // class: BorrowerRecord
```

This leaves us with the Publication class. As it acts as a base class for the Book and Journal classes, it implements common features pertaining to the architecture. It follows that it implements the Comparable interface (allowing its descendants to be compared on the same basis) and holds a reference to its borrower. This results in the following outline code:

```
package librarysubsystem;
public abstract class Publication implements Comparable {
    // ----- Operations ----------
    public void    attachBorrower(BorrowerRecord aBorrower) {
        theBorrower = aBorrower;
    } // method: attachBorrower

    public void    detachBorrower() {
        theBorrower = null;
    } // method: detachBorrower

    public final int    compareTo(Object obj) {
        //...
    } // method: compareTo

    public final boolean    equals(Object obj) {
        //...
    } // method: equals

    public final int    hashCode(){
        //...
    } // method: hashCode

    // ...
    // ----- Relations ----------
    private BorrowerRecord    theBorrower;
} // class: Publication
```

6.2.4.2 Method code

For the Library class all that remains is to complete its operation signatures and methods. They are essentially unchanged from chapter 4. Of course we must remember when naming operations that the Library refers to Publications rather than Books but that is the only difference. For example, we have getPublicationsIterator rather than getBooksIterator.

Similarly there are no significant changes to the BorrowerRecord class except for the trivial renaming of operations, e.g. getBorrowedPublicationsIterator not getBorrowedBooksIterator.

The Publication class has the common features of Books and Journals as well as having architectural responsibilities. This gives the following outline code:

```
package librarysubsystem;
    public abstract class Publication implements Comparable {
        // ----- Operations ----------
        protected    Publication(String aTitle, int aCatalogueNumber) {
```

```
      theTitle = aTitle;
      theCatalogueNumber = aCatalogueNumber;
      theBorrower = null;
   } // method: Publication
   protected    Publication() {
      this("",0);
   } // method: Publication
   public int    getCatalogueNumber() {
      return theCatalogueNumber;
   } // method: getCatalogueNumber
   public String    getBorrowerName(){
      String name = "";
       if(theBorrower != null) {
         name = theBorrower.getName();
       }
      return name;
   } // method: getBorrowerName
   public BorrowerRecord    getBorrower(){
      return theBorrower;
   } // method: getBorrower
   public String    toString() {
      return theCatalogueNumber + " : " + theTitle;
   } // method: toString
   // ...

   // ----- Attributes ----------
   private String    theTitle;
   private final int    theCatalogueNumber;
   // ----- Relations ----------
   private BorrowerRecord    theBorrower;
 } // class:Publication
```

Notice that even though it is **abstract** its constructors are **protected** to minimize
(even further) any chance of misuse by other classes that are not descendants.

When it comes to the Book and Journal classes we find that the majority of the work
has been done in the Publication class. All they have to do is to hold any additional
attributes, supply a constructor and redefine toString for use by the Action class. This
leaves us with:

```
package librarysubsystem;

public final class Book extends Publication {

   // ----- Operations ----------
   public Book(String aTitle, int aCatalogueNumber, String anAuthor) {
      super(aTitle, aCatalogueNumber);
```

```
        theAuthor = anAuthor;
    } // method: Book

    public Book() {
        this("", 0, "");
    } // method: Book

    public String    toString() {
        return super.toString() + " by " + theAuthor;
    } // method: toString

    // ----- Attributes ----------
    private String    theAuthor;
} // class: Book
```

and

```
    package librarysubsystem;

    public final class Journal extends Publication {

        public Journal(String aTitle, int aCatalogueNumber, String anEditor,
                        DateIF aDateOfPublication)
        {
            super(aTitle, aCatalogueNumber);
            theEditor = anEditor;
            theDateOfPublication = aDateOfPublication;
        } // method: Journal

        public Journal() {
            this("", 0, "", new DateImp());
        } // method: Journal

        public String    toString() {
            return super.toString() + " edited by " + theEditor + " on " + theDateOfPublication;
        } // method: toString

        // ----- Attributes ----------
        private String theEditor;
        private DateIF theDateOfPublication;
} // class: Journal
```

as the complete listings for the Book and Journal classes. The interface DateIF and the DateImp class are those described earlier in chapter 5. Notice how brief the class declarations are and the extent to which they reuse methods defined by the superclass. This is typical of a good object-oriented design. Hopefully we accomplish more and more in each iteration with less and less effort. This is a major benefit that should not be underestimated.

Finally, the Action class is updated to reflect the use-cases developed in section 6.2.1. However, its overall structure is unchanged from the case study of chapter 4. The only

real difference is that its methods recognize that Publications are borrowed and
returned to the Library. A typical method is:

```
// class Action
public void    addOneBook() {
    // Get book details from the user.
    ConsoleIO.out.print("\t" + "Enter the title >>> ");
    String title = ConsoleIO.in.readLine();
    ConsoleIO.out.print("\t" + "Enter the author >>> ");
    String author = ConsoleIO.in.readLine();
    ConsoleIO.out.print("\t" + "Enter the catalogue number >>> ");
    int catalogueNumber = ConsoleIO.in.readInt();
    //
    // Add the book to the library.
    theLibrary.addOnePublication(new Book(title, catalogueNumber, author));
    //
    // Display the outcome
    ConsoleIO.out.println("\n\t" + theLibrary.getStatus() );
} // method: addOneBook
```

Full listings of the classes developed in this section are to be found with the supplied
software (model Lib6_1.uml).

6.2.4.3   Testing

As in the chapter 4 case study we could make use of **private** operations such as
testUseCases and testErrorConditions to help automate the burden of testing. We
might even use a commercially available testing tool (http://www.junit.org). However,
we leave this as an exercise to the reader and assume that manual testing is sufficient for
the purpose of this discussion. Of course we should use the test cases documented in
section 6.2.1 as the basis for any tests. An outline of the Application run method is:

```
// class Application
public void   run() {
    // Create a unified interface (facade) for the Library
    Action action = new Action("Books-R-Us");
    //
    // Get and process the user's choice
    String choice = "";
    do {
        // Get the user's selection
        choice = this.getSelection();
        ConsoleIO.out.println();
        // Action the user's choice
        if(choice.equals("0")) {
            action.close();
        } else if(choice.equals("1")) { // Register one // borrower.
```

```
        action.registerOneBorrower();
    }
    //
    // One selection and the corresponding actions for each use-case
    // ...
    } else {
        action.unknownSelection();
    }
  } while(choice.equals("0") == false);
```

} // method: run

Assuming that two Books and a Journal have been added to the Library a typical successful testing outcome is:

```
0: Quit
1: Register one borrower
2: Display borrowers
3: Add one book
4: Add one journal
5: Load publications from file
6: Display publications available for loan
7: Display publications on loan
8: Lend one publication
9: Return one publication

    Enter the choice >>> 6

Books-R-Us :3 book(s) :0 borrower(s)

    Publications available for loan
      1: Java by Ken

      2: Basic by Sally

      3: UML edited by John on 17/2/2003
```

It clearly shows that we have been able use the polymorphic effect. The list of publications available for loan include both books and journals as shown in the report. We also believe that the model is a sensible one. Therefore we have achieved the aims of this iteration.

6.2.5   Reconcile model diagrams

There are no actions that we need take in this activity. However, it is always useful to make this decision explicit.

6.3   Iteration 2

This iteration has only one aim. It is that to demonstrate that we can significantly improve the previous design by using the full power of the object-oriented paradigm. In particular the design should be able to accommodate new requirements without the necessity for substantial changes. We must avoid having to redesign the system or recode large parts of it each time we conduct an iteration. Therefore the challenge is to develop a design by refactoring it so that is stable in the face of change.

6.3.1   Establish use-cases

As this iteration does not result from new requirements there are no changes to be made to the use-case diagram from the previous iteration.

6.3.2   Analysis

At this point we consider the design from a different perspective to that taken in the previous iteration. If we think in more general terms about the objects that populate the design then there is a good chance of success. For example, the Library can be viewed as a lender that just happens to lend Books and Journals. Perhaps in the future it will lend other items such as compact discs or videos. We may even have other kinds of lenders, e.g. a car hire company or a video shop.

This train of thought leads us to introduce an interface LenderIF. Our intention is that it should specify the essence of what it is to be a lender, not just a Library. Similarly an interface LoanItemIF specifies what it is to be a loan item, not just a Book or Journal.

This may appear to be a small change but it is extremely significant. Anticipating the design that follows, we intend that any class that implements LenderIF can be viewed as a lender. We are not restricted to the Library class. Similarly any class that implements LoanItemIF can be viewed as a loan item. Again we are not restricted to a single class such as Book. This gives us the element of flexibility we desire.

6.3.3   Design

It might seem sensible to implement LenderIF and LoanItemIF with problem domain-specific objects such as the Library, Book and Journal. However, we take a different approach and try to stay more general in our thinking. Therefore we implement each interface with an **abstract** class. The first is LenderImp that implements LenderIF and the second is LoanItemImp that implements LoanItemIF.

There is a zero-to-many aggregation relationship between a lender and loan items inferred from the system specification. This means that we can establish an initial architectural framework as shown in figure 6.7.

Notice that the Library, Publication, Book and Journal classes are absent. It is a general framework that can be applied to any lender or loan item. Our intention is to specialize the LenderImp and LoanItemImp classes when we need to "instantiate" a particular application. For example, the Library class is a specialization of LenderImp and Publication is a specialization of LoanItemImp in this case study. Figure 6.8 illustrates the relationships between the interfaces and classes concerned.

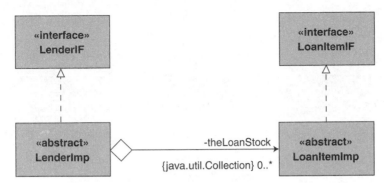

Figure 6.7   *An initial architectural framework*

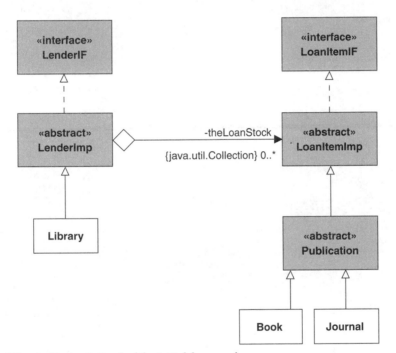

Figure 6.8   *An "instantiation" of the initial framework*

It is important to realize that the problem domain-specific Library and Publication classes have no responsibility for the architecture. This is all taken care of by the **abstract** classes of the framework. It is this fact that goes a long way towards giving us the resilience to change that we desire.

To complete this part of the design we focus on the BorrowerRecord class introduced in chapter 4. The approach used for lenders and loan items leads us to a BorrowerRecordIF interface implemented by an **abstract** BorrowerRecordImp class. We infer aggregation relations between the BorrowerRecordImp, LenderImp and LoanItemImp classes from the system specification.

A suitably modified architectural framework and its "instantiation" is shown in figures 6.9(a) and 6.9(b).

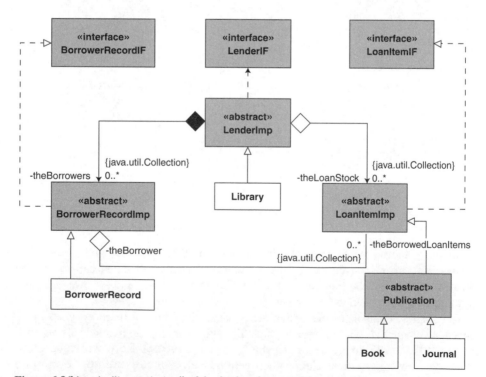

Figure 6.9(a)   *The final architectural framework*

Figure 6.9(b)   *An "instantiation" of the final architectural framework*

Although it may appear rather complex at first sight, it is in fact quite straightforward. Essentially it consists of three layers. The top layer has the interfaces that specify the minimum behaviour that any lender, loan item or borrower record must have. The middle layer implements the architectural relationships that exist between them. Finally, the bottom layer has the problem domain-specific objects that populate our design.

6.3.4   Implementation

Clearly the new design affects the manner in which we implement the final system. One approach is to implement the first two layers to produce a framework that can be reused in subsequent iterations. In keeping with our decision to partition code into Java packages, it seems sensible to locate them in a package named loansubsystem.

The implementation of the bottom layer can be treated as a separate activity. As before we can locate the objects developed in a librarysubsystem package. Our intention is that it will make use of the interfaces and abstract classes exported by the loansubsystem package. The relationship between the two packages is shown in figure 6.10.

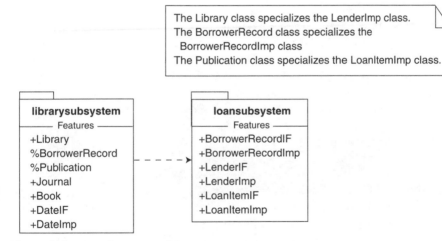

Figure 6.10   *A package-view of the system*

6.3.4.1   Architectural framework code

During this activity we establish the detailed nature of the three interfaces LenderIF, LoanItemIF and BorrowerRecordIF. After some thought, and using our previous experiences, we conclude that a lender must be able to:

- register a borrower
- add a loan item
- record the loan and return of a loan item

- provide a means of displaying its borrowers and loan items
- provide a textual representation for display purposes and
- provide status information

Translating this into Java we have:

```
package loansubsystem;

public interface LenderIF {
    // ----- Operations ----------
    public abstract void    registerOneBorrower(String aBorrowerName);
    public abstract void    addOneLoanItem(LoanItemIF aLoanItem);
    public abstract void    lendOneLoanItem(int aCatalogueNumber,
                                            String aBorrowerName);
    public abstract void    returnOneLoanItem(int aCatalogueNumber);
    public abstract Iterator    getBorrowersIterator();
    public abstract Iterator    getLoanItemsIterator();
    public abstract String    toString();
    public abstract String    getStatus();
} // interface: LenderIF
```

Similarly a loan item must be able to:

- attach and detach a borrower
- provide its catalogue number
- provide its borrower and its borrower name
- provide a textual representation for display purposes

These requirements map to:

```
package loansubsystem;
public interface LoanItemIF {
    // ----- Operations ----------
    public abstract void    attachBorrower(BorrowerRecordIF aBorrower);
    public abstract void    detachBorrower();
    public abstract int    getCatalogueNumber();
    public abstract BorrowerRecordIF    getBorrower();
    public abstract String    getBorrowerName();
    public abstract String    toString();
} // interface: LoanItemIF
```

Finally, a borrower record must be able to:

- attach and detach a loan item
- provide its name
- provide a means of displaying its loan items and
- provide a textual representation for display purposes

giving:

package loansubsystem;

public interface BorrowerRecordIF {

    // ----- Operations ----------
    public abstract void   attachLoanItem(LoanItemIF aLoanItem);
    public abstract void   detachLoanItem(LoanItemIF aLoanItem);
    public abstract Iterator   getLoanItemsIterator();
    public abstract String   getName();
    public abstract String   toString();

} // interface: BorrowerRecordIF

If we now focus on the second layer of the design then we implement each interface with a corresponding **abstract** class. They are of course located in the loansubsystem package.

As might be expected, the code for the LoanItemImp class that implements LoanItemIF is taken more or less directly from code previously developed for the Publication class of the previous iteration. The major difference is that formal parameters and some attributes are interface and not class declarations. This gives us the flexibility to refer to any object whose class implements the corresponding interface.

An outline of the LoanItemImp class is:

package loansubsystem;

public abstract class LoanItemImp **implements** Comparable, LoanItemIF {

    // ----- Operations ----------
    public   LoanItemImp(**int** aCatalogueNumber) {
      theCatalogueNumber = aCatalogueNumber;
      theBorrower = **null**;
    } // method: LoanItemImp

    public final void   attachBorrower(loansubsystem.BorrowerRecordIF aBorrower) {
      theBorrower = (BorrowerRecordImp) aBorrower;
    } // method: attachBorrower

    public final void   detachBorrower() {
      theBorrower = **null**;
    } // method: detachBorrower

    public final   loansubsystem.BorrowerRecordIF getBorrower() {
      return theBorrower;
    } // method: getBorrower

    public String   toString() {
      return String.valueOf(theCatalogueNumber) + " : ";
    } // method: toString

    // getCatalogueNumber, equals, compareTo and hashCode are as in
    // previous versions except that in method compareTo the cast is to a
    // LoanItemIF not a Book or Publication.
    // ...

```
// ----- Attributes ----------
private final int   theCatalogueNumber;
// ----- Relations ----------
private   BorrowerRecordImp theBorrower;
} // class: LoanItemImp
```

Notice how we have imposed the requirement that all loan items must have a catalogue number represented as an **int**. It acts as the basis for any comparisons that are made in the equals, compareTo and hashCode methods. Also we control the redefinition of a method in a subclass by including or omitting the **final** qualifier.

Similarly, the code for the BorrowerRecordImp class is based on the BorrowerRecord class of iteration 1. Its outline is:

```
package loansubsystem;

import java.util.*;

public abstract class BorrowerRecordImp implements Comparable, BorrowerRecordIF {
   // ----- Operations ----------
   public   BorrowerRecordImp(String aName) {
      theName = aName;
      theBorrowedLoanItems = new LinkedList();
   } // method: BorrowerRecordImp

   public final void   attachLoanItem(loansubsystem.LoanItemIF aLoanItem) {
      aLoanItem.attachBorrower(this);
      theBorrowedLoanItems.add(aLoanItem);
   } // method: attachLoanItem

   public final void   detachLoanItem(loansubsystem.LoanItemIF aLoanItem) {
      theBorrowedLoanItems.remove(aLoanItem);
      aLoanItem.detachBorrower();

   } // method: detachLoanItem

   public String   toString() {
      int borrowedItemsCount = theBorrowedLoanItems.size();
      return theName + " : " + borrowedItemsCount + " loan item(s)";
   } // method: toString

   // getName, getBorrowedLoanItemsIteratorequals, compareTo and hashCode are as in
   // previous versions except that in method compareTo the cast is to a
   // BorrowerRecordIF not a BorrowerRecord.

   // ...

   // ----- Attributes ----------
      private String   theName;
   // ----- Relations ----------
      private java.util.Collection   theBorrowedLoanItems;   // of LoanItemImp
} // class: BorrowerRecordImp
```

As with the LoanItemImp class, notice that we have imposed the requirement that all borrower records must have a name represented as a String. It is used for all comparisons of borrower records.

When it comes to the LenderImp class we use the Library class of iteration 1 as our guide. Virtually all of the code that results is the same except that there are appropriate interface and not class declarations. For example, in the method for addOneLoanItem we have:

```
// class LenderImp
public final void   addOneLoanItem(LoanItemIF aLoanItem) {
    int  aCatalogueNumber = aLoanItem.getCatalogueNumber();
    LoanItemIF  foundLoanItem = this.getLoanItem(aCatalogueNumber);
    // ...
} // method: LenderImp
```

Notice that getLoanItem returns a reference to a LoanItemIF. This means that no matter what class of object it actually refers to we know that it must conform to the LoanItemIF interface. Therefore we can send it the message getCatalogueNumber later in the method body. Many other examples of this approach are to be found in the code for this iteration.

An important point to understand is that when a reference to an interface is declared, only messages corresponding to operations advertised by that interface can be sent through that reference. Even though the class of the object referenced advertises other operations, messages corresponding to them cannot be sent.

We can use this fact to our advantage by treating an object of some arbitrary class as one that conforms to a specific interface. No matter what class it belongs to, we know a subset of its operations that can be used, i.e. those advertised by the interface. In a sense we view it as an interface object not an object of the class it actually belongs to. This allows us to write code that is generally applicable.

When implementing the LenderImp class, the only real problem lies in the method for registerOneBorrower. The difficulty is that it has the responsibility for creating each borrower record the lender holds. Unfortunately it cannot do so as it has no knowledge of what class the borrower record actually belongs to. The reason is that it is problem domain-specific. One solution is to defer this decision by introducing an **abstract** operation createBorrower declared as:

```
// class LenderImp
private abstract BorrowerRecordIF   createBorrower(String aBorrowerName);
```

The intention is that some concrete subclass (Library) with the knowledge of what class the borrower record belongs to (BorrowerRecord) actually implements this operation. We can then make use of it in registerOneBorrower as follows:

```
// class LenderImp
public final void   registerOneBorrower(String aBorrowerName) {
    BorrowerRecordIF  foundRecord = this.getBorrowerRecord(aBorrowerName);
    //
    if(foundRecord == null) {
        BorrowerRecordIF  borrowerRecord = this.createBorrower(aBorrowerName);
```

```
    boolean result = theBorrowers.add(borrowerRecord);
    // ...
} // method: registerOneBorrower
```

We have achieved the apparently impossible by creating an object without knowing what class it belongs to! This is because createBorrower returns a reference to an object that implements BorrowerRecordIF. It does not matter that its method is to be found in a subclass of LenderImp not LenderImp itself.

This subtle solution to a difficult problem is an example of a design pattern. In fact the method registerOneBorrower is a *Template Method* design pattern that uses a *Factory Method* design pattern createBorrower. Chapter 8 expands upon this important field of study.

Full listings of the interfaces and classes developed in this section are to be found with the supplied software (model Lib6_2.uml).

6.3.4.2   Problem domain code

Having established a reusable architectural framework we can now focus on those classes that are specific to the problem domain. They are the classes Publication, Book, Journal, Library and BorrowerRecord previously identified. All being well, we should find their implementation is quite straightforward.

Starting with the BorrowerRecord class we discover that there is no work to do! All that is required is to specialize the **abstract** class BorrowerRecordImp then implement a constructor using superclass methods. The resulting code is as follows:

```
package librarysubsystem;

import loansubsystem.*;

public class BorrowerRecord extends BorrowerRecordImp {

    // ----- Operations ----------
    public   BorrowerRecord(String aBorrowerName) {
       super(aBorrowerName);
    }
}  // class: BorrowerRecord
```

In keeping with our earlier decision the code is located in the librarysubsystem package.

For the Library class we intend giving it a name. This leads to a parameterized and default constructor that initialize the name attribute. We must also redefine toString to reflect the fact that a Library has a name. All that remains is to define the method for createBorrower inherited from the superclass BorrowerRecordImp by creating a BorrowerRecord. The code for the Library class is:

```
package librarysubsystem;

import loansubsystem.*;

public final class Library extends LenderImp {

    // ----- Operations ----------
    public   Library(String aName) {
       super();
```

```
      theName = aName;
  }
  public   Library() {
    this("");
  }
  public BorrowerRecordIF    createBorrowerRecord(String aBorrowerName) {
    return new BorrowerRecord(aBorrowerName);
  }
  public String    toString() {
    return theName +   " : " + super.toString();
  }
  // ----- Attributes ----------
  private String    theName;
} // class Library
```

Finally we have the Publication class hierarchy. Again it does not present any major difficulties. We concern ourselves only with the characteristics of Publications, Books and Journals. The fact that they are loan items is irrelevant. This has all been taken care of by the architectural framework.

The role of the Publication class is to act as a common protocol for all of its descendants. It also has the title and catalogue number attributes common to all Publications. As no instances of it are required it is qualified as **abstract** and has package scope only. There is a default and a parameterized constructor to initialize attributes. However, they are **protected** to avoid the possibility of misuse by classes that are not descendants. We also redefine toString from the superclass LoanItemImp and allow it to be redefined by descendants. The resulting code for the Publication class is:

```
package librarysubsystem;

import loansubsystem.*;

abstract class Publication extends LoanItemImp {
    // ----- Operations ----------
    protected    Publication(String aTitle, int aCatalogueNumber) {
      super(aCatalogueNumber);
        theTitle = aTitle;
    }
    protected    Publication() {
      this("", 0);
    }
    public String    toString() {
      return super.toString() + theTitle;
    }
    // ----- Attributes ----------
    private String    theTitle;
} // class Publication
```

The concrete class Book is a specialization of the **abstract** Publication class and is available outside the package. It has one additional attribute, initialized by default and parameterized constructors, that represents the author of a Book. The redefined method for toString reflects this additional attribute. The code for the Book class is:

```
package librarysubsystem;

public final class Book extends Publication {
    // ----- Operations ----------
    public   Book(String aTitle, int aCatalogueNumber, String anAuthor) {
      super(aTitle, aCatalogueNumber);
      theAuthor = anAuthor;
    }
    public   Book() {
      this("",0, "");
    }
    public String    toString() {
      return super.toString() + " by " + theAuthor;
    }
    // ----- Attributes ----------
    private String    theAuthor;
}   // class Book
```

Notice just how brief the code is even though a Book can be loaned, borrowed, displayed and held in a collection.

The concrete class Journal is very similar to Book. The only real difference is that it has two additional attributes. The first is the editor and the second the date of publication. This gives the following code for the Journal class:

```
package librarysubsystem;

public class Journal extends Publication {
    // ----- Operations ----------
    public Journal(String aTitle, int aCatalogueNumber, String anEditor,
                   DateIF aDateOfPublication)
    {
      super(aTitle, aCatalogueNumber);
      theEditor = anEditor;
      theDateOfPublication = aDateOfPublication;
    }
    public   Journal() {
      this("", 0, "", new DateImp(1,1,2002));
    }
    public String    toString() {
      return super.toString() + " edited by " + theEditor + " on " + theDateOfPublication;
    }
```

```
// ----- Attributes ----------
private String    theEditor;
private DateIF    theDateOfPublication;
}
```

As with iteration 1, we have used the DateIF interface and DateImp class discussed in chapter 5 for the date of publication of a Journal. The Action and Application classes from the previous iteration are unchanged. Full listings of the classes developed in this section are to be found with the supplied software (model Lib6_2.uml).

6.3.4.3 Testing

During this activity there is no need to introduce any new tests as there are no new requirements. However, it is vital that all of the tests conducted in the previous iteration give exactly the same results. This is in fact the case.

6.3.5   Reconcile model diagrams

As with the previous iteration there are no explicit actions that need be taken. However, it is worth noting that Java code is also part of the documentation for the system. In fact there is a school of thought (*Agile Programming*) that concludes that it is the most important documentation we have. Therefore it is important to consider this when taking part in this activity. For example, we make extensive use of javadoc comments (see http://java.sun.com/j2se/javadoc).

6.3.6   A review of iteration 2

The aim of this iteration is to make the design more general and resilient to change by using the full power of the object-oriented paradigm. It is hard to look into the future! However, to help us decide just how successful we have been, we can explore some requirements that we anticipate will arise.

First of all, it is probable that different kinds of Publication will be introduced. For example, there may be Ordinance Survey maps held by the Library. This presents no real problem as all we have to do is to introduce an OSMap class as a specialization of Publication. Figure 6.11 illustrates.

Similarly if there is a requirement to specialize the Book or Journal classes then the impact on the design would be much the same. Better still, exactly the same conclusions would apply to the specialization of the Library or BorrowerRecord classes.

The ease with which we are able to cope with this type of change lies in the fact that the classes we introduce are treated as examples of a superclass, e.g. a Publication. It makes no difference that they belong to a subclass. The polymorphic effect permits messages sent through a superclass reference to be received by a subclass object.

It is also likely that there may be loan items that are not publications. For example, we might have compact disks. Again this requirement presents no real difficulty. We

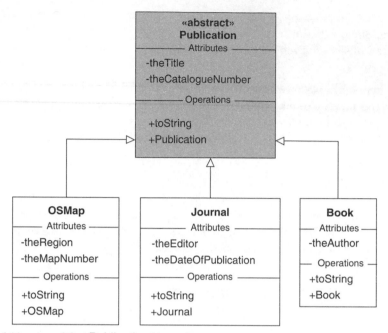

Figure 6.11 *A modified Publication hierarchy*

might introduce an **abstract** class Recording analogous to the Publication class. It would in turn be specialized into a CompactDisk class just as the Book class was specialized from Publication. Figure 6.12 illustrates a modified LoanItemImp hierarchy.

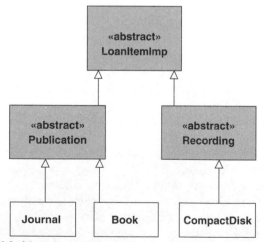

Figure 6.12 *A modified LoanItemImp hierarchy*

The reason why this kind of change is so straightforward is that the Recording class specializes LoanItemImp. The framework code references all loan items as LoanItemImps and so it makes no difference to what subclass they actually belong. As with the previous change it is the polymorphic effect at work. Specializations of LenderImp and BorrowerRecordImp would have a similar minimal impact on the design.

One serious difficulty that might arise is that the Recording class may already be a specialization of some other class. For example, it might be a kind of MusicItem. As Java does not support multiple inheritance we cannot specialize it from LoanItemImp. As this is the kind of problem that will inevitably occur again it is best to find a generally applicable solution.

We can introduce a LoanItemImpAdapter class that is a specialization of LoanItemImp. Our intention is that it should act as a replacement for the Recording (or indeed the Publication) class. As it does not implement any methods it is **abstract**. All it does is to relay messages to an aggregate component instead. In fact the only message of interest corresponds to the toString operation used in displays by the Action class. (In more realistic examples there would of course be many more operations.) Therefore we introduce an interface DisplayIF that advertises toString. Any aggregate component of the LoanItemImpAdapter must implement it.

Figure 6.13 shows the changes introduced into the revised architecture. They have no effect on the rest of the design. Essentially we can choose to use the adapter or not as circumstances dictate.

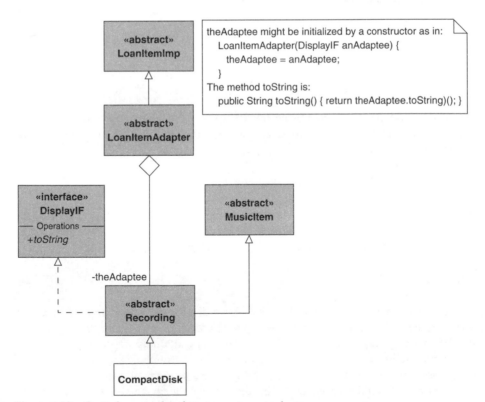

Figure 6.13   *Part of a revised architecture using an adapter*

The approach is known as the *Adapter* design pattern and it is applicable to other parts of the design should it prove to be necessary. It is discussed in more detail in chapter 8.

Consider what happens if there is a requirement to display only the books available for loan. The problem is that we have decided that a book is just a publication that we view as a loan item. It might be a book, journal or even a compact disk. The crucial point is we do not distinguish between them. Clearly books and journals are the same or they are different: they cannot be both. Therefore whatever we do is going to be a compromise unless we radically change our design.

One pragmatic solution is to discover the class of each loan item in the library and then take the appropriate action. Typical code that uses this approach is:

```
Iterator  iter = theLibrary.getLoanItemsIterator();

while(iter.hasNext()) {
    LoanItemIF loanItem = (LoanItemIF)  iter.next();
    // ...
    if(loanItem.getBorrower() == null) {
        String name = loanItem.getClass().getName();
        if(name.equals("Book")) {
            // Display this loan item
        }
        // ...
    }
}
// ...
```

We might also have considered using the instanceof operator in place of the method calls getClass and getName. However, we prefer the latter as they give the true type of the loan item. The statement:

```
String name = loanItem.getClass().getName();
```

uses the method getClass (defined in Object) to return a Class object produced by the compiler. It is interrogated by getName (defined in Class) for the String that is the class name. It is then compared with the class name that interests us with:

```
if(name.equals("Book")) ...
```

and the appropriate action taken. In this case we display the loan item, which is now known to be a Book. In other words we have abandoned the polymorphic effect temporarily. Clearly this approach should be used with caution as we must populate our code with statements to identify classes by name. Apart from being against the spirit of object orientation, it could easily lead to difficulties during maintenance.

If it turns out that this approach is not suitable then the conclusion is that this architectural framework cannot be used in its current form. This is not really the fault of the designers, as one of the assumptions they made (loan items are treated uniformly) is no longer valid. The solution is to learn from this experience and if possible redesign the framework to accommodate this type of change. For example, the LenderImp class might have a collection of collections each of which holds a particular kind of loan item. However, this is for future iterations. On balance we consider this iteration to be successful.

6.4   Iteration 3

In all the previous versions of the library application we have supported a rudimentary mechanism whereby we could load the initial library stock with details read from a file. The stock details could then be augmented by additional books and journals but this new information was not reflected back to the data file. This version of the library application is given support for making its data persist. A *persistence mechanism* provides data storage between separate executions of an application. A number of strategies are available to us, from simple interfaces that persistent class objects must implement, to complex schemes integrated into database technologies. In this chapter we consider the serialization mechanism supported by Java.

A shortened lifecycle that finishes at the implementation of persistence to the application is applied.

6.4.1   Establish use-cases

As there are no new requirements for this iteration there is no need to develop a new use-case diagram. However, we add explicit start-up and close-down use-cases to the one shown in figure 6.2. A suitable note attached to each helps clarify their purpose. Figure 6.14 illustrates.

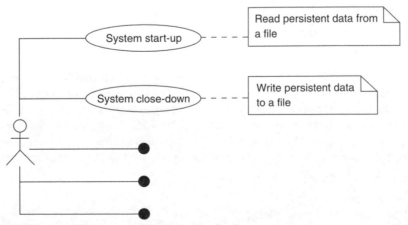

Figure 6.14   *A modified use-case diagram: iteration 3*

6.4.2   Analysis

Object serialization is a mechanism for saving the state of an object as a byte sequence in a disk file. The serialization mechanism supports re-creation of the Java object from this persistent file store. Thus separate activations of an application can retain an image of its objects.

To persist an object in Java the object's class must implement the Serializable interface. This is a *tagging interface* with no features and simply indicates that objects of that class can persist. A class diagram in which the Book class from chapter 4 is made persistent is shown in figure 6.15.

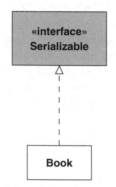

Figure 6.15   *Implementing the Serializable interface*

If the Book class extends the Publication class as in this chapter, and all Publications are to be made persistent, then we arrange for the **abstract** class Publication to implement the Serializable interface. This means that the Book and Journal classes remain unchanged.

For simplicity, if we focus on just the Book class then we have:

```
import java.io.*;
// ...
public final class Book implements Comparable, java.io.Serializable {
   // All features are unchanged from the case study of chapter 4
} // class: Book
```

To persist a Book object we use the ObjectOutputStream class. An ObjectOutputStream object is wrapped around a byte stream object of the class FileOutputStream to persist the object's byte sequence. This is an example of the *decorator pattern* described in chapter 8. The following code illustrates how this approach operates to persist a single Book object. Other than opening and closing the persistent file, we need only execute the method writeObject defined in ObjectOutputStream.

```
public class MainSave {
   // ----- Operations ----------
   public static void   main(String[] args) {
      Book bk = new Book( ... );                    // make a Book object
```

```
      //
      try {
         FileOutputStream fos = new FileOutputStream("Book.ser"); // open persistent file
         ObjectOutputStream oos = new ObjectOutputStream(fos);   // wrap with decorator
         oos.writeObject(bk);                                     // now save instance
         oos.close();                                             // close persistent file
      } catch(IOException ex) {
      // report error
      }
   } // method: main
}    // class: MainSave
```

Restoring the object during a subsequent execution of the application is equally simple. The readObject method call from the ObjectInputStream class is all we require. Since this method is used to restore any serialized object, then we must cast it to the correct type. We then have a Book object containing the same state information as the original.

```
public class MainLoad {
   // ----- Operations ----------
   public static void        main(String[] args) {
      try {
         FileInputStream fis = new FileInputStream("Book.ser");   // open persistent file
         ObjectInputStream ois = new ObjectInputStream(fis);      // wrap with decorator
         Book bk = (Book)ois.readObject();                        // restore the instance
         ois.close();                                             // close persistent file
         Console.out.println("Title: " + bk.getTitle());         // confirm all is ok
      } catch(IOException ex) {
      // report error
      } catch(ClassNotFoundException ex) {
      // report error
      }
   } // method: main
}    // class: MainLoad
```

We should note that when we serialize an object then not only that object's state is made to persist, but also all persistent objects referenced by it. Thus if our model is a Library object with its stock comprising many Book objects, then when we persist the Library object then so too do all the associated Book objects. Similarly, when we restore the single Library object, as in the second illustration above, then the Book objects are also restored as the library stock.

6.4.3   Design

Applying this revision to iteration 2 of this case study is relatively straight-forward. First, we arrange that the classes LoanItemImp, LenderImp and BorrowerRecordImp

implement the Serializable interface so that their objects may persist. We must do the
same to the DateImp class in the librarysubsystem package. Figure 6.16 illustrates.
The only other changes are made to the Action class as described in section that
follows.

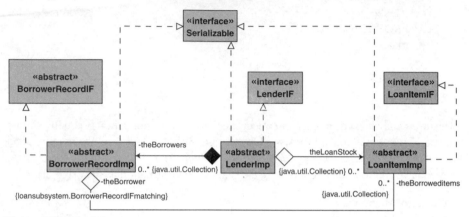

Figure 6.16   *Serialization*

Observe how in this class diagram all the implementation classes have been made to
implement the Serializable interface. We might have considered having all the interface classes extend from Serializable. The implication of this latter choice is that
all implementations must persist. It may be that in some circumstances this is not
appropriate.

6.4.4   Implementation

Again there is relatively little work to do in this activity. The only significant change is
made to the Action class as shown below. Specifically, we arrange that at program
close-down the application's state is written as serialized data to the persistent file store.
At program start-up we recover this state by reading the same file. Respectively, we capture this in the close method and in the constructor to the Action class.

```
package librarysubsystem;

import java.io.*;
import textio.*;
import java.util.*;

public class Action  {

    // ----- Operations ----------
    public    Action(String aLibraryName) {
        //
        //    Restore the persistent application objects, if any
        File file = new File(PERSISTENT_FILENAME);
```

```
   if(file.exists()) {
     try {
        FileInputStream fis = new FileInputStream(PERSISTENT_FILENAME);
        ObjectInputStream ois = new ObjectInputStream(fis);
        theLibrary = (Library)ois.readObject();
        ois.close();
     } catch(IOException ex) {
        System.out.println("Error reading persistent store");
     } catch(ClassNotFoundException ex) {
        System.out.println("Error restoring persistent application objects");
     }
   } else
     theLibrary = new Library(aLibraryName);
}  // method: Action

public void    close() {
   //
   //   Save the persistent application objects
   try {
     FileOutputStream fos = new FileOutputStream(PERSISTENT_FILENAME);
     ObjectOutputStream oos = new ObjectOutputStream(fos);
     oos.writeObject(theLibrary);
     oos.close();
   } catch(FileNotFoundException ex) {
     System.out.println("Cannot create persistent file store");
   } catch(IOException ex) {
     System.out.println("Error writing to persistent file store");
   }
   ConsoleIO.out.println("\n\t" + "SYSTEM CLOSING" + "\n");
}  // method: close

   // ...
   // ----- Attributes ----------
   private static final String    PERSISTENT_FILENAME = "Library.ser";

   // ----- Relations ----------
   private Library    theLibrary;
}  // class: Action
```

Full listings for this iteration can be found in the software provided under Lib6_3.uml.

6.4.5 Testing

As with the previous iteration we do not engage in extensive testing during this activity. All that is required is to demonstrate that the software behaves in the same manner as iteration 1 and that it supports persistence. This is indeed the case therefore we consider the aim of this iteration to have been met.

6.5 Summary

1. Scenarios represent different paths through a use-case. Scenarios have basic and alternate flow of events through a use-case.
2. Use-cases can have include relationships and extend relationships. They simplify use-cases by sharing some common functionality.
3. Specialization and the use of the polymorphic effect can radically simplify our designs and implementation code. The code is expressed in terms of **abstract** superclasses rather than specific concrete classes.
4. The full power of the object-oriented paradigm lets our designs accommodate new requirements without the necessity for substantial changes.
5. The full power of the object-oriented paradigm includes class specialization, object substitution, the polymorphic effect and designs expressed in terms of interfaces.
6. An architectural framework is a general solution that can be instantiated for a particular domain-specific application.
7. A persistence mechanism provides data storage between separate executions of an application. Object serialization in Java is a mechanism for saving the state of an object as a byte sequence and from which the object's state can be restored.

6.6 Exercises

1. Develop documentation for the "Display publications on loan" and "Return one publication" use-cases. The "Add one book" use-case discussed in section 6.2.1 should act as a guide.
2. Develop test-cases for the "Display publications on loan" and "Return one publication" use-cases. The "Add one book" test-case discussed in section 6.2.1 should act as a guide.
3. In the Action class from iteration 1 (model Lib6_1.uml) the methods displayPublicationsAvailableForLoan and displayPublicationsOnLoan need to determine if a publication is on loan to a borrower. Introduce a method isOnLoan for this purpose:

   ```
   // class Publication
   ...boolean isOnLoan() { ... }
   ```

4. Use examples from this chapter to illustrate the value of **final** methods and **final** classes.
5. What changes would be made to the analysis and design of iteration 1 if there were several different kinds of borrower? For example, there might be a limit on the number of publications that can be borrowed by a borrower. Perhaps some borrowers have privileged borrowing rights that permit them to borrow more publications. A similar situation might apply to the length of time that a publication can be borrowed for without incurring a fine.
6. Consider the impact on the analysis and design of iteration 1 if there were a requirement that some borrowers can borrow books and journals but others can only borrow books.

7. We have not shown an archiectural relationship between two interface classes. Explain why and suggest an alternative. See, for example, theBorrower relation of the class LoanItemImp of section 6.3.4.1.

8. (a) Using the class diagram of figure 6.5 as your guide, extend iteration 1 (model Lib6_1.uml) by introducing music CDs as loan items. You can assume that each CD is given a catalogue number, a title and the name of the artist.

   (b) Now extend iteration 2 (model Lib6_2.uml) using the class diagram of figure 6.12 as your guide.

   (c) Discuss the benefits of each approach.

9. (a) In the second iteration (model Lib6_2.uml) a framework was established to represent a lender of arbitrary loan items supplied to arbitrary borrowers. Use this framework to create an application in which a company hires cars and light vans to registered borrowers. You can assume that the hire company has several cars and light vans each of which have a unique registration number. They are hired out to registered customers each of whom has a unique name. Cars and light vans have a model name and a year of registration. However cars also have an engine capacity and light vans their maximum load capacity.

   (b) Comment on the difficulty this exercise. Does it justify building the framework in the first place?

10. (a) Revisit exercise 8 and using the third iteration (model Lib6_3.uml) as your starting point, introduce a LoanItemAdapter class as shown in figure 6.13.

   (b) Discuss the benefits of this approach as compared to those taken in exercise 8.

   (c) Use your solution to illustrate the benefits of Java's serialization capability.

7

Graphical User Interfaces

The final version of the library case study from chapter 4 and in subsequent chapters introduced the class Action to separate the concerns of user interaction from the other application classes. This separation permits us to offer various user interfaces without disrupting the domain model. In this chapter we evolve the library application from a system driven from the command line to one that uses a graphical interface.

In this chapter we also meet some new OO features, details of which are given in later chapters. Some are simple extensions to our existing Java knowledge and are easily assimilated. Some we have now detailed, such as class specialization, and the reader is invited to revisit the latter parts of the preceding two chapters. This chapter is not a definitive guide to the Swing library as it is a very large topic and is beyond the scope of this book. The reader is directed to the references Elliott (2002) and Topley (1999).

Throughout this chapter emphasis is given to the design of our library application, both as it presently exists and as we make the necessary revisions. Further, we also consider the architectural design of the Java Swing library to gain a better understanding of how to deploy it. The Swing library also serves as an illustration of an industrial strength framework building on the discussions from the previous chapter.

7.1   Overview of Swing

A graphical Java application is developed using the Swing class library. This is a large and complex library consisting of over 300 classes and interfaces. The software engineers that developed it made full use of leading edge technologies such as *design patterns*, and this further complicates its usage. A design pattern is a recognized solution for a particular type of problem. In this chapter we will have some direct and indirect contact with these patterns and present sufficient material to understand and exploit them. A full discussion is, however, reserved until the next chapter.

Many classes in the Swing library represent the familiar components that appear in a graphical application. These include buttons, menus, text fields, etc. For example, figure 7.23 is an application dialog with three labelled text fields at the top and two buttons at the foot. The user is expected to enter information into the text fields then select the button to accept or cancel the action. Much of our work in this chapter either involves specializing a component class to give it application-specific features, or associating a handler object with it so that some action is taken when, for example, a button is pressed or a menu item selected. This makes the learning curve a little less steep than it might first appear.

Obviously, we will not meet all the classes in the Swing library. However, the overall architecture of the library means there is a significant amount of similarity between the Swing components. Hence the illustrations presented can be the foundations for further study.

The similarity between the Swing components arises from the extensive use of specialization. Many components are specializations of the JComponent class. This **abstract** class carries most of the common behaviour associated with graphical components. For example, this class defines whether a component is visible, and if so renders itself on to the screen. Most components are rendered as a member of some other component referred to as its parent as in a whole/part relationship described in the preceding chapters. Note this is not to be confused with a parent/child relationship as found in specialization. For example, a button component may appear on a parent dialog component. The JComponent class implements the notion of aggregate subcomponents with the aggregate as the parent. The JComponent class also defines the location of a component, and is usually in terms of its position relative to the parent component.

The JComponent class is then the root of a specialization hierarchy that extends to various concrete components that might be used in a graphical application. For example, a dialog may have a text field into which the user enters some value. The class JTextField represents such a component. Alongside the field there may be a simple label documenting the purpose of the text field. A label is obtained using the class JLabel that is an immediate descendant of JComponent (see figure 7.1).

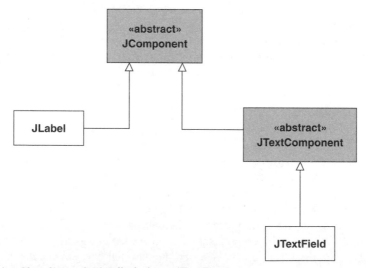

Figure 7.1   *Class hierarchy for JLabel and JTextField*

Most of the other graphic components are developed in this manner, specializing directly (as for JLabel) or indirectly through other **abstract** classes (as is the case for the class JTextField). The following table lists some of the components we introduce.

Component	Description
JButton	An implementation of a "push" button. Buttons can be decorated with text or a graphic icon.
JFrame	A JFrame is a top-level window with a captioned title and a border. A frame is used to represent the application.
JLabel	A JLabel is a display area for a short length text string. Labels are often used alongside input text areas to document their purpose.
JMenuBar	A JMenuBar represents a menu-bar as might appear at the top of a frame.
JPanel	A JPanel is a general-purpose container frequently used to group other components.
JTextArea	A JTextArea is a multi-line area for presenting plain text.
JTextField	A JTextField is a component that permits the editing of a single line of text.

Since the Swing class JComponent extends (is a specialization of) the class Container (see figure 7.2), Swing components can act as containers for other Swing components. These subcomponents are the *child components* (the parts) of their *parent component* (the whole) and are included in a parent using the method add from the class Container. The relative position of a child within a parent is determined by a *layout manager* (see later) associated with the parent container.

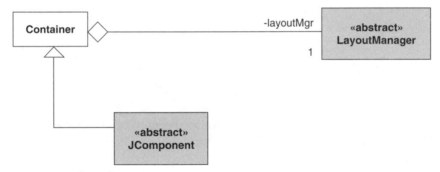

Figure 7.2 *The Container class hierarchy*

7.2   Rebuilding the library case study

This chapter emphasizes the iterative style of development introduced from chapter 2. Recognize that all the development work has been done by the designers of the Java Swing library. All we are required to do is to make use of it. Therefore, we produce a series of versions that successively augment the preceding version until we produce the final product. At each iteration we will set an objective so that we have a purpose that we can measure our achievement against and provide a harness for any testing.

In chapter 6, the final version of the library case study carefully separated off the input/output (IO) mechanisms from the domain model for the application. Thus the classes Library, Book, Journal and BorrowerRecord provide toString methods to give access to the state of the objects. None of them perform any IO. All the IO is provided by the methods in the Action class and the Application class.

This separation between the domain application model and its IO mechanism is inspired by the *model-view-controller* (MVC) design pattern. The model, as already discussed, represents the actual application state. The view element of the MVC is concerned with the visualization (Action class) of the data while the controller is concerned with user input that may result in state changes to the data (Application class and Action class). Figure 7.3 illustrates.

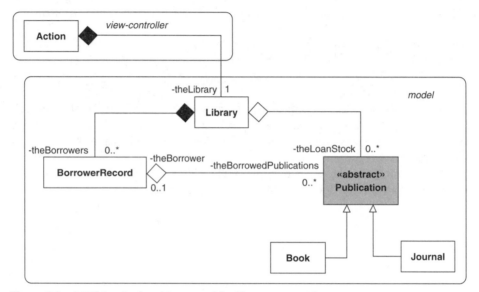

Figure 7.3   *MVC inspired architecture of the library case study*

Note that we have ignored the framework developed in chapter 6 and have presented a conceptual representation of the model (see figure 6.9(b) for the full architecture). This permits us to focus on the primary model classes.

The MVC is a common architecture to use when developing graphical user interfaces (GUIs). In fact, it has been used in the development of the Swing components. For example, the JButton class acts as the view-controller for a button while a separate class implements its model. This separation of model from view-controller permits us to replace the Action class with Swing classes to give our application a modern user interface.

In this first version we demonstrate how to introduce a graphical window that represents the application's interface. The parameterized class constructor for JFrame takes a single String argument representing the title for the application that is presented in the caption bar. Having created such an object, we merely have to make it visible. In previous chapters we have made the run method of the Application class responsible for the program's overall control. Now it will reside in the GUI classes that replace class Action. For this reason method main now creates the primary GUI object instead of the Application object as shown in program 7.1.

Program 7.1 A first frame (model Lib7_1.uml)

```
import javax.swing.*;
public class Main {
   // ----- Operations ----------
   public static void   main(String[] args) {
      JFrame frame = new JFrame("Library");
      frame.setVisible(true);
   }   // method: main
}   // class: Main
```

When we run this program a new window appears at the top left of the screen as shown by figure 7.4. It is a fully featured window that can be dragged, opened, iconized, and closed. When the window is resized, we see it has a caption bar with the title "Library", a system menu to the left of the caption, and minimize, maximize and close buttons to the right of the caption. Strictly, the application does not close correctly. After having apparently closed the application with this button, when we return to the command window we must issue a CONTROL-C to reinstate the command prompt. We shall see in program 7.4 how to correct this.

Figure 7.4 *The library application window*

To bring the window up with some start-up size we merely have to set the window bounds. The method **setBounds** has four arguments representing, respectively, the X and Y coordinates for the upper left of the window, the width and the height. All are measured in pixel values. Program 7.2 shows how this is achieved and figure 7.5 is its output.

Program 7.2 Sizing the frame (model Lib7_2.uml)

```
import javax.swing.*;
public class Main {
   // ----- Operations ----------
   public static void   main(String[] args) {
      JFrame frame = new JFrame("Library");
      frame.setBounds(0, 0, 400, 300);
      frame.setVisible(true);
   }   // method: main
}   // class: Main
```

Perhaps a final improvement we can make to the program start-up procedure is to locate the main application window in the centre of the screen. The Toolkit class has the

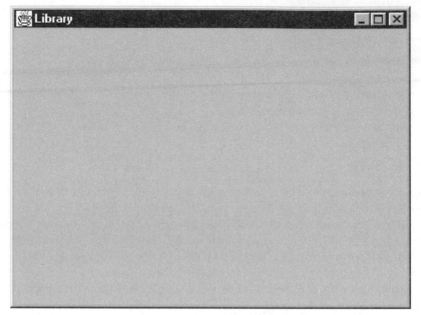

Figure 7.5 *Frame with an initial start-up size*

method getScreenSize to obtain the size of the screen. The Dimension value returned by this method encapsulates the width and height of the screen. We can access these values directly with the attributes width and height or by using the accessor methods getWidth and getHeight. Thereafter, some simple arithmetic centres the frame as demonstrated in program 7.3.

Program 7.3   Centring the application window (model Lib7_3.uml)

```
import librarysubsystem.LibraryFrame;
public class Main {
    public static void   main(String[] args) {
        LibraryFrame frame = new LibraryFrame("Library");
    }   // method: main
}   // class: Main
```

and:

```
package librarysubsystem;
import javax.swing.*;
import java.awt.*;
public class LibraryFrame extends JFrame {
    // ----- Operations ----------
    public   LibraryFrame(String aCaption) {
        super(aCaption);
```

```
    Dimension screen = Toolkit.getDefaultToolkit().getScreen Size();
    int width = screen.width * 3 / 4;
    int height = screen.height * 3 / 4;
    this.setBounds(screen.width / 8, screen.height / 8, width, height);
    this.setVisible(true);
  }   // method: LibraryFrame
}   // class: LibraryFrame
```

In this version we have moved all of the frame building logic into a specialized LibraryFrame class (see figure 7.6). The class LibraryFrame extends the superclass JFrame. The class constructor for LibraryFrame first invokes the superclass JFrame constructor with the statement **super**(caption). It then sets its bounds and position and then makes itself visible.

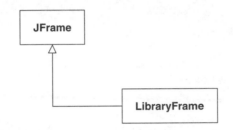

Figure 7.6   *Specialized JFrame class*

7.3 Events

Graphical applications are described as *event driven*. They await some user activity then spring into action to service that event. For example, when the user moves the mouse across a component, the mouse's movement is reflected by the motion of the mouse cursor on the screen. Equally, when the user selects a menu from the menu bar, a popup list of menu items is displayed from which the user may select some service request. Both illustrations exhibit how a system is effectively inactive until the user does something. When the user moves the mouse or selects a menu then an event occurs, stimulating the system, perhaps ultimately leading to a state change in the domain model.

In the Java event model different types of events are represented by objects of different event classes. For example, mouse movements are represented by objects of the class MouseEvent. Similarly, selecting a menu item is represented by an object of the class ActionEvent. The event object produced as a result of an event occurring is made available to the program code that processes the event, the so-called *event handler*. The event object has various attributes describing aspects of the event. For example, a MouseEvent object has the co-ordinates of the position of the mouse when the mouse event occurred. Figure 7.7 shows an extract of the event class hierarchy.

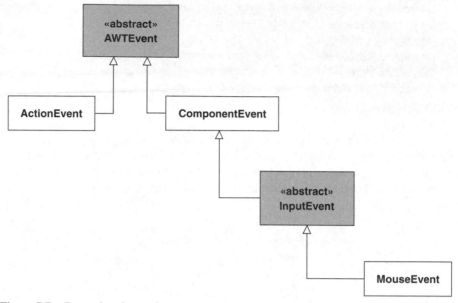

Figure 7.7   *Event class hierarchy*

The Java event model is based on the notion of an *event listener*. A listener is an object that is interested in being advised of events. When an event is generated by an event source, the source notifies all its listener objects by calling a specified method and passing to it the appropriate event object. For the source to be able to call a specific method in a listener object, the listener must implement a particular method protocol as defined by a corresponding listener interface. For example, for some classes XListener and YListener to operate as listeners for ActionEvents they must implement the ActionListener interface as shown by figure 7.8. To implement the interface, each subclass must have a definition for the actionPerformed method.

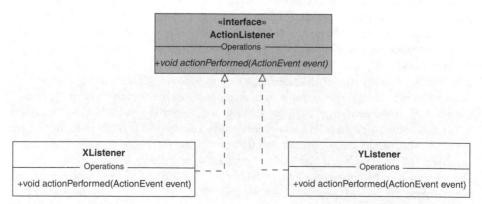

Figure 7.8   *ActionListener hierarchy*

An object that generates events is referred to as the *event source* and is some specialized JComponent object. An event source object maintains a list of listener objects that wishes to be notified of events generated by the source. For example, figure 7.9 shows two action event listeners associated with our LibraryFrame. When a source object generates an action event it calls the actionPerformed method on all action listeners registered with it. The listener object will specify those events with which it is interested. A listener object listening for only ActionEvent events, for example, will not hear any MouseEvent events.

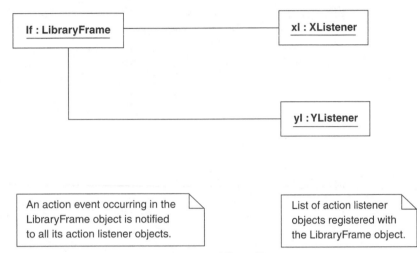

Figure 7.9   *Action listeners registered with the LibraryFrame*

This set of listeners can be dynamically updated with various methods to add and remove listeners for an event source object as shown in figure 7.10. For the event type ActionEvent there are the methods addActionListener and remove ActionListener on a component. These listener objects will then be notified by the source object of any ActionEvents. Figure 7.10 shows the particular methods to add and remove action listeners.

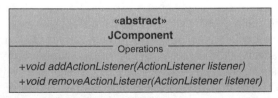

Figure 7.10   *Adding and removing action event listeners*

Figure 7.11 shows a sequence diagram that traces the behaviour of our LibraryFrame object. Consider that we have classes XListener and YListener that describe action listeners as in figure 7.8. The LibraryFrame object lf creates an instance of each of these listener classes then registers each with itself. The latter is achieved by lf sending itself the message addActionListener(xl) and addActionListener(yl). Then when the user generates an event through, for example, pressing a button or selecting a menu item, the LibraryFrame object lf operates as the event source and notifies the action listener objects xl and yl through the actionPerformed message.

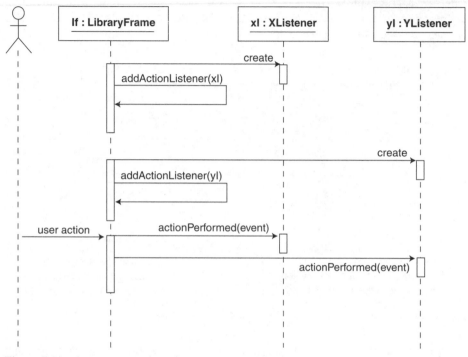

Figure 7.11   *Registering and notifying action listeners*

The above illustration is described in terms of action events. An identical scheme applies were we interested in, say, mouse events. First, we develop one or more listener classes that implement the MouseListener interface. We then register objects of these listener subclasses with the event source. When a particular mouse event occurs then the event source notifies its mouse listener objects. An object of the class MouseEvent is created with the details of the event (e.g. mouse position at time of the event) and is passed as the parameter to the listener's event handler (the equivalents of xl and yl).

In this fourth version of the library case study we include the code to properly close the application when the user selects close from the system menu or from the close button. To do so we register a WindowListener object with our LibraryFrame component. This listener object is from the class LibraryFrameClosing that implements WindowListener. WindowListener is an interface with a protocol that includes seven **abstract** methods such as the one we are interested in, windowClosing (this is discussed further in chapter 8). The Swing library also provides classes that fully implement the interface with each redefined method having empty method bodies. These are generally known as adapter classes, and for this example is the WindowAdapter class.

We then specialize from the WindowAdapter class and only implement the handler that we are interested in. Figure 7.12 shows this arrangement.

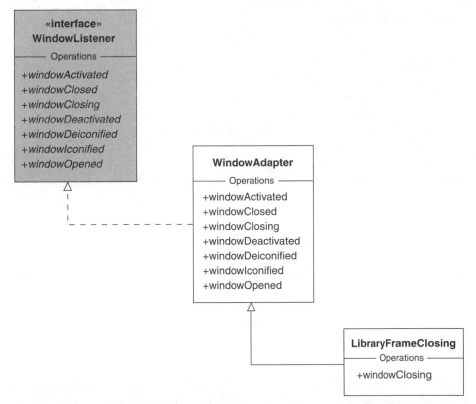

Figure 7.12 *WindowListener hierarchy*

In our library application we are interested in the user closing the application. To correctly handle this situation we register an object of the LibraryFrameClosing class as a *window listener*. The class LibraryFrameClosing extends the class WindowAdapter (which implements the interface WindowListener) and provides an implementation for the one method windowClosing. The method simply calls the method exit from the class System to close the application. The full listing for the class LibraryFrame is given in program 7.4.

Program 7.4   LibraryFrame with an inner class event handler (model Lib7_4.uml)

```
package librarysubsystem;
import java.awt.*;
import javax.swing.*;
```

```
public class LibraryFrame extends javax.swing.JFrame {

// ----- Operations ----------
public   LibraryFrame(String aCaption) {
    super(aCaption);
    Dimension screen = Toolkit.getDefaultToolkit().getScreenSize();
    int width = screen.width * 3 / 4;
    int height = screen.height * 3 / 4;
    this.setBounds(screen.width / 8, screen.height / 8, width, height);
    this.setVisible(true);
    //   Proper close-down procedure.
    this.addWindowListener(new LibraryFrameClosing());
} // method: LibraryFrame
// ----- Inner classes ----------
public class LibraryFrameClosing extends java.awt.event.WindowAdapter {

    // ----- Operations ----------
    public void   windowClosing(java.awt.event.WindowEvent event) {
        System.exit(0);
    } // method: windowClosing
} // class: LibraryFrameClosing
} // class: LibraryFrame
```

First, note the final statement in the LibraryFrame class constructor:

this.addWindowListener(new LibraryFrameClosing());

Here, the LibraryFrame object sends a message to itself. The message is addWindowListener and the parameter is an instance of the class LibraryFrameClosing, the event handling class. Note how the latter is an example of an *inner class*, a class defined within the LibraryFrame class. For the present, there is nothing significant in doing it this way and we could just as easily presented LibraryFrameClosing as a normal class, but inner classes were introduced into Java to provide support for event handling code. The full effect of this we shall see shortly.

7.4   Menu bar

Most graphical applications support a *menu bar* located along the top of the application window. The menu bar carries a number of drop-down *menus* each of which has a set of *menu items*. The user selects one of these items to have some functionality performed by the application.

A basic menu is assembled with the classes JMenuBar, JMenu and JMenuItem. The class JMenuBar is used to create a menu bar object that is attached to the top of the main application frame window. The menu bar is populated with a number of menus. In figure 7.13 the menus are File, Options and Help. Each menu is represented by an instance of the class JMenu. When the user selects any such menu a menu popup containing one or more menu items (from the class JMenuItem) or menu item separators (a bar dividing groups of menu items) is displayed.

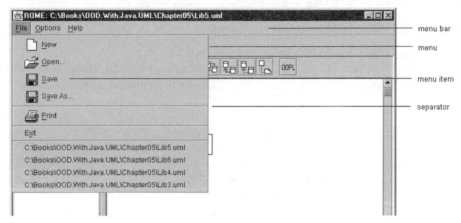

Figure 7.13   *Elements of a menu*

Once again, all the work is done in the class LibraryFrame with the **private** support method assembleMenuBar. The attribute theMenuBar is initialized with the statement:

theMenuBar = **new** JMenuBar();

and attached to the frame with:

this.setJMenuBar(theMenuBar);   // inherited from JFrame

Menus and menu items are prepared with code of the form:

```
// Prepare the File menu ...
JMenu fileMenu = new JMenu("File");
fileMenu.setMnemonic('F');
```

```
// ... and the Exit menu item
theFileExitAction = // ... see later ...
JMenuItem fileExit = fileMenu.add(theFileExitAction);
fileExit.setMnemonic('x');
```

```
// Place the menu on the frame.
theMenuBar.add(fileMenu);
this.setJMenuBar(theMenuBar);
```

The first statement uses the JMenu constructor to establish a menu labelled "File". The penultimate statement adds this menu object to the menu bar with the add method from class JMenuBar. Note the calls to the method setMnemonic to associate shortcuts for both the menu and the menu item.

The remaining statements assemble a single menu item, adding it to the menu. The add method of the class JMenu is prepared to accept an action object as a parameter. The action object is used to create and return a JMenuItem object. This can then be configured in a number of ways. Here, we set the mnemonic "x" for this menu item. The interesting code for this is the single statement:

```
JMenuItem fileExit = fileMenu.add(theFileExitAction);
```

the details of which are discussed in the next section.

The ordering of menu items in a menu is determined by the order in which the items are added to the menu. A similar arrangement applies to the menus on the menu bar.

7.4.1 Responding to a menu selection

Menus and menu items usually have shortcuts by which the user can activate the required service without having to use the mouse. This way an application that makes extensive use of the keyboard, such as a word processor, does not require the user to leave the keyboard to make a menu selection with the mouse. Typically, a menu has one of the characters in the menu name underlined (for example, File) that is used in conjunction with the ALT key to invoke the popup. In a similar way menu items can also be annotated. When the menu is visible, the highlighted character of the menu item activates that service. Thus ALT + F might open the File menu and ALT + N might select the New menu item (see figure 7.13). The method setMnemonic in the classes JMenu and JMenuItem achieve this effect. The parameter is a single char representing the shortcut. Thus we have:

```
fileMenu.setMnemonic('F');
```

Attaching the menu to the frame does exactly that and no more. We have to include more code to our solution to handle the menu selections made by the user. Selecting a menu item causes the menu item to generate an ActionEvent. To process it we must register a listener object with the menu item for this type of event. In section 7.3 we indicated how this is achieved.

According to figure 7.11 we need to create an object of a subclass that implements ActionListener, the listener interface for these types of events. We then register this event handler object with our menu item object using the method addActionListener. However, the JMenu class supports an alternative to this. Rather than adding an object instance of the class JMenuItem to a JMenu object, we can add an instance of the specialized ActionListener object using an overloaded version of the method add. The add method returns a newly created JMenuItem that can be configured as shown in the code listing above for preparing menus and menu items. This effect is described by the sequence diagram in figure 7.14.

Here, the LibraryFrame object creates an action listener object of the class FileExitAction (subclassed from AbstractAction). This action object is then added to the JMenu object that creates a JMenuItem object with that action listener registered as its event handler. The newly created JMenuItem object is then delivered as the return value for the add operation.

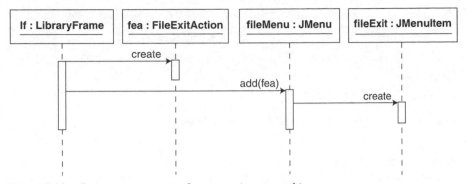

Figure 7.14   *Creating a menu item from an action event object*

Following our discussions from section 7.3 FileExitAction would implement the interface ActionListener (see figure 7.8). However, if we subclass from the class AbstractAction then the menu item that is created can be given some text or an icon or both. The relevant class hierarchy is shown in figure 7.15. Note how the Action interface is a specialization of the ActionListener interface. The **abstract** class AbstractAction is responsible for introducing labels and icons that decorate the objects created from actions. The figure also shows that the FileExitAction merely has to provide a suitable constructor and a definition for the actionPerformed method. The code for this is given in program 7.5.

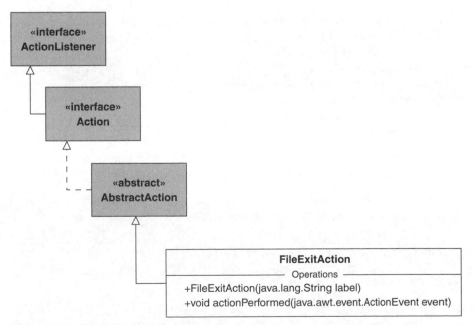

Figure 7.15   *FileExitAction class hierarchy*

Program 7.5   The LibraryFrame class with two inner classes
(model Lib7_5.uml)

```java
package librarysubsystem;

import java.awt.*;
import javax.swing.*;

public class LibraryFrame extends javax.swing.JFrame {

// ----- Operations ----------
   public LibraryFrame(String aCaption) {
     super(aCaption);
     this.assembleMenuBar();
     Dimension screen = Toolkit.getDefaultToolkit().getScreenSize();
     int width = screen.width * 3 / 4;
     int height = screen.height * 3 / 4;
     this.setBounds(screen.width / 8, screen.height / 8, width, height);
     this.setVisible(true);
     //   Proper close-down procedure.
     this.addWindowListener(new LibraryFrameClosing());
   }   // method: LibraryFrame

   private void    assembleMenuBar() {
     //   Prepare the File menu ...
     JMenu fileMenu = new JMenu("File");
     fileMenu.setMnemonic('F');
     //   ... and the Exit menu item.
     JMenuItem fileExit = fileMenu.add(theFileExitAction);
     fileExit.setMnemonic('x');
     //   Add the menu to the menu bar.
     theMenuBar.add(fileMenu);
     //   Then place the menu on the frame.
     this.setJMenuBar(theMenuBar);
   }   // method: assembleMenuBar

// ----- Attributes ----------
   private javax.swing.JMenuBar    theMenuBar = new JMenuBar();

   private FileExitAction    theFileExitAction = new FileExitAction("Exit");

// ----- Inner classes ----------
   public class LibraryFrameClosing extends java.awt.event.WindowAdapter {
     // ----- Operations ----------
     public void    windowClosing(java.awt.event.WindowEvent event) {
       LibraryFrame.this.theFileExitAction.actionPerformed(null);
     }   // method: windowClosing

   }   // class: LibraryFrameClosing

   public class FileExitAction extends javax.swing.AbstractAction {
```

```
// ----- Operations ----------
public    FileExitAction(String aLabel) {
   super(aLabel);
}    // method: FileExitAction

public void    actionPerformed(java.awt.event.ActionEvent event) {
   System.exit(0);
}    // method: actionPerformed
}    // class: FileExitAction
}    // class: LibraryFrame
```

In the windowClosing method of LibraryFrameClosing we see one of the key features of inner classes. An object instance of the inner class LibraryFrameClosing has effectively a reference to the object instance of the outer class, here LibraryFrame. Further, the inner class instance forms a closely coupled relationship with its outer class instance such that it has access to the attributes and operations of the outer class object, even to those that have private visibility. Hence the statement:

LibraryFrame.**this**.theFileExitAction.actionPerformed(**null**);

references the private attribute theFileExitAction of an instance of the enclosing class LibraryFrame. Strictly, the qualifier LibraryFrame.**this** is not required where there is no ambiguity. However, it is a useful piece of documentation that reminds us that we are referring to an object of the outer class. This object executes the actionPerformed method of the FileExitAction class, closing the application. Since that method does not use its parameter we can safely pass the **null** actual parameter value. Here, we ensure that whether the user closes the application through the system menu or this new F̲ile + E̲xit menu, the same closedown code is executed. For example, if our application required files to be properly closed, then we could place all the necessary code in the actionPerformed method of the inner class FileExitAction.

Figure 7.16 now shows our updated application. The user has selected the File menu. The Exit menu item is highlighted when the mouse is placed across it.

7.5 Application menus

Using the scheme introduced in the preceding section we can start to populate the LibraryFrame class with a number of inner classes that provide event handlers for the various services our application must support. Rather than attempt this in one major increment we shall satisfy ourselves that our approach is correct by first including two such inner classes. Of course, we aim to support the same functionality achieved with the final version in chapter 6, but for the present we shall establish the necessary architecture to support lending and returning a publication. Here we emphasize setting up the required architecture, the actionPerformed methods simply issue messages to the system console. The true code for these and the other services follows in the next iteration.

Following on from program 7.5, we introduce two new inner classes Lend PublicationAction and ReturnPublicationAction. As above, both are specializations of the AbstractAction class and present implementations of the actionPerformed

Figure 7.16   *The library application with its* File *menu and* Exit *menu item*

method and will, at the next version, provide the service described by its class name. Additionally, we include a new menu entitled Application with these action objects used in the creation of the corresponding menu items. Program 7.6 presents the relevant elements of the LibraryFrame class. Only the details for the LendPublicationAction and the ReturnPublicationAction classes are shown.

Program 7.6   Inner classes for lending and returning books (model Lib7_6.uml)

```
package librarysubsystem;

import java.awt.*;
import javax.swing.*;
import java.util.*;

public class LibraryFrame extends javax.swing.JFrame {

    // ----- Operations ----------
    private void assembleMenuBar() {
        //   Prepare the File menu ...
        JMenu fileMenu = new JMenu("File");
        fileMenu.setMnemonic('F');
        //   ... and the Exit menu item.
        JMenuItem fileExit = fileMenu.add(theFileExitAction);
        fileExit.setMnemonic('x');
        //   Prepare the Application menu ...
        JMenu applicationMenu = new JMenu("Application");
        applicationMenu.setMnemonic('A');
```

```
    //   ... and its various menu items.
    JMenuItem lendPublication = applicationMenu.add(theLendPublicationAction);
    JMenuItem returnPublication = applicationMenu.add(theReturnPublicationAction);
    //   Add the menus to the menu bar.
    theMenuBar.add(fileMenu);
    theMenuBar.add(applicationMenu);
    //   Then place the menu on the frame.
    this.setJMenuBar(theMenuBar);
  }  // method: assembleMenuBar

// ...

// ----- Attributes ----------
private javax.swing.JMenuBar    theMenuBar = new JMenuBar();

private FileExitAction    theFileExitAction = new FileExitAction("Exit");

private LendPublicationAction    theLendPublicationAction =
                              new LendPublicationAction("Lend publication");

private ReturnPublicationAction    theReturnPublicationAction =
                              new ReturnPublicationAction("Return publication");

// ----- Inner classes ----------
public class LendPublicationAction extends javax.swing.AbstractAction {

  // ----- Operations ----------
  public    LendPublicationAction(String aLabel) {
    super(aLabel);
  }  // method: LendPublicationAction

  public void actionPerformed(java.awt.event.ActionEvent event) {
      System.out.println("Lend publication action");
  }  // method: actionPerformed
}  // class: LendPublicationAction

public class ReturnPublicationAction extends javax.swing.AbstractAction {

  // ----- Operations ----------
  public    ReturnPublicationAction(String aLabel) {
    super(aLabel);
  }  // method: ReturnPublicationAction

  public void    actionPerformed(java.awt.event.ActionEvent event) {
    System.out.println("Return publication action");
  }  // method: actionPerformed
}  // class: ReturnPublicationAction

// ...

}   // class: LibraryFrame
```

When we execute this version of the application we gain some real confidence that we are progressing in the correct direction. Now we have a second menu entitled Application with two menu items Lend publication and Return publication. When, for example, the former is selected the message "Lend publication action" appears in the command console. Now we simply have to develop a series of inner classes, one for each of the services originally provided by the methods in the Action class from chapter 6. Like that Action class we reintroduce the Library and the other classes as the model to underpin the MVC controller that we have developed. An equivalent class diagram for this is given in figure 7.17 (see also the comments concerning figure 7.3). We note here the similarity with figure 7.3 with the Action class now replaced with a LibraryFrame. The latter has also absorbed the functionality of Application run method.

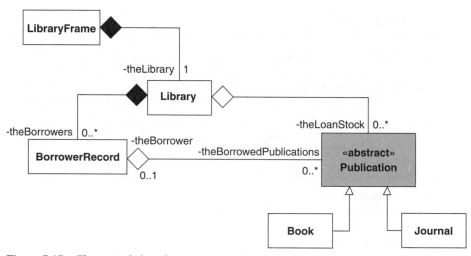

Figure 7.17   *The revised class diagram*

Our primary objective with the final version in chapter 6 was to successfully separate the view and controller from the model. In our graphical application we must construct a replacement for the view that combines with our model and controller elements. An assembly for the view component is described by the class diagram shown in figure 7.18.

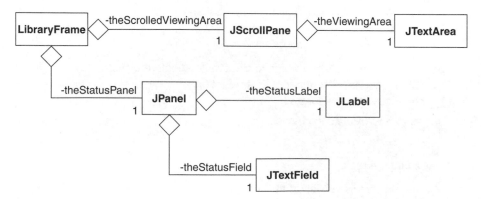

Figure 7.18   *The view component*

The principal elements are the JTextArea, the JLabel and the JTextField. The JTextArea, with role name theViewingArea, is a multi-line display area where all program output is presented. We shall arrange for this component to be read-only so that the user cannot change its content. Further, we wrap it in a Swing JScrollPane so that it has horizontal and vertical scrollbars. The JLabel and JTextField are used to present any status information from the application. Recollect from chapter 6 that each action produced a message representing the status of that action. Such messages are routed to this JTextField. The JLabel is simply a static piece of text that decorates the field, describing its purpose. The two are assembled into a single component by adding them to a JPanel. The effect is shown in figure 7.19.

Figure 7.19   *Appearance of the application*

Strictly, the ScrollPane and JPanel are added to the *content pane*, as part of the JFrame. Both the content pane and the JPanel need to be associated with layout managers that will determine how their subcomponents are presented. We set the JPanel to have a FlowLayout manager that left justifies its subcomponents, so that the label and the text field appear horizontal and adjacent to each other. For the content pane of the LibraryFrame we use a BorderLayout manager that partitions the content pane into five regions described as NORTH, SOUTH, EAST, WEST and CENTER. We place the viewing area into the central region and the status panel in the southern region. If any region is missing, the centre expands outward to occupy it. The relevant parts of the LibraryFrame are given below:

```
public class LibraryFrame extends javax.swing.JFrame {

   // ----- Operations ----------
   public   LibraryFrame(String aCaption) {
      super(aCaption);
      //   Populate the frame with its sub-components.
      Container contentPane = this.getContentPane();
      contentPane.setLayout(new BorderLayout());
      contentPane.add(theScrolledViewingArea, BorderLayout.CENTER);
```

```java
    contentPane.add(theStatusPanel, BorderLayout.SOUTH);
    this.assembleMenuBar();
    Dimension screen = Toolkit.getDefaultToolkit().getScreenSize();
    int width = screen.width * 3 / 4;
    int height = screen.height * 3 / 4;
    this.setBounds(screen.width / 8, screen.height / 8, width, height);
    this.setVisible(true);
    //   Proper close-down procedure.
    this.addWindowListener(new LibraryFrameClosing());
    theLibrary = new Library("Napier");
    }   // method: LibraryFrame

    // ...

    // ----- Attributes ----------

    private javax.swing.JTextArea theViewingArea = new JTextArea();
    {   // initialization block
       theViewingArea.setEditable(false);
    }
    private javax.swing.JScrollPane    theScrolledViewingArea =
       new JScrollPane(theViewingArea);

    private javax.swing.JLabel    theStatusLabel = new JLabel("Status");

    private javax.swing.JTextField    theStatusField = new JTextField(60);

    private javax.swing.JPanel    theStatusPanel =
       new JPanel(new FlowLayout(FlowLayout.LEFT));

    {   // initialization block
       theStatusPanel.add(theStatusLabel);
       theStatusPanel.add(theStatusField);
    }
}   // class: LibraryFrame
```

Note also the use of initialization blocks. These are code blocks executed along with the initialization of the class attributes. Here we use them to further initialize the attributes following that provided by their class constructors. As an illustration, once we have created theViewingArea we then set it so it operates as read-only.

Development of the action event handlers mirrors the original code in the Action class from chapter 6. The changes we have to make revolve around displaying the results and obtaining information from the user. For those handlers primarily concerned with producing output displays we use the append message of the JTextArea class. This method simply adds new text on to the end of the multi-line text area. See the handler for the class DisplayPublicationsAvailableForLoanAction in program 7.7 for examples of this coding. User input is obtained by using various dialogs from the class JOptionPane. A number of simple dialogs are provided by this class, including the input dialogs provided by the showInputDialog method. An example of its use is given in the AddBookAction class.

Program 7.7   Restoration of all the actions (model Lib7_7.uml)

```
package librarysubsystem;
import java.awt.*;
import javax.swing.*;
import java.util.*;
import textio.*;
public class LibraryFrame extends javax.swing.JFrame {
    // ----- Operations ----------
    private void assembleMenuBar() {
        // Prepare the File menu ...
        JMenu fileMenu = new JMenu("File");
        fileMenu.setMnemonic('F');
        // ... and the Exit menu item.
        JMenuItem fileExit = fileMenu.add(theFileExitAction);
        fileExit.setMnemonic('x');
        // Prepare the Application menu ...
        JMenu applicationMenu = new JMenu("Application");
        applicationMenu.setMnemonic('A');
        // ... and its various menu items.
        JMenuItem registerBorrower = applicationMenu.add(theRegisterBorrowerAction);
        JMenuItem displayBorrowers = applicationMenu.add(theDisplayBorrowersAction);
        JMenuItem addBook = applicationMenu.add(theAddBookAction);
        JMenuItem addJournal = applicationMenu.add(theAddJournalAction);
        JMenuItem loadPublications = applicationMenu.add(theLoadPublicationsAction);
        JMenuItem publicationsForLoan = applicationMenu.add
            (theDisplayPublicationsAvailableForLoanAction);
        JMenuItem publicationsOnLoan = applicationMenu.add
            (theDisplayPublicationsOnLoanAction);
        JMenuItem lendPublication = applicationMenu.add(theLendPublicationAction);
        JMenuItem returnPublication = applicationMenu.add(theReturnPublicationAction);
        // Add the menus to the menu bar.
        theMenuBar.add(fileMenu);
        theMenuBar.add(applicationMenu);
        // Then place the menu on the frame.
        this.setJMenuBar(theMenuBar);
    }   // method: assembleMenuBar

    // ...

    // ----- Attributes ----------
    private AddBookAction    theAddBookAction = new AddBookAction("Add book");
    private DisplayPublicationsAvailableForLoanAction
        theDisplayPublicationsAvailableForLoanAction
        = new DisplayPublicationsAvailableForLoanAction("Display publications for loan");
    // ...
```

```java
// ----- Inner classes ----------
public class AddBookAction extends javax.swing.AbstractAction {
    // ----- Operations ----------
    public    AddBookAction(String aLabel) {
        super(aLabel);
    }   // method: AddBookAction

    public void actionPerformed(java.awt.event.ActionEvent event) {
        //    Get book details from the user.
        String title = JOptionPane.showInputDialog(LibraryFrame.this,
            "Enter the book title", "Book details", JOptionPane.QUESTION_MESSAGE);
        String catalogueString = JOptionPane.showInputDialog(LibraryFrame.this,
            "Enter the catalogue number","Book details",
                JOptionPane.QUESTION_MESSAGE);
        String author = JOptionPane.showInputDialog(LibraryFrame.this,
            "Enter the author name", "Book details",
                JOptionPane.QUESTION_MESSAGE);
        int catalogueNumber = Integer.parseInt(catalogueString);
        //
        // Add the book the the library.
        LibraryFrame.this.theLibrary.addOneLoanItem(new Book(title,
                                                    catalogueNumber, author));
        //
        // Display the outcome.
        LibraryFrame.this.theStatusField.setText(LibraryFrame.this.theLibrary.getStatus());
    }   // method: actionPerformed

}   // class: AddBookAction

public class DisplayPublicationsAvailableForLoanAction extends
                javax.swing.AbstractAction {

    // ----- Operations ----------
    public    DisplayPublicationsAvailableForLoanAction(String aLabel) {
        super(aLabel);
    } // method: DisplayPublicationsAvailableForLoanAction

    public void    actionPerformed(java.awt.event.ActionEvent event) {
        // Display information about the Library
    LibraryFrame.this.theViewingArea.append("\n" + theLibrary);
        //
        // Display information about the publications available for loan
        LibraryFrame.this.theViewingArea.append("\n\t" + "Publications available for loan");
        boolean publicationFound = false;
        Iterator iter = theLibrary.getLoanItemsIterator();
        //
        while(iter.hasNext()) {
            Publication pub = (Publication) iter.next();
```

```
            if(pub.getBorrower() == null) {
               LibraryFrame.this.theViewingArea.append("\n\t\t" + pub);
               publicationFound = true;
            }
         }
         if(publicationFound == false)
            LibraryFrame.this.theViewingArea.append("\n\t\t" + "None");
         //
         // Display outcome
         LibraryFrame.this.theStatusField.setText("");
      } // method: actionPerformed
   } // class: DisplayPublicationsAvailableForLoanAction
} // class: LibraryFrame
```

In class DisplayPublicationsAvailableForLoanAction, the actionPerformed method simply appends the library name and publication details to theViewingArea component. At the end of the method we use the message setText to change the content of theStatusField.

The actionPerformed method in the AddBookAction class prompts the user for the individual book details, then adds a new Book object to the library stock. The method showInputDialog has four parameters: parent component, message, title and message type. The parent component identifies the Swing component that acts as the parent while the dialog is visible. Here, it is the LibraryFrame object. The message is the prompt that advises what information is sought from the user. The title is the text that decorates the dialog caption. The message type is a symbolic constant describing the type of dialog, here a questioning dialog. An example of this kind of dialog is shown in figure 7.20.

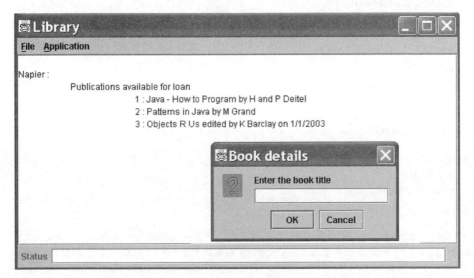

Figure 7.20 *The application with some output and a dialog requesting a book title*

7.6 Application buttons

In this final iteration we aim to provide the user with a simpler interface comprising on-screen buttons labelled with the applications services. In effect they provide shortcuts to the items on the Application menu. In keeping with the theme of this chapter we make one incremental change to that in program 7.7 which leaves the program unchanged but gives us a better structure to include the action buttons.

The code for the method assembleMenuBar shown in program 7.7 is not especially inspirational. It is simply a list of statements to create the menus and attach the menu items. One useful revision is to place the action objects and the menu details into a data structure. Here we have chosen an array, but see also appendix E for alternatives. We can then replace our existing code with new code to traverse each array to build the menus and their menu items. First, let us construct an array with the details for the File menu. In this case it is the menu name and the action object for its single menu item:

```
private Object[]   theFileMenu = {
   "File",
   theFileExitAction
};
```

Note how the array has a String and a FileExitAction object and so we declare it as an array of Objects. That way the array can reference any type other than the primitives. In a similar manner we have the Application menu:

```
private Object[]   theApplicationMenu = {
   "Application",
   theLoadPublicationsAction,
   theAddBookAction,
   theAddJournalAction,
   theDisplayPublicationsAvailableForLoanAction,
   theDisplayPublicationsOnLoanAction,
   theRegisterBorrowerAction,
   theLendPublicationAction,
   theReturnPublicationAction
};
```

Now we place references to these two arrays into a third array:

```
private Object[] theMenus = {
   theFileMenu,
   theApplicationMenu
};
```

These structures are described by figure 7.21. Two one-dimensional tables are established to hold the menus. Each is declared as an array of Objects. The first entry is a String object acting as the menu name. The remaining entries are the action objects that act as the handlers for the menu items. The third array is an array with references to the menu arrays. This third array then acts as the root of the structure from which all the menus can be assembled.

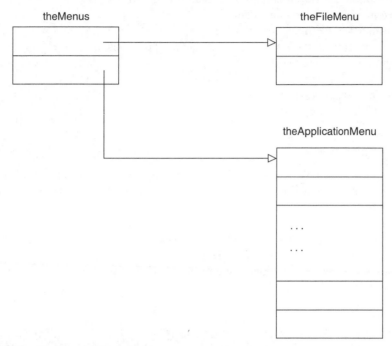

Figure 7.21   *Structure of the menus*

With these simple changes we can now reprogram the **assembleMenuBar** method. It iterates through the elements of **theMenus** passing each member to another auxiliary method **assembleMenu**. The formal parameter to this new method is an array of Objects, reflecting that it is a heterogeneous container. The first element in the array is the menu name and the remaining are the action objects. For such an array the **assembleMenu** method creates the menu from the first element and the menu items from the remaining action elements. Program 7.8 shows the necessary details.

Program 7.8   Revised menu assembly (model Lib7_8.uml)

```
public class LibraryFrame extends JFrame {
    // ----- Operations ----------
    private void    assembleMenuBar() {
        int menusLength = theMenus.length;
        for(int k = 0; k < menusLength; k++)
            this.assembleMenu((Object[ ])theMenus[k]);
        this.setJMenuBar(theMenuBar);
    } // method: assembleMenuBar
```

```
    private void   assembleMenu(Object[ ] menuEntries) {
      int menuLength = menuEntries.length;
      // Peel off the first item as the menu name.
      JMenu menu = new JMenu((String)menuEntries[0]);
      // Remainder are the menu items.
      for(int k = 1; k < menuLength; k++) {
        AbstractAction action  = (AbstractAction)menuEntries[k];
        menu.add(action);
      }
      theMenuBar.add(menu);
    }   // method: assembleMenu
    // ...

} // class: LibraryFrame
```

These code revisions have no impact on the appearance and execution of the program. The changes, however, permit us to readily incorporate a toolbar of buttons corresponding to the Application menu items. Many applications have a toolbar positioned immediately below the menu. Such toolbars have small buttons with iconic representations for the services they provide. For our example we use large buttons labelled the same as the Application menu items. We arrange the buttons vertically in a panel to the left of the view area where the output is displayed. The appearance is now as demonstrated in figure 7.22.

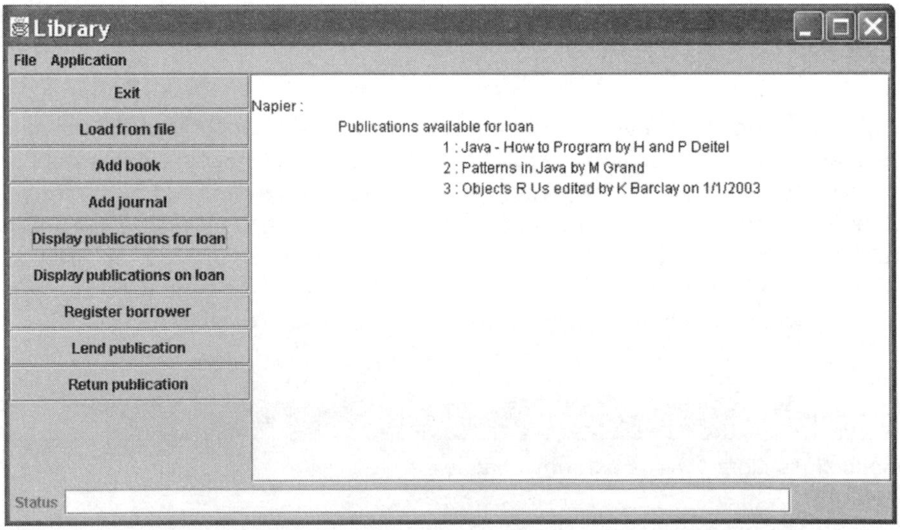

Figure 7.22   *Button toolbar*

The new panel arrangement is simple to achieve. Recollect that the frame's content pane was given a BorderLayout manager. We simply add the button panel to the left (WEST) of the view panel. The panel itself is made from the Box class that uses a

BoxLayout manager. This layout manager is suitable for preparing its components as a row of horizontal members or a column of vertical members. Program 7.9 highlights the relevant new code.

Program 7.9   A panel of buttons (model Lib7_9.uml)

```
public class LibraryFrame extends JFrame {

   private void   assembleMenu(Object[ ] menuEntries) {
      int menuLength = menuEntries.length;
      // Peel off the first item as the menu name.
      JMenu menu = new JMenu((String)menuEntries[0]);
      // Remainder are the menu items.
      for(int k = 1; k < menuLength; k++) {
         AbstractAction action = (AbstractAction)menuEntries[k];
         menu.add(action);
         JButton button = new JButton(action);
         theToolBar.add(button);
         button.setPreferredSize(BUTTONSIZE);
         button.setMinimumSize(BUTTONSIZE);
         button.setMaximumSize(BUTTONSIZE);
      }
      theMenuBar.add(menu);
   }   // method: assembleMenu

   // ...

   // ----- Attributes ----------

   private javax.swing.JMenuBar   theMenuBar = new JMenuBar();
   private javax.swing.Box   theToolBar = Box.createVerticalBox();

   private static final java.awt.Dimension   BUTTONSIZE = new Dimension(200, 30);
      // ...
} // class: LibraryFrame
```

Observe how the private method **assembleMenu** simultaneously prepares both the menus and the buttons. A **JButton** object is created from the action, decorated with the same text as the corresponding menu item. Each button is added to **theToolBar**, an object of the **Box** class.

Finally, we reinstate the serialization mechanism. In version 3 from chapter 6 the saving was implemented in the close method of the Action class and the restoration of objects was realized in the Action class constructor. With our graphical version of the application the corresponding location for this code is the event handler for FileExitAction and the LibraryFrame class constructor.

Program 7.10   Serialization (model Lib7_10.uml)

```java
public class LibraryFrame extends javax.swing.JFrame {

  // ----- Operations ----------
  public   LibraryFrame(String aCaption) {
    super(aCaption);
    //   Populate the frame with its sub-components.
    Container contentPane = this.getContentPane();
    contentPane.setLayout(new BorderLayout());
    contentPane.add(theStatusPanel, BorderLayout.SOUTH);
    contentPane.add(theToolBar, BorderLayout.WEST);
    contentPane.add(theScrolledViewingArea, BorderLayout.CENTER);
    this.assembleMenuBar();
    Dimension screen = Toolkit.getDefaultToolkit().getScreenSize();
    int width = screen.width * 3 / 4;
    int height = screen.height * 3 / 4;
    this.setBounds(screen.width / 8, screen.height / 8, width, height);
    this.setVisible(true);
    //   Proper close-down procedure.
    this.addWindowListener(new LibraryFrameClosing());
    //   Restore the persistent application objects.
    File file = new File(PERSISTENT_FILENAME);
    if(file.exists()) {
      try {
        FileInputStream fis = new FileInputStream(PERSISTENT_FILENAME);
        ObjectInputStream ois = new ObjectInputStream(fis);
        theLibrary = (Library)ois.readObject();
        ois.close();
      } catch(IOException ex) {
          theStatusField.setText("Error reading persistent store");
      } catch(ClassNotFoundException ex) {
          theStatusField.setText("Error restoring application objects");
      }
    } else
        theLibrary = new Library("Napier");
} // method: LibraryFrame

// ...

// ----- Inner classes ----------
public class FileExitAction extends javax.swing.AbstractAction {

// ----- Operations ----------
  public   FileExitAction(String aLabel) {
    super(aLabel);
  }   // method: FileExitAction

  public void   actionPerformed(java.awt.event.ActionEvent event) {
```

```
    try {
      FileOutputStream fos
          = new FileOutputStream(LibraryFrame.this.PERSISTENT_FILENAME);
      ObjectOutputStream oos = new ObjectOutputStream(fos);
      oos.writeObject(LibraryFrame.this.theLibrary);
      oos.close();
    } catch(IOException ex) {
      LibraryFrame.this.theStatusField.setText("Cannot open persistent file");
    }
    System.exit(0);
  }   // method: actionPerformed
}   // class: FileExitAction
}   // class: LibraryFrame
```

7.7   Dialogs

Obtaining the book and journal details is somewhat pedestrian since we use separate
input dialogs. Normally, the user would expect a single dialog requesting each value. To
achieve this we introduce specialized JDialog classes. For example, the BookDialog
class gives rise to the input dialog as shown in figure 7.23.

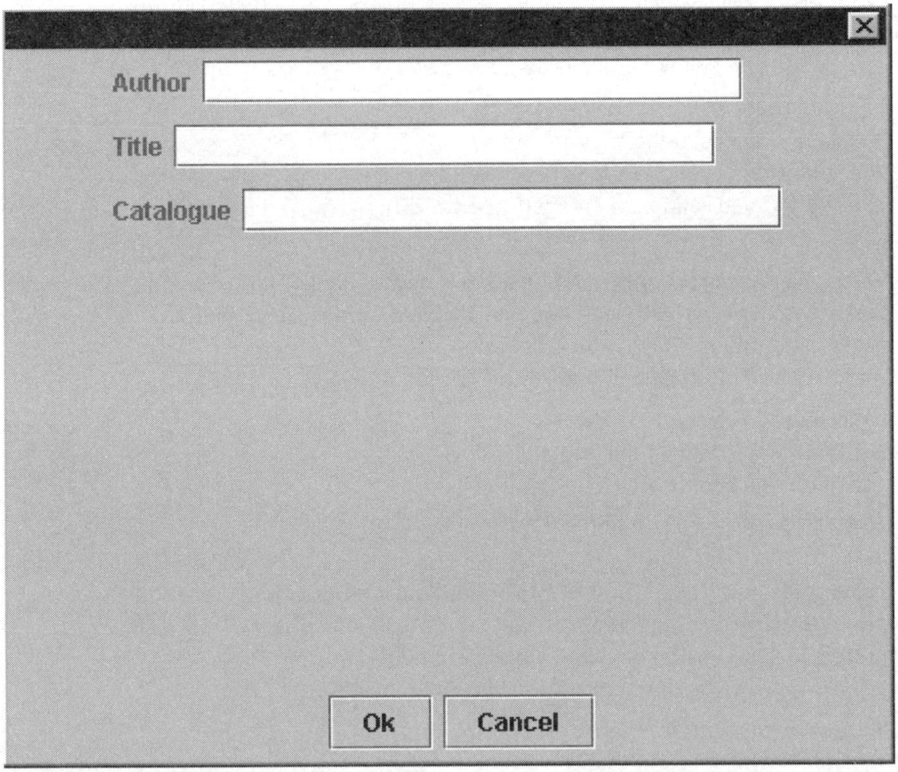

Figure 7.23   *Book details dialog*

The BookDialog class assembles this dialog, populating it with the three text fields and associated labels at the top, and the two buttons at the lower edge. The JDialog class is a component with a content pane that is given a BorderLayout manager. The constructor for the class BookDialog will place a panel in its centre to carry the three text fields and a buttons panel at the SOUTH position. The code is:

```
// class BookDialog
public   BookDialog(JFrame frame) {
   super(frame, "Book details", true);
   this.setSize(500, 400);
   //   Build the sub-components.
   Container contentPane = this.getContentPane();
   contentPane.setLayout(new BorderLayout());
   contentPane.add(theDetailsPanel, BorderLayout.CENTER);
   contentPane.add(theButtonsPanel, BorderLayout.SOUTH);
   this.setResizable(false);
   this.setLocationRelativeTo(frame);
}   // method: BookDialog
```

The details panel is assembled with a series of declarations and initializations (the code commentary has, for brevity, been removed):

```
private javax.swing.JLabel    theAuthorLabel = new JLabel("Author");
private javax.swing.JTextField    theAuthorTextField = new JTextField(30);
private javax.swing.JPanel    theAuthorPanel =
                new JPanel(new FlowLayout(FlowLayout.LEFT));
{   // initialization block
   theAuthorPanel.add(theAuthorLabel);
   theAuthorPanel.add(theAuthorTextField);
   theAuthorPanel.setMaximumSize(PANELSIZE);
}
private javax.swing.JLabel    theTitleLabel = new JLabel("Title");
private javax.swing.JTextField    theTitleTextField = new JTextField(30);
private javax.swing.JPanel    theTitlePanel
                = new JPanel(new FlowLayout(FlowLayout.LEFT));
{   // initialization block
   theTitlePanel.add(theTitleLabel);
   theTitlePanel.add(theTitleTextField);
   theTitlePanel.setMaximumSize(PANELSIZE);
}
private javax.swing.JLabel    theCatalogueLabel = new JLabel("Catalogue");
private javax.swing.JTextField    theCatalogueTextField = new JTextField(30);
private javax.swing.JPanel    theCataloguePanel
                = new JPanel(new FlowLayout(FlowLayout.LEFT));
{ // initialization block
   theCataloguePanel.add(theCatalogueLabel);
```

```
    theCataloguePanel.add(theCatalogueTextField);
    theCataloguePanel.setMaximumSize(PANELSIZE);
}
private javax.swing.Box   theDetailsPanel = Box.createVerticalBox();
{ // initialization block
    theDetailsPanel.add(theAuthorPanel);
    theDetailsPanel.add(theTitlePanel);
    theDetailsPanel.add(theCataloguePanel);
    theDetailsPanel.add(Box.createVerticalGlue());
}
```

An inner class OkCancelAction describes the event handling associated with the two buttons. For either button we make the dialog invisible. If the user presses the "Ok" button we set the property theUserAction of the class BookDialog to the value JOptionPane.OK_OPTION that identifies this button. The value JOptionPane.CANCEL_OPTION records that the "Cancel" button was selected.

```
// ----- Inner classes ----------
public class OkCancelAction implements java.awt.event.ActionListener {
    // ----- Operations ----------
    public void   actionPerformed(java.awt.event.ActionEvent event) {
        BookDialog.this.setVisible(false);
        if(event.getActionCommand().equals("Ok"))
            BookDialog.this.theUserAction = JOptionPane.OK_OPTION;
        else
            BookDialog.this.theUserAction = JOptionPane.CANCEL_OPTION;
    }   // method: actionPerformed
}   // class: OkCancelAction
```

When the user dismisses the dialog, the method getUserAction from the BookDialog class allows us to determine the choice made. If the user dismissed the dialog with the "Ok" button then the three accessor methods getTitle, getCatalogueNumber and getAuthor give us the values entered by the user. Program 7.11 shows the revisions made to the inner class AddBookAction. Similar work can give us a dialog for the user details.

Program 7.11   Using the dialog (model Lib7_11.uml)

```
public class LibraryFrame extends JFrame {
    // ...
    // ----- Inner classes ----------
    public class AddBookAction extends javax.swing.AbstractAction {
        // ----- Operations ----------
        public   AddBookAction(String label) {
            super(label);
        } // method: AddBookAction
```

```
public void actionPerformed(java.awt.event.ActionEvent event) {
    BookDialog dialog = new BookDialog(LibraryFrame.this);
    dialog.setVisible(true);
    if(dialog.getUserAction() == JOptionPane.OK_OPTION) {
        String title = dialog.getTitle();
        int catalogueNumber = dialog.getCatalogueNumber();
        String author = dialog.getAuthor();
        //
        // Add the book to the library.
        LibraryFrame.this.theLibrary.addOneBook(new Book(title, catalogueNumber,
                                                          author));
        //
        // Display the outcome
        LibraryFrame.this.theStatusField.setText(LibraryFrame.this.theLibrary.getStatus() );
    } else
            LibraryFrame.this.theStatusField.setText("No Book added");
    } // method: actionPerformed
} // class: AddBookAction
// ...
} // class: LibraryFrame
```

7.8   Summary

1. The **abstract** class JComponent is the root class for many of the graphic components.
2. Components can include other sub-components in a parent/child arrangement. A **LayoutManager** is responsible for positioning the children of a parent component.
3. The model-view-controller design pattern is a significant feature of the architecture of the Swing classes. The view-controller is responsible for the presentation and the interaction with a graphical component. The model element represents the state information for the component.
4. Events in Swing are represented by objects of different event classes. When an event occurs, an event object is passed to an event handler. The event object carries the data associated with the event.
5. The Java event model is based on the notion of event listeners. When an event is generated by an event source, the source notifies all its listener objects by calling a fixed method and passing to it the appropriate event object. For the source to be able to call a specific method in a listener object, the listener object must implement a particular method protocol as defined by a corresponding listener interface.
6. Inner classes are frequently used to realize event listeners.

7.9   Exercises

1. Use the API documentation to construct a class hierarchy for the class JFrame. Why does this hierarchy not include the class JComponent? What terms are used

to distinguish the class JFrame from the concrete subclasses of JComponent? Give an explanation for these terms.

2. What class of object is used to arrange GUI components in a Container?

3. Use the API documentation to construct a class diagram that is rooted in the interface LayoutManager. Briefly summarize the layout schemes of the various layout managers.

4. What method is used to specify the layout manager for a component? Name three concrete layout managers.

5. Is BorderLayout the default layout manager for a content pane? When using a BorderLayout what is the maximum number of components can it display? Do we have to add GUI components in a particular order to a BorderLayout?

6. In a manner similar to that shown in figure 7.1, present a class hierarchy for the class JButton by using the information in the API documentation. The class AbstractButton has an architectural attribute of the class ButtonModel. Incorporate this class into the class diagram including its associated class hierarchy. Explain the role for this second group of classes.

7. Explain the difference between the terms model, view and controller. How is the MVC deployed in Swing for the class JButton? Find how it is used in the class JTextField. How is it deployed in the library case study?

8. Explain the difference between the terms "event driven", "event object" and "event handler".

9. The methods addActionListener and removeActionListener are used to associate and disassociate action listener objects with a JComponent. What are the names and signatures for the corresponding operations for mouse events? How did we do the same for window events in the library case study?

10. How do we ensure that an action listener object can be sent the correct message by an event source?

11. Give an outline of how events are handled in Swing by considering what happens when a mouse click occurs over some component. In particular identify how other components are advised of this occurrence. For a component to be informed how do we determine which of its methods we should execute as the handler? What type of object is passed to that handler to give details of the event.

12. Program 7.4 introduced inner classes to act as event handler objects. Go to the Java website that follows and checkout some details for these constructs.
http://java.sun.com/products/jdk/1.1/docs/guide/innerclasses/spec/inner-classes.doc.html

13. Develop a class diagram for the classes JMenu and JMenuItem. What in this architecture permits a menu to have sub-menus?

14. What is the rationale for adding ActionListener objects to both JMenu objects and to JToolBar objects?

15. The BookDialog class is used to obtain the book author, title and catalogue number. It also has event handlers for the Ok and Cancel buttons. The JournalDialog class achieves much the same for a journal. What might be done to reduce the amount of common code? Use model Lib7_11.uml and make the necessary changes.

16. Using program 7.11 as the base code, augment this application with two additional use-cases:
   • a list of borrowers that have no books on loan and
   • a list of borrowers that have more than some specified number of books on loan.

8

Design Patterns

No book on object-oriented development would be complete without a discussion of software *design patterns*. They first became popular with the seminal book Gamma (1994) and have subsequently attracted a considerable following among software professionals. "Patterns are ways to describe best practice, good design, and capture experience in a way that it is possible for others to reuse this experience" (http://www.hillside.net). Their aim is to build a body of knowledge to support design and implementation in a form that can be promulgated to software developers. Our aim is to illustrate their use and to bring a greater understanding of OOAD to the reader.

8.1 Delegation

Delegation is at the heart of many design patterns. Therefore a review of delegation is a useful introduction to this chapter. The essential idea is that when an object receives a message it can choose to implement none or part of the method. This is because it has the option of making use of a *delegate*. The class diagram of figure 8.1 and the accompanying outline Java code illustrate a delegator implementing operation_1 but using a delegate in the implementation of operation_2 and operation_3.

Figure 8.1   *A delegator with a delegate*

```
public class Delegator {
   // ----- Operations ----------
   public Delegator(Delegate aDelegate) {
      theDelegate = aDelegate;
   } // method: Delegator
   public void operation_1( ) {
      //...
   } // method: operation_1
```

```
      public void operation_2( ) {
        //...
        theDelegate.operation_4();
      } // method: operation_2

      public void operation_3( ) {
        theDelegate.operation_5();
      } // method: operation_3

        // ----- Attributes ----------
        private Delegate theDelegate;

        // ...

} // class: Delegator


public class Delegate {
    // ----- Operations ----------
    public void operation_4( ) {
      //...
    } // method: operation_4

    public void operation_5( ) {
      //...
    } // method: operation_5

    // ...

} // class: Delegate


public class Application {
    // ----- Operations ----------
    public void run () {
      Delegator delegator = new Delegator(new Delegate(...));
      //
      delegator.operation_1();   // Implemented directly by the delegator.
      delegator.operation_2();   // Implemented in part by the delegator and the delegate.
      delegator.operation_3();   // Implemented by the delegate.

    } // method: run

} // class: Application
```

Recall that when using the Java programming language:

- specialization is implemented by class (interface) inheritance
- inheritance is static in the sense that it is a compile-time phenomenon
- all instances of a class behave in the same manner
- an object cannot change its class and
- encapsulation may be broken if **protected** (or **package**) access to a parent's features is used.

By way of contrast:

- delegation is implemented by object composition
- the delegate can be changed at run-time therefore delegation may be dynamic if required
- delegation uses public services therefore it respects encapsulation.

To illustrate delegation we can use an amended version of the example from chapter 5. Consider a software company, employing programmers each of which have a unique payroll number. We can assume that a programmer gets paid a basic weekly salary of 100 pounds. When a new programmer joins the staff he is assigned to a more experienced programmer called a mentor. The mentor is paid a bonus of 10 per cent for this duty. The company management also recognizes that some mentors are exceptional programmers and designates them as gurus. They receive a 20% bonus. We can model the relationship between the programmer, mentor and guru as one of specialization.

As indicated in figure 8.2, the **Programmer** class acts as the base class holding the payroll number and basic salary as **protected** attributes. Its **getSalary** method is redefined in the descendant **Mentor** and **Guru** classes to reflect their enhanced salary. Similarly the **toString** method is redefined to support the display of the objects.

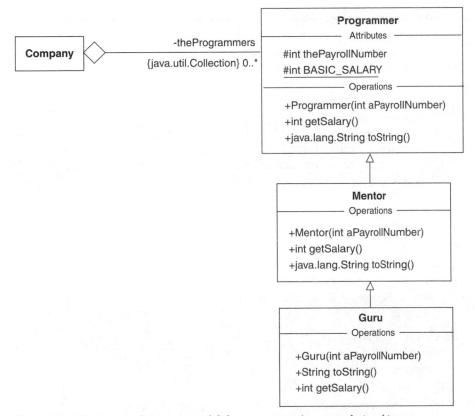

Figure 8.2 *Using specialization to model the programmer/mentor relationship*

Polymorphic substitution allows the Company to maintain a collection of Programmer, Mentor and Guru objects.

To calculate the salary bill, the Company accumulates a total by iterating over the collection of Programmers requesting the salary of each object it encounters. Of course the object in question could be a Programmer, Mentor or Guru. Each responds according to the method defined for getSalary in its class. Outline Java code for these classes is shown in program 8.1.

Program 8.1   Using specialization (model Prog8_1.uml)

```java
import textio.*;
public class Application {
    // ----- Operations ----------
    public void    run() {
        //
        // Create a new organisation
        Company co = new Company("The Object University");
        //
        // Hire some new employees
        co.hireProgrammer(new Programmer(123));
        co.hireProgrammer(new Mentor(456));
        co.hireProgrammer(new Guru(789));
        //
        // Display all the employees
        co.displayProgrammers();
        //
        // Obtain and print the wage bill
        ConsoleIO.out.println("The salary bill is: " + co.getSalaryBill());
        ConsoleIO.out.println();
    }   // method: run
}   // class: Application

public class Programmer {
    // ----- Operations ----------
    public    Programmer(int aPayrollNumber) {
        thePayrollNumber = aPayrollNumber;
    }   // method: Programmer

    public int    getSalary() {
        return BASIC_SALARY;
    }   // method: getSalary
    public String    toString() {
        return "Programmer with payroll number: " + thePayrollNumber;
    }   // method: toString
```

```
// ----- Attributes ----------
protected int   thePayrollNumber;
protected static final int   BASIC_SALARY = 100;
}   // class: Programmer

public class Mentor extends Programmer {
  // ----- Operations ----------
  public   Mentor(int aPayrollNumber) {
    super(aPayrollNumber);
  }   // method: Mentor

  public int   getSalary() {
    return BASIC_SALARY + (int)(BASIC_SALARY * 0.1);
  }   // method: getSalary

  public String   toString() {
    return "Mentor with payroll number: " + thePayrollNumber;
  }   // method: toString
} // class: Mentor

public class Guru extends Mentor {
  // ----- Operations ----------
  public Guru(int aPayrollNumber) {
    super(aPayrollNumber);
  }   // method: Guru

  public String   toString() {
    return "Guru with payroll number: " + thePayrollNumber;
  }   // method: toString

  public int   getSalary() {
    return BASIC_SALARY + (int)(BASIC_SALARY * 0.2);
  }   // method: getSalary
}   // class: Guru
```

Note that for the sake of clarity we have omitted the methods equals, compareTo and hashCode where they would normally be expected. Similarly the method qualifier **final** is often omitted. The Application (created in a class Main with method main as normal) produces an output as follows:

Staff list for The Object University:

```
  Programmer with the payroll number: 123
  Mentor with the payroll number: 456
  Guru with the payroll number: 789
```
The salary bill is: 330

The difficulty with this solution arises when there is a need for a mentor or guru to revert back to his normal programming job! We can easily make the situation worse by introducing the requirement that a programmer adopts other roles. For example, he might also be an administrator. In this case specialization is not appropriate. The reason is that we must change the class of a particular object, e.g. from Mentor to Programmer. A better alternative is to use a delegate to represent the role that an employee adopts. This approach is shown in the class diagram of figure 8.3. Note that we have decided that a Guru is just a special kind of Programmer not a Mentor. This prevents the specialization hierarchy becoming unnecessarily deep.

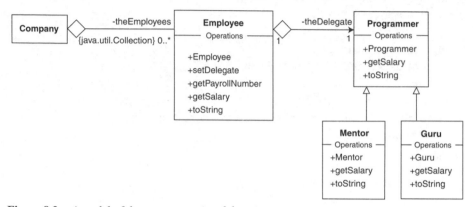

Figure 8.3   *A model of the company using delegation*

Now when the Company accumulates a salary bill it does so by asking each Employee for its salary. The difference is that the Employee uses a Programmer delegate to do so. The toString method is treated similarly. As before we specialize the Programmer into a Mentor or Guru. The crucial point is that the Company only has Employee objects but each may have a different role. It depends on the class of the delegate. The Java code for this example is given in program 8.2.

Program 8.2   Using a delegate (model Prog8_2.uml)

```
import textio.*;

public class Application {
    // ----- Operations ----------
    public void    run() {
        Company co = new Company("The Object University");
        Employee emp = null;
        //
        emp = new Employee(123);
        emp.setDelegate(new Programmer());
        co.hireEmployee(emp);
        //
        emp = new Employee(456);
        emp.setDelegate(new Mentor());
```

```
      co.hireEmployee(emp);
      //
      co.displayEmployees();
      ConsoleIO.out.println("The salary bill is: " + co.getSalaryBill());
      ConsoleIO.out.println();
      //
      emp = co.getEmployee(123);
      emp.setDelegate(new Mentor());
      //
      emp = co.getEmployee(456);
      emp.setDelegate(new Guru());
      //
      co.displayEmployees();
      ConsoleIO.out.println("The salary bill is: " + co.getSalaryBill());
      ConsoleIO.out.println();
   }   // method: run
}   // class: Application

import textio.*;
import java.util.*;
public class Company {
   // ----- Operations ----------
   public   Company(String aName) {
      theName = aName;
      theEmployees = new ArrayList();
   }   // method: Company
   public void   hireEmployee(Employee anEmployee) {
      theEmployees.add(anEmployee);
   }   // method: hireEmployee
   public void   displayEmployees() {
      ConsoleIO.out.println();
      ConsoleIO.out.println("Staff list for " + theName);
      ConsoleIO.out.println();
      Iterator iter = theEmployees.iterator();
      while(iter.hasNext()) {
         Employee emp = (Employee)iter.next();
         ConsoleIO.out.println("\t" + emp);
      }
      ConsoleIO.out.println();
   }   // method: displayEmployees
   public int   getSalaryBill() {
      int totalBill = 0;
      Iterator iter = theEmployees.iterator();
      while(iter.hasNext()) {
         Employee emp = (Employee)iter.next();
```

```java
        totalBill += emp.getSalary();
    }
    return totalBill;
}   // method: getSalaryBill

public Employee    getEmployee(int aPayrollNumber) {
    Iterator iter = theEmployees.iterator();
    while(iter.hasNext()) {
        Employee emp = (Employee)iter.next();
        if(aPayrollNumber == emp.getPayrollNumber())
            return emp;
    }
    return null;
}   // method: getEmployee

// ----- Attributes ----------
private String    theName;
private java.util.Collection    theEmployees;    // of Employee
}   // class: Company

public class Employee {
    // ----- Operations ----------
    public    Employee(int aPayrollNumber) {
        thePayrollNumber = aPayrollNumber;
        theDelegate = null;
    }   // method: Employee

    public void    setDelegate(Programmer aProgrammer) {
        theDelegate = aProgrammer;
    }   // method: setDelegate

    public int    getPayrollNumber() {
        return thePayrollNumber;
    }   // method: getPayrollNumber

    public int    getSalary() {
        return theDelegate.getSalary();
    }   // method: getSalary

    public String    toString() {
        return "Employee with payroll number: " + thePayrollNumber +
                " that is a " + theDelegate;
    }   // method: toString

    // ----- Attributes ----------
    private int    thePayrollNumber;
    private Programmer    theDelegate;
}   // class: Employee

public class Programmer {
    // ----- Operations ----------
```

```
   public Programmer() {
   }   // method: Programmer
   public int   getSalary() {
      return BASIC_SALARY;
   }   // method: getSalary
   public String   toString() {
      return "Programmer";
   }   // method: toString
   // ----- Attributes ----------
   protected static final int   BASIC_SALARY = 100;
}   // class: Programmer

public class Mentor extends Programmer {
   // ----- Operations ----------
   public Mentor() {
      super();
   }   // method: Mentor
   public int   getSalary() {
      return BASIC_SALARY + (int)(BASIC_SALARY * 0.1);
   }   // method: getSalary
   public String   toString() {
      return "Mentor";
      }   // method: toString
}   // class: Mentor

public class Guru extends Programmer {
   // ----- Operations ----------
   public   Guru() {
      super();
   }   // method: Guru
   public int   getSalary() {
      return BASIC_SALARY + (int)(BASIC_SALARY * 0.2);
   }   // method: getSalary
   public String   toString() {
      return "Guru";
   }   // method: toString
}   // class: Guru
```

The Application produces the following output:

Staff list for The Object University:

Employee with payroll number: 123 that is a Programmer
Employee with payroll number: 456 that is a Mentor

The salary bill is: 210

Staff list for The Object University:

  Employee with payroll number: 123 that is a Mentor
  Employee with payroll number: 456 that is a Guru

The salary bill is: 230

8.2   Interface

Although apparently rather simple, the *interface design pattern* is probably the most
widely used design pattern of all. Recall from the discussions of chapters 5 and 6 that
we frequently establish a relationship with the base class of a class hierarchy. The intention is that an object belonging to any of its specialized classes can act as a substitute.
In fact a typical example is that shown in figure 8.2. Clearly this is a useful approach but
it can sometimes be too restrictive. The main problem that we face is that all of the
objects involved must belong to the same hierarchy. This may not always be appropriate.

A solution is to replace the base class with an interface. This is a class that provides
only a specification of its services. It allows us to design to a specification only and not
to a specification and its implementation. In order to conform to an interface a class
must implement all of its services, i.e. supply a method body for each operation. There
is no requirement for conformant classes to have any relationship with each other. They
need only implement the same interface. Crucially they can belong to different class
hierarchies. It is this fact that gives us the flexibility we need.

As is shown in the class diagram in figure 8.4 the Programmer, Mentor, and Guru
classes are now free to have separate implementations. Their only commonality is that
they must implement the Role interface.

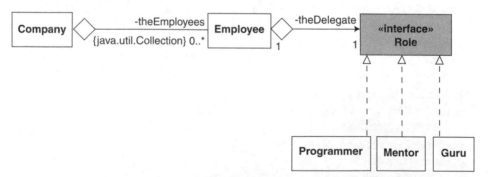

Figure 8.4   *A model of the company using specialization, delegation and an interface*

Corresponding Java code is given in program 8.3. The classes Application and
Company are unchanged from the previous example giving:

Program 8.3   An interface delegate (model Prog8_3.uml)

```
public class Employee {
  // ----- Operations ----------
```

```
    public   Employee(int aPayrollNumber) {
        thePayrollNumber = aPayrollNumber;
        theDelegate = null;
    }   // method: Employee

    public void   setDelegate(Role aRole) {
        the Delegate = aRole;
    }   // method: setDelegate

    public int   getPayrollNumber() {
        return thePayrollNumber;
    }   // method: getPayrollNumber

    public int   getSalary() {
        return theDelegate.getSalary();
    }   // method: getSalary

    public String   toString() {
        return "Employee with payroll number: " + thePayrollNumber + "that is a" +
            theDelegate;
    } // method: toString

    // ----- Attributes ----------
    private int        thePayrollNumber;
    private Role    theDelegate;
} // class: Employee


public interface Role {

    // ----- Operations ----------
    public abstract int   getSalary();
    public abstract String   toString();
    public static final int    BASIC_SALARY = 100;
} // interface: Role


public class Programmer implements Role {

    // ----- Operations ----------
    public   Programmer() {
    } // method: Programmer

    public int   getSalary() {
        return Role.BASIC_SALARY;
    }   // method: getSalary

    public String   toString() {
        return "Programmer";
    } // method: toString
} // class: Programmer


public class Mentor implements Role {

    // ----- Operations ----------
```

```
    public   Mentor() {
    } // method: Mentor

    public int   getSalary() {
       return Role.BASIC_SALARY + (int)(Role.BASIC_SALARY * MULTIPLIER);
    } // method: getSalary

    public String   toString() {
       return "Mentor";
    } // method: toString

    // ----- Attributes ----------
    private static final double   MULTIPLIER = 0.1;
} // class: Mentor

public class Guru implements Role {

    // ----- Operations ----------
    public   Guru() {
    } // method: Guru

    public int   getSalary() {
       return Role.BASIC_SALARY + (int)(Role.BASIC_SALARY * MULTIPLIER);
    } // method: getSalary

    public String   toString() {
       return "Guru";
    } // method: toString

    // ----- Attributes ----------
    private static final double   MULTIPLIER = 0.2;
} // class: Guru
```

The Application produces an output as follows:

Staff list for the Object University

Employee with payroll number: 123 that is a Programmer
Employee with payroll number: 456 that is a Mentor

The salary bill is 210

Staff list for the Object University

Employee with payroll number: 123 that is a Mentor
Employee with payroll number: 456 that is a Guru

The salary bill is 230

The interface pattern is widely used in the Java API. For example, when using the Swing class library (see chapter 7) there is a common requirement to create an application that processes events. Typically the programmer must ensure that on notification of an event such as a mouse click on a **JComponent**, some specific action should take place. The Swing library provides a **Listener** interface for this purpose.

In fact there are several **Listener** interfaces: one for each kind of event that might occur. For example, there is a **MouseListener** interface that is used to handle mouse events. To be a MouseListener a class must implement the MouseListener interface:

```
public interface MouseListener {
    // ----- Operations ----------
    public abstract void    mouseClicked(MouseEvent e);
    public abstract void    mouseEntered(MouseEvent e);
    public abstract void    mouseExited(MouseEvent e);
    public abstract void    mousePressed(MouseEvent e);
    public abstract void    mouseReleased(MouseEvent e);
} // interface: MouseListener
```

To connect a specific Listener to a JComponent there are a number of operations provided by the JComponent class. For example, there is the operation addMouseListener (MouseListener listener). Typically we might have:

```
import java.awt.*;
import java.awt.event.*;
public class Demo extends JFrame {
    // ...
    private class    FrameListener implements MouseListener {
        // ----- Operations ----------
        public void    mouseClicked(MouseEvent e) {
            // ...
        } // method: mouseClicked

        public void    mouseEntered(MouseEvent e) {
            // ...
        } // method: mouseEntered

        public void    mouseExited(MouseEvent e) {
            // ...
        } // method: mouseExited

        public void    mousePressed(MouseEvent e) {
            // ...
        } // method: mousePressed

        public void    mouseReleased(MouseEvent e) {
            // ...
        } // method: mouseReleased
    } // class: FrameListener
    // ...
    this.addMouseListener(new FrameListener());
    // ...
} // class: Demo
```

and

```
public class Main {
   // ----- Operations ----------
   public static void main (String[] args) {
      Demo demo = new Demo();
      demo.setVisible(true);
      // ...
   } // method: main
} // class: Main
```

8.3   Iterator

It has been demonstrated in the preceding chapters the usefulness of the collections provided by the Java Collections Framework. However, each element must be held privately
to prevent client abuse. Therefore each collection must provide some way of accessing
its elements. The *iterator design pattern* is used to do so.

This design pattern defines an interface that declares methods for sequentially accessing
the objects in the collection. Each collection provides clients with an object that implements this interface. Clients that access the elements in a collection do so through the interface and remain independent of the class that actually implements it. An example from
program 8.2 will suffice.

```
public class Company {
   // ----- Operations ----------
   public   Company(String aName) {
      theEmployees = new ArrayList();
      // ...
   } // method: Company
   public int   getSalaryBill() {
      int totalBill = 0;
      Iterator iter = theEmployees.iterator();
      while(iter.hasNext()) {
         Employee emp = (Employee)iter.next();
         totalBill += emp.getSalary();
      }
      return totalBill;
   } // method: getSalaryBill
   // ...
   // ----- Attributes ----------
   private java.util.Collection   theEmployees;   // of Employee
   // ...
} // class: Company
```

The Company uses an ArrayList to hold Employee references. In the method
getSalaryBill it must visit each Employee in turn and request its salary. It does this by

declaring iter, a reference to an Iterator initialized with a reference to the object returned by the iterator method of the ArrayList. It can then be sent the message hasNext to control a while loop and next to access each reference in the ArrayList. The Object reference returned must be cast to an Employee so that the message getSalary can be sent.

The important points for us to appreciate is that Iterator is an interface defined in the java.util package as:

```
public interface Iterator {
    // ----- Operations ----------
    public abstract boolean   hasNext();
    public abstract Object    next();
    public abstract void      remove(Object obj);
    // ...

}
```

and that the method iterator in ArrayList returns an implementation of it. The detail of how it does so is of no concern to us. Note that any other collection that provides an Iterator could be used with minimal changes to our code.

8.4   Adapter

Sometimes a class may present the wrong set of services to a client. By using the *adapter pattern* we can easily remedy this rather than undertake extensive modifications. All that happens is that the adaptee is a delegate of the adapter. This allows an Adapter class to advertise those services expected by a client but to use an adaptee to implement them. The class diagram of figure 8.5 illustrates the approach.

Figure 8.5   *Using a delegate to design an adapter*

When an Adapter is sent a message corresponding to operations 1, 2 or 3 it uses operations 4, 5 and 6 respectively in the Adaptee. This means that it can present the services expected by a client and use another object to implement them. The Action class of Chapter 4 is based on an Adapter. In this context it is also known as the *façade design pattern* as it acts as a front end to its delegate, the Library.

An adapter is also useful when there are too many services on offer. Often the adapter is a specialization of the adaptee. Typically the adapter provides default implementations of the adaptee's operations. Further specializations of the adapter can redefine them as is necessary. However, an interface can be used as the adaptee. In this case the adapter just implements the adaptee. Figure 8.6 and the corresponding Java code fragments illustrate the approach.

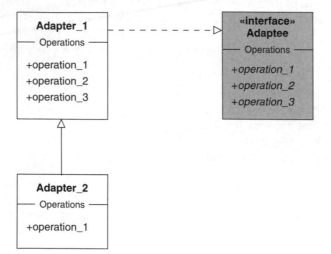

Figure 8.6   *Using specialization to design an adapter*

public interface Adaptee {
    // ----- Operations ----------
    public abstract void   operation_1();
    public abstract void   operation_2();
    public abstract void   operation_3();
} // interface: Adaptee

public class Adapter_1 **implements** Adaptee {
    // ----- Operations ----------
    public void operation_1() {
        // ...
    } // method: operation_1
    public void operation_2() {
        // ...
    } method: operation_2
    public void operation_3() {
        // ...
    } // method: operation_3
} // class: Adapter_1

public class Adapter_2 **extends** Adapter_1 {
    // ----- Operations ----------
    public void operation_1() {
        // ...
    } // method: operation_1
} // class: Adapter_2

```
public class Application {
    // ----- Operations ----------
    public void run () {
        Adaptee firstAdapter = new Adapter_1(...);
        firstAdapter.operation_1();    // Default implementation
        firstAdapter.operation_2();    // Default implementation
        firstAdapter.operation_3();    // Default implementation
        //
        Adaptee secondAdapter = new Adapter_2(...);
        secondAdapter.operation_1();   // Special implementation
        secondAdapter.operation_2();   // Default implementation
        secondAdapter.operation_3();   // Default implementation
    } // method: run
} // class: Application
```

There are many examples of this kind of adapter in the Java API. For example, there is an abstract adapter class for each kind of event listener interface. All that each class does is to provide empty methods. This means that the programmer need only extend the adapter and provide methods that are relevant to the application. Crucially there is no need to even know about the others. This simplifies an otherwise tedious task.

For example, the abstract WindowAdapter class provides seven default methods for the WindowListener interface described in figure 7.12. An application is expected to specialize it and redefine one or more of its methods as shown in program 7.4.

8.5   Singleton

The *singleton design pattern* is concerned with object creation. In many applications we find a requirement that no more than one instance of a particular class should exist. For example, throughout this text we have assumed a single instance of the class Application. However, with a default constructor we can create any number of instances as in:

```
public class Main {
    // ----- Operations ----------
    public static void main(String[] args) {
        Application app1 = new Application();
        Application app2 = new Application();
        // ...
    } // method: main
} // class: Main
```

In part, this illustration points the way to achieving our objective. That is, the Application class is given programmer-defined default or parameterized constructors that do not have public visibility. Without a public constructor clients cannot create any instances. Note that if there is no declared constructor for a class, then Java assumes a default constructor with public visibility.

Additionally, if a single instance is to be guaranteed, then the class must have some operation, which we shall refer to as a *factory operation*, to create the singleton. Hence we are hiding the class constructor behind the factory operation.

In Java we realize the *singleton pattern* in the Application class with a **static** method representing the factory that guarantees that only one instance is available. This method has access to the attribute that refers to the unique instance. The attribute representing the instance references an initialized object. The class is declared as in program 8.4.

Program 8.4   Singleton design pattern (model Prog8_4.uml)

```java
public class Application {
    // ----- Operations ----------
    public static Application getApplication() {
        return theApplication;
    } // method: getApplication

    private Application() {
        // No initialization required
    } // method: Application

    public void run() {
        // ...
    } // method: run

    // ----- Attributes ----------
    private static final Application    theApplication = new Application();
} // class: Application
```

Class Main now appears as:

```java
public class Main {
    // ----- Operations ----------
    public static void      main(String[] args) {
        Application app = Application.getApplication();
        app.run();
    } // method: main
} // class: Main
```

8.6   Visitor

Consider program 8.2 from section 8.1. The Company class has the operation displayEmployees which presents a list of staff members:

```java
// class Company
    public void displayEmployees() {
        ConsoleIO.out.println();
        ConsoleIO.out.println("Staff list for " + theName);
```

```
      ConsoleIO.out.println();
      Iterator iter = theEmployees.iterator();
      while(iter.hasNext()) {
         Employee emp = (Employee)iter.next();
         ConsoleIO.out.println("\t" + emp);
      }
      ConsoleIO.out.println();
} // method: displayEmployees
```

Consider also the operation getSalaryBill to compute the total cost of the salaried employees:

```
// class Company
   public int getSalaryBill() {
      int totalBill = 0;
      Iterator iter = theEmployees.iterator();
      while(iter.hasNext()) {
         Employee emp = (Employee)iter.next();
         totalBill += emp.getSalary();
      }
      return totalBill;
} // method: getSalaryBill
```

We immediately identify a common logic fragment in both examples, namely, the need to iterate across the collection of Employees and to perform some action against each. In the operation displayEmployees the action is to display the details of each Employee. In the operation getSalaryBill the action is to form a running total of the salaries of each Employee.

The *visitor pattern* lets us separate out the code to traverse a possibly complex structure of objects from the processing that is performed against each object. While our collection of Employee objects is relatively straightforward, we see from the two illustrative methods that we mix in traversal with action code. It is often the case that it is better to separate these two concerns.

The separation we desire can be achieved by locating action operations in a separate object called a *visitor*. This visitor object is passed to each element of the structure (here, an Employee object) as the structure is traversed. The architecture of the visitor classes is shown in figure 8.7.

The EmployeeVisitor is an interface with the abstract operation visit:

```
public interface EmployeeVisitor {
   // ----- Operations ----------
   public abstract void visit(Employee emp);
} // interface: EmployeeVisitor
```

The EmployeeDisplayVisitor class provides an implementation for the visit operation. The method simply displays the Employee object referenced by the parameter supplied by a client. This is, of course, the action we associate when displaying the collection of employees.

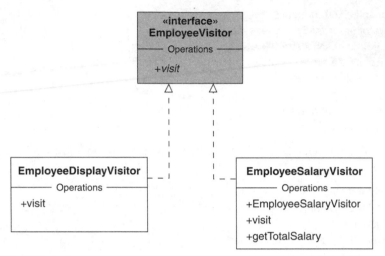

Figure 8.7   *Visitor classes*

```
public class EmployeeDisplayVisitor implements EmployeeVisitor {
    // ----- Operations ----------
    public void visit(Employee emp) {
        ConsoleIO.out.println("\t" + emp);
    } // method: visit
} // class: EmployeeDisplayVisitor
```

The EmployeeSalaryVisitor also implements the visit operation but the method adds the salary of the Employee to a private attribute. Additionally the class provides an accessor for this **private** attribute and a constructor to initialize it.

```
public class EmployeeSalaryVisitor implements EmployeeVisitor {
    // ----- Operations ----------
    public    EmployeeSalaryVisitor() {
        theTotalSalary = 0;
    } // method: EmployeeSalaryVisitor

    public void    visit(Employee emp) {
        theTotalSalary += emp.getSalary();
    } // method: visit

    public int    getTotalSalary() {
        return theTotalSalary;
    } // method: getTotalSalary

    // ----- Attributes ----------
    private int    theTotalSalary;
} // class: EmployeeSalaryVisitor
```

The Company class is no longer responsible for displaying or accumulating a salary bill. However, it must provide a method accept that iterates over its collection of

Employees giving an EmployeeVisitor access to each in turn. Of course different EmployeeVisitors carry out different actions on each Employee. We can now revise the Company class from program 8.4 in the outline code that follows. The other classes are unchanged.

Program 8.5 Using a visitor (model Prog8_5.uml)

```
public class Company {
  // ----- Operations ----------
  public void   displayEmployees() {
    ConsoleIO.out.println();
    ConsoleIO.out.println("Staff list for " + theName);
    ConsoleIO.out.println();
    //
    EmployeeDisplayVisitor displayVisitor = new EmployeeDisplayVisitor();
    this.accept(displayVisitor);
    ConsoleIO.out.println();
  } // method: displayEmployees
  public int   getSalaryBill() {
    EmployeeSalaryVisitor salaryVisitor = new EmployeeSalaryVisitor();
    this.accept(salaryVisitor);
    return salaryVisitor.getTotalSalary();
  } // method: getSalaryBill
  private void   accept(EmployeeVisitor visitor) {
    Iterator iter = theEmployees.iterator();
    while(iter.hasNext()) {
      Employee emp = (Employee)iter.next();
      visitor.visit(emp);
    }
  } // method: accept
  // ...
  // ----- Attributes ----------
  private String   theName;
  private java.util.Collection   theEmployees; // of Employee
} // class: Company
```

The two original methods for displayEmployees and getSalaryBill now create an appropriate visitor object then call the private accept method. This accept method iterates across the Employee objects and requests the visitor object to visit each. Of course, different visitor objects have different behaviours when they visit an Employee.

8.7   Observer

In object-oriented designs there is a common need for one or more objects (the observers) to be informed when the state of another (the subject) changes. This gives rise to the *observer pattern* as illustrated in figure 8.8.

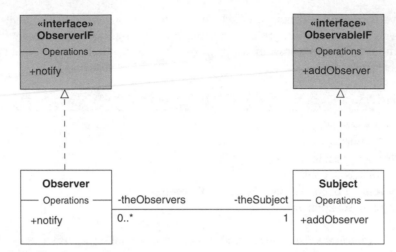

Figure 8.8 *The observer design pattern*

To be an observer a class must implement the interface ObserverIF. The intention is that the subject can inform an observer that its state has changed. This is done by the subject sending the message notify. The method for notify consists of the actions the observer actually takes on notification. To be a subject a class must implement the interface ObservableIF. The intention is that observers can be registered with the subject for notification with the message addObserver.

The observer design pattern is so widely used that it is supported by the Java API. There is an interface Observer (corresponding to ObserverIF in figure 8.8):

public interface Observer {

    public abstract void update(Observable obj, Object arg);

} // interface: Observer

It specifies that a class must implement the method update to be an observer. The subject makes use of it when notifying its observers of a change to its state.

There is also the class Observable (corresponding to Subject in figure 8.8):

public class Observable {

    public void addObserver(Observer obj) { ... }
    protected void setChanged() { ... }
    public void notifyObservers(Object obj) { ... }
    // ...

} // class: Observable

It takes care of the implementation details associated with being a subject. For example, we can register an observer with addObserver, note that the subject has changed with setChanged and inform its observers with notifyObservers. All that we need do is to extend Observable to represent an object that is to be observed.

As a simple example, consider an application in which there are students each with a name and a matriculation number that can be changed. If a student's matriculation number is changed then administrators must be informed. Alternatively if a student's name is changed then lecturers must be informed. We can make each student Observable by extending the Observable class. The Lecturers and Administrators become Observers by implementing the Observer interface. Program 8.6 illustrates the approach.

Program 8.6   A simple observer example (model Prog8_6.uml)

```
import java.util.Observable;

public class Student extends java.util.Observable {
    // ----- Operations ----------
    public   Student(String aName, int aMatriculationNumber) {
        theName = aName;
        theMatriculationNumber = aMatriculationNumber;
    } // method: Student

    public void   setName(String aName) {
        theName = aName;
        this.setChanged();
        this.notifyObservers("name");
    } // method: setName

    public void   setMatriculationNumber(int aMatriculationNumber) {
        theMatriculationNumber = aMatriculationNumber;
        this.setChanged();
        this.notifyObservers("matriculation number");
    } // method: setMatriculationNumber

    public String   toString() {
        return theName + " with matriculation number " + theMatriculationNumber + "\n";
    } // method: toString

    // ----- Attributes ----------
    private String theName;
    private int   theMatriculationNumber;
} // class: Student

import java.util.Observable;
import textio.*;

public class Lecturer implements java.util.Observer {
    // ----- Operations ----------
    public Lecturer(String aName) {
        theName = aName;
    } // method: Lecturer

    public void   update(Observable obj, Object arg) {
        if(arg.equals("name"))
```

```
            ConsoleIO.out.println(this + " knows that " + obj + " has changed its " + arg);
        else
            ConsoleIO.out.println(this + " cannot handle changes to " + arg);
    } // method: update

    public String   toString() {
        return "Lecturer: " + theName + "\n";
    } // method: toString

    // ----- Attributes ----------
    private String theName;
} // class: Lecturer


import java.util.Observable;
import textio.*;

public class Administrator implements java.util.Observer {
    // ----- Operations ----------
    public Administrator(String aName) {
        theName = aName;
    } // method: Administrator

    public String   toString() {
        return "Administrator: " + theName + "\n";
    } // method: toString

    public void    update(Observable obj, Object arg) {
        if(arg.equals("matriculation number"))
            ConsoleIO.out.println(this + " knows that " + obj + " has changed its " + arg);
        else
            ConsoleIO.out.println(this + " cannot handle changes to " + arg );
    } // method: update
    // ----- Attributes ----------
    private String theName;
} // class: Administrator

public class Application {
    // ----- Operations ----------
    public void run() {
        Student s1 = new Student("John", 123);
        //
        s1.addObserver(new Lecturer("Ken"));
        s1.addObserver(new Administrator("Peter"));
        //
        s1.setName("Alec");
        s1.setMatriculationNumber(456);
    } // method: run
} // class: Application
```

Notice the methods setName and setMatriculationNumber in the Student class. They change the appropriate attribute, record that the recipient has changed with setChanged then notify its observers with notifyObservers. If the change is not recorded then notifyObservers does nothing. As both setChanged and notifyObservers are inherited from the superclass Observable their implementation is of no concern to us. However, we expect that the Observable class maintains a collection of Observer references.

As the Lecturer and Administrator classes implement the Observer **interface** there is an update method in each. It is this method that is called by notifyObservers. The first parameter is a reference to the Observable object that has changed. If necessary the update method can make use of the second parameter to respond accordingly.

Finally, the Application makes use of the addObserver method to register a Lecturer and an Administrator as Observers. As with setChanged and notifyObservers it is inherited by the Student class from Observable.

The output produced is:

```
Administrator: Peter
  cannot handle changes to name
Lecturer: Ken
  knows that Alec with matriculation number 123
  has changed its name
Administrator: Peter
  knows that Alec with matriculation number 456
  has changed its matriculation number
Lecturer: Ken
  cannot handle changes to matriculation number
```

A second example is the Swing class library that supports the implementation of event-driven designs. An important part of its design uses a variation of the observer pattern. The class JComponent is a base class describing any kind of visual component such as a panel or a text field. Subclass instances of this JComponent class can act as both subjects and observers. Therefore a text field operating as a subject (an event source) might inform a panel, operating as an observer that some new data is available for display.

8.8   Template method

Recall that in the second iteration of the case study of chapter 6 (section 6.3.4.2) we encountered a serious problem with the implementation of the LenderImp class. The role of one of its methods, registerOneBorrower was to define the logic for registering a borrower. It also had to create a borrower record held by the lender. Unfortunately the necessary information to create a borrower record was not available as the detailed nature of the borrower is problem domain-specific. For example, it might be a library borrower with a name, address and telephone number or a video shop borrower with a name and registration number. How do we define a method when some of the information required to do so is only known by a subclass? One solution is the *Template Method* design pattern.

The template method pattern uses an **abstract** method in the implementation of a method. The subtle point is that the superclass method is **final** and the **abstract** method is defined by a subclass. This means that we can define a method in a superclass and expect a subclass to complete it without altering the ordering of its statements.

In the case study we know that registerOneBorrower considers the borrower to be an object belonging to a class that implements the BorrowerRecordIF interface. We also know that it must have a name. Therefore we can declare:

protected abstract BorrowerRecordIF createBorrower(String aBorrowerName);

in the LenderImp class. It can then be used by its registerOneBorrower method as in:

```
// class LenderImp
    public final void registerOneBorrower(String aBorrowerName) {
        BorrowerRecordIF foundRecord = this.getBorrowerRecord(aBorrowerName);
        // ...
        if(foundRecord == null) {
        BorrowerRecordIF borrowerRecord = this.createBorrower(aBorrowerName);
        boolean result = theBorrowers.add(borrowerRecord);
        // ...
        }
} // method: registerOneBorrower
```

A problem domain-specific class such as Library or VideoShop can then define it as:

```
// class Library
    protected final BorrowerRecordIF createBorrower(String aBorrowerName) {
        String address;
        String telephoneNo;
        // Get the values for address and telephoneNo – perhaps using a dialog box
        // ...
            return new BorrowerRecord(aBorrowerName, address, telephoneNo);
    } // method: createBorrower
```

or assuming that RentalRecord implements BorrowerRecordIF:

```
// class VideoShop
    protected final BorrowerRecordIF createBorrower(String aBorrowerName) {
        String registrationNo;
        // Get the values for registrationNo – perhaps using a dialog box
        // ...
        return new RentalRecord(aBorrowerName, registrationNo);
} // method: createBorrower
```

8.9   Abstract factory

A framework is a generic architecture that can be customized for a particular application. Their use in object-oriented systems has become widespread. Because of evolving requirements they need to be malleable, so that they may adapt to change. For example, a framework may need to be portable across many different hardware platforms. Often this leaves the framework designer with the rather difficult problem of creating an object without knowing about its concrete implementation. In fact we encountered such an example in the preceding section with the creation of a borrower record in the LenderImp class.

The *abstract factory pattern* solves this kind of problem by delegating construction of a concrete class object to an appropriate subclass. Figure 8.9 illustrates the architecture that is often used.

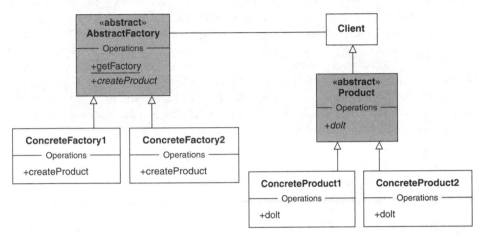

Figure 8.9   *The abstract factory pattern*

The AbstractFactory class defines an **abstract** method createProduct for creating instances of the required concrete class (such as ConcreteProduct1 or ConcreteProduct2). A concrete factory such as ConcreteFactory1 or ConcreteFactory2 implements that interface. In class ConcreteFactory1, for example, the method createProduct creates a new object of the class ConcreteProduct1. The actual concrete factory object is supplied by the **static** method getFactory in class AbstractFactory. A client calls this method to get the concrete factory for creating the various concrete products. In this case it is ConcreteProduct1 or ConcreteProduct2. For example, a client may have code such as:

```
Product myProduct = AbstractFactory.getFactory().createProduct();
myProduct.doIt();
```

The client code does not know (or need to know) what concrete class myProduct actually references. This is probably fixed when the system is initialized and might change from one hardware platform to another. Note that for simplicity the concrete factories shown create only one concrete product. Typically they might create a family of concrete products.

The Swing class library provided by the Java API is a classic example of a framework that makes use of this pattern to help make it portable. Its Toolkit class is an abstract factory used to create objects that work with the native windowing system. The concrete factory it uses for a given hardware platform is determined by initialization code. So that it can be accessed by other classes, a **static** method getDefaultToolkit is provided. In addition it has **abstract** operations such as createFrame and createButton that are defined by the concrete factory. Swing makes use of code similar in nature to that shown above.

8.10   Decorator

Perhaps the most elegant of the design patterns is the *Decorator design pattern*. It achieves the apparently impossible by allowing the programmer to dynamically add functionality to an object. Objects of the same class can have different run-time behaviours! Decorator achieves its effect by using a wrapping (or chaining) technique.

Consider what happens if we change the company scenario from section 8.2 so that an employee is basically a programmer, but he is expected to adopt other roles. For example, a programmer might become a mentor or an administrator. The "twist" that we introduce is that a programmer can adopt combinations of these roles in some arbitrary fashion. For example, a programmer might also be a mentor, or an administrator or both at various times. One obvious way to achieve this is to have a class for every possible combination of roles. For example, we might have the classes:

Programmer,
ProgrammerAndMentor,
ProgrammerAndAdministrator and
ProgrammerAndMentorAndAdministrator.

However, this would lead to an explosion of classes. If we expect to add other roles later in the development of the software it is even less attractive. Finally it does not give us the run-time flexibility we need as it is based on a static class hierarchy.

A better approach to the problem is shown in the class diagram of figure 8.10.

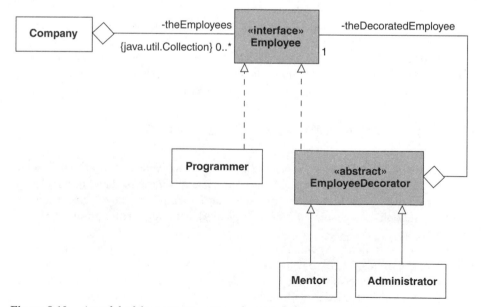

Figure 8.10   *A model of the company using a decorated class*

As might be expected from previous discussions, the design has the concrete classes Programmer, Mentor and Administrator. They correspond to each role an Employee might adopt.

However, the EmployeeDecorator class requires some explanation. It has an aggregate component named theDecoratedEmployee that acts as a delegate. Normally we say that it decorates it. By polymorphic substitution this delegate can belong to any class that implements the Employee interface. For example, it could be a Programmer. However, the delegate could also be an EmployeeDecorator, i.e. a Mentor or Administrator. The important point to understand is that in this case the object concerned decorates an Employee. As before this could be a Programmer or an EmployeeDecorator. This gives us the flexibility we require to decorate an Employee as necessary.

For example, as all of these classes define the getSalary method we might have the collaboration diagram as shown in figure 8.11.

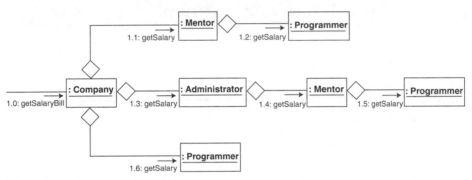

Figure 8.11   *Representative collaborations*

It shows:

- a Mentor that decorates a Programmer
- an Administrator that decorates a Mentor that decorates a Programmer and
- an undecorated Programmer

When each is sent the message getSalary, it is relayed to each decorated Employee as appropriate.

The basic approach is to create a Programmer and decorate it as appropriate. Therefore we might have:

Employee programmer_1 = **new** Programmer(123);

and

Employee programmer_2 = **new** Programmer(456);
Employee teamLeader = **new** Mentor(programmer_2);

To avoid the confusion of referencing an object when it is undecorated and decorated (programmer_2 above and theDecoratedEmployee in DecoratedEmployee respectively) it is normal for the decorated object to be created during the constructor call of the decorating object. Therefore we have:

Employee teamLeader = **new** Mentor(**new** Programmer(456));

to create a Mentor that decorates a Programmer. This approach continues with more complex decorations to give:

Employee projectLeader = **new** Administrator(**new** Mentor(**new** Programmer(912)));

to give an Administrator that decorates a Mentor that decorates a Programmer.

Note that as a decorated object carries the same data as its decorator, we strive to make a class such as EmployeeDecorator as lightweight as possible, i.e. with a minimum number of attributes.

The Java code from program 8.7 illustrates the decorator design pattern in more detail.

Program 8.7   Using a decorator (model Prog8_7.uml)

```
import textio.*;

public class Application {
  // ----- Operations ----------
  public void run() {
    Company co = new Company("The Object University");
    //
    Employee programmer = new Programmer(123);
    co.hireEmployee(programmer);
    //
    Employee teamLeader = new Mentor(new Programmer(456) );
    co.hireEmployee(teamLeader);
    //
    Employee documentationLeader = new Administrator(new Programmer(789));
    co.hireEmployee(documentationLeader);
    //
    Employee projectLeader = new Administrator(new Mentor(new Programmer(912)));
    co.hireEmployee(projectLeader);
    //
    co.displayEmployees();
    ConsoleIO.out.println("The salary bill is: " + co.getSalaryBill());
    ConsoleIO.out.println();
  } // method: run
} // class: Application

public interface Employee {
  // ----- Operations ----------
  public abstract int   getPayrollNumber();
  public abstract int   getSalary();
```

```
    public abstract String    toString();
} // interface: Employee

public class Programmer implements Employee {
    // ----- Operations ----------
    public    Programmer(int aPayrollNumber) {
        thePayrollNumber = aPayrollNumber;
    } // method: Programmer

    public int    getPayrollNumber() {
        return thePayrollNumber;
    } // method: getPayrollNumber

    public int    getSalary() {
        return BASIC_SALARY;
    } // method: getSalary

    public String    toString() {
        return " Programmer with payroll number: "  + this.getPayrollNumber();
    } // method: toString

    // ----- Attributes ----------
    private int    thePayrollNumber;
    private static final int    BASIC_SALARY = 100;
} // class: Programmer

public abstract class EmployeeDecorator implements Employee {
    // ----- Operations ----------
    public    EmployeeDecorator(Employee anEmployee) {
        theDecoratedEmployee = anEmployee;
    } // method: EmployeeDecorator

    public int    getPayrollNumber() {
        return theDecoratedEmployee.getPayrollNumber();
    } // method: getPayrollNumber

    public int    getSalary() {
        return theDecoratedEmployee.getSalary();
    } // method: getSalary

    public String    toString() {
        return theDecoratedEmployee.toString();
    } // method: toString

    // ----- Attributes ----------
    private Employee    theDecoratedEmployee;
} // class: EmployeeDecorator

public class Mentor extends EmployeeDecorator {
    // ----- Operations ----------
```

```java
public Mentor(Employee anEmployee) {
    super(anEmployee);
} // method: Mentor

public int    getSalary() {
    return super.getSalary() + (int)(super.getSalary() * MULTIPLIER);
} // method: getSalary

public String    toString() {
    return " Mentor and " + super.toString();
} // method: toString

// ----- Attributes ----------
private static final double    MULTIPLIER = 0.1;
} // class: Mentor

public class Administrator extends EmployeeDecorator {

// ----- Operations ----------
public    Administrator(Employee anEmployee) {
    super(anEmployee);
} // method: Administrator

public int    getSalary() {
    return super.getSalary() + (int)(super.getSalary() * MULTIPLIER);
} // method: getSalary

public String    toString() {
    return " Administrator and " + super.toString();
} // method: toString

// ----- Attributes ----------
private static final double    MULTIPLIER = 0.2;
} // class: Administrator
```

The output produced is:

Staff list for the Object University

Programmer with payroll number: 123
Mentor and Programmer payroll number: 456
Administrator and Programmer payroll number: 789
Administrator and Mentor and Programmer payroll number: 912

The salary bill is: 462

The input/output classes of the Java API are strongly influenced by the decorator design pattern. For example, a FileOutputStream object references a text file and is created with a statement such as:

```java
FileOuputStream fos = new FileOutputStream("Library.txt");
```

However, the ObjectOutputStream class supports object serialization discussed in section 6.4.2 and has a constructor that permits it to decorate a FileOutputStream object.

A typical example is:

```
ObjectOutputStream oos = new ObjectOutputStream(new
    FileOutputStream("Library.ser"));
Book bk = new Book (...);
oos.writeObject(bk);
```

The crucial point is that a FileOutputStream object is made to behave like an ObjectOutputStream object with no modification to the FileOutputStream class.

8.11   Summary

1. Patterns are ways to describe best practices, good designs, and capture experience in a way that it is possible for others to reuse this experience.
2. Specialization and delegation are widely used in object-oriented systems. Both provide powerful ways of reusing code. Delegation can often be used in place of specialization, offering flexibility at run-time.
3. The use of interfaces can increase the flexibility we seek. An interface provides only a specification of its services. In order to conform to an interface a class must implement all of its services, i.e. supply a method body for each operation. Crucially the classes can belong to different class hierarchies.
4. The adapter design pattern is used to introduce a class with the required set of services that is realized by another class that has the wrong set of services for a client.
5. The singleton design pattern guarantees that no more than one instance of a particular class exists in a program. The singleton also offers a common point of access to the object throughout the application.
6. The visitor pattern is used to separate the code to traverse a possible complex structure of objects from the processing that is performed against each object.
7. The template method pattern lets us fix the ordering of steps in an algorithm but lets subclasses vary the details of the separate steps.
8. The abstract factory method delegates the construction of concrete class objects to an appropriate subclass. This way the client code need not know what actual concrete object it references.
9. The decorator pattern is used to dynamically add new functionality to an object.
10. Many of these design patterns have been incorporated into the Java API.

8.12   Exercises

1. The final model from the previous chapter (program 7.11, Lib7_11.uml) included a class BookDialog used to obtain the details of a Book. In the actionPerformed method of the inner class AddBookAction of LibraryFrame a new BookDialog is created and made visible. The user then enters the required information and selects either the Ok or Cancel buttons. The dialog is then dismissed and if Ok has been selected then the input values are obtained and used to initialize a newly created Book object.

Object creation is one of the most expensive operations for a Java program to perform. Consider using the singleton design pattern so that only one BookDialog object is ever created. You will probably also require operations to empty the dialog controls so that new data can be entered. Otherwise when the dialog appears it will show the data from the preceding usage.

2. The ROME case tool is a large, complex Java program. It should come as no surprise that it was designed before it was implemented. Therefore there are several UML diagrams that were used in the development of ROME. For example, there is a class diagram for the class diagrammer it supports.

We expect this diagram to be populated with classes that refer to model elements such as a class symbol or an association symbol. For example, figure 8.12 is a simplified class diagram for the class diagrammer used in ROME. The **abstract** class ModelElement represents any model element that can appear in a diagram. The **abstract** class NodeElement represents any kind of node while the **abstract** class RelationElement represents any kind of relation between two nodes. The concrete classes ClassSymbol, NoteSymbol, AggregationSymbol, AssociationSymbol and SpecializationSymbol represent themselves.

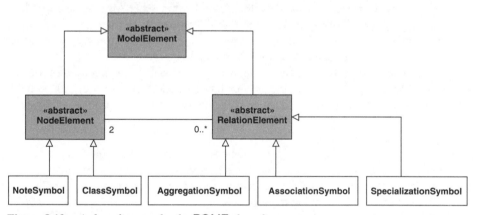

Figure 8.12   *A class diagram for the ROME class diagrammer*

This may appear odd and perhaps in some sense recursive. However, there is no contradiction. It is just a class diagram for a software system. The fact that it is a class diagrammer makes no difference. You might like to reflect on the fact that ROME is being used to develop a model of the next version of itself.

We understand that an object diagram is a concrete example of the class diagram. Therefore an object diagram corresponding to figure 8.12 is a class diagram. Use a selection of class diagrams from this book to check on its accuracy. Remember that it is somewhat simplified.

3. Introduce the concrete class PackageSymbol that is an aggregate of any number of other modelling elements into the class diagram of the previous exercise. You should arrive at the structure for the composite design pattern. Visit http://www.hillside.net for a discussion on this pattern. Given that a class may include further inner classes with their own relationships, then how should ClassSymbol be revised? Compare the composite and decorator patterns.

4. Figure 8.12 implies that as a class symbol is dragged by the user then all **RelationElements** associated with it must be informed so that visually they retain the appearance of being connected. What design pattern might be used to ensure this occurs? Outline how it would be introduced into figure 8.12.

5. Figure 8.13 is a simplified class diagram for the primary user-interface panels used in ROME.

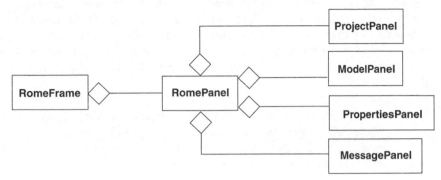

Figure 8.13   *The primary user-interface panels from ROME*

For example, the class **ProjectPanel** represents the panel at the upper left for the model's project tree (see also appendix B).

When the user selects a new diagram from the project panel tree then the model panel responds by rendering the diagram represented by the tree node. Equally when the user selects a class symbol in the model panel then the properties panel presents the features of the chosen class.

What design pattern is appropriate to achieve this effect? Reflect on the Java documentation for the classes **PropertyEvent** and **PropertyChangeListener** then consider how they would be deployed for this purpose.

6. In chapters 4, 6 and 7 we have developed software used to support a librarian. Unfortunately in the light of new requirements it requires extensive modification. Rather than adopt an ad-hoc code-based approach, it has been decided to redesign it making use of design patterns wherever appropriate. It is anticipated that this should minimize the effort required to accommodate future changes as well as bringing an element of "best practice". It is also expected that any changes to any existing classes should be minimal.

The new requirements are as follows:

Requirement #1
There should be a guarantee that only one library is created.

Requirement #2
There is a need to track the borrowing habits of frequent borrowers. If a borrower has more than four publications out on loan then from that point on the details of his borrowings should be recorded. The librarian can ask at any time for the borrowing/returning history of all such borrowers.

All transactions for a borrower should be sensibly grouped together, e.g. loans followed by returns. However, the groupings should be ordered by the borrower's name.

Similarly the system should allow specific borrowers to be tracked no matter how many publications they have out on loan. The borrowing history of any one of these borrowers can be requested by the librarian.

Requirement #3
From time to time a publication will not be available for lending. For example, the librarian may have to have its cover refurbished. Our intention is not to remove it from the loan stock even though it cannot be borrowed. When the details of such a publication are displayed then a short message to the effect that it cannot be borrowed should also be displayed. Clearly it should not already be out on loan. However, it should also be possible for it to revert back to being a normal borrowable publication.

7. When testing software it is often useful to be able to follow the thread of control i.e. to trace the sequence of method calls. This is especially true with object-oriented systems where there is often extensive message passing. At its simplest, this might involve a method printing its name before starting normal execution. For example, when testing the library case study we might have an output such as:

```
// ...
method: Book.toString
method: Publication.toString
1: Java by Ken
// ...
```

It shows that displaying the details of a book involves the method toString in the class Book invoking the method toString in the class Publication.

An obvious implementation strategy is to add a print statement to each method. However, it has the major disadvantage that normal code is populated with testing code. Ultimately it must be deleted or commented out when testing is complete. Use of a boolean variable alleviates the problem but it still leaves us with code that is more complex than it should be. If there is a requirement that we should be able to switch testing code on and off for selected classes, then our difficulties are exacerbated. In the worst case our testing code could introduce errors to the methods we are trying to test.

One solution is to separate testing behaviour from normal behaviour. Our intention is to have no testing code in the body of a method under test. This implies that we need to be able to change an object's behaviour. We expect the same object to behave differently when it is under test and when it is not.

After some thought we come to the conclusion that the decorator design pattern will help us determine the most useful approach to take. For example, testing code might be located in decorator methods. Normal code would then be located in corresponding methods in its delegate. Explore this idea using the Publication hierarchy in iteration 11 (model Lib7_11.uml) from chapter 7.

Case Study: A Final Review

In section 2.1 it was pointed out that there might be occasions on which a design and its implementation could be refactored. Such an occasion arose in section 4.4 with the introduction of the Action class and later in section 6.3 with the development of a framework. On both occasions the intention was to aid further development of the case study. However, it is probable that future changes will not involve the original developers. Bearing this in mind, we take a last opportunity to review the case study.

9.1   Refactoring

A useful working definition of refactoring is:

To make changes to the internal structure of software so that it is easier to understand and modify without changes in its observable behaviour.

Refactoring often depends on individual experience and expertise. However, it has been documented in a manner similar to the design patterns described in chapter 8. For example, there is a catalogue of named refactorings (Fowler 1999) each of which is a distillation of the hard-won experience of experts. Interested readers should consult the reference cited.

A golden rule when refactoring is to make relatively small changes, as it is surprisingly easy to introduce errors! Therefore when refactoring the case study we take the most recent version (iteration 11 from chapter 7) as our starting point and apply changes to it iteratively.

As with previous iterations there is no question of abandoning the desire to develop the case study in a controlled manner. Therefore we must be sure that all of its documentation is updated where necessary. To bring the necessary focus to each refactoring it should also have a stated aim.

Clearly the integrity of the system should not be compromised when refactoring. Therefore testing becomes a critical activity. It is essential that we demonstrate that the system has exactly the same behaviour before and after refactoring. Although it would be more convenient and efficient to automate the testing process, it is sufficient for our purposes to undertake manual testing based on the test cases from section 6.2.1. Interested readers should refer to more advanced sources (Fowler 1999, http://www.junit.org).

9.2   Iteration 1

The aim of this iteration is to reduce the perceived complexity of the system. Obviously the easier a system is to understand the easier it is to make changes. It is worth noting that the developers have not intentionally made it too complex. There is no question of attributing blame. It is a fact of life that just like anyone else software engineers learn from experience. Sometimes when working intensively on a part of a large, complex system it is hard to keep the bigger picture in mind. In this iteration we stand back and take a more objective view.

Recall from section 7.2 that the architecture of the case study is inspired by the model-view-controller (MVC) architecture and the use of a framework. Anyone trying to understand the case study (perhaps with the intention of maintaining it) would benefit from being aware of this. For example, the Library class does not show an explicit architectural relationship with its Publications. This is because it specializes the framework class LenderImp that does. Given that the Library class is located in a different package from the framework classes (see figure 6.9(a)) this fact might not be readily understood. Therefore any steps we can take to make this clearer would be advantageous.

Before proceeding some important points to remember are:

- The domain model is application-specific and should have no responsibility for input/output. In the case study it is the Library with its Publications and BorrowerRecords that constitute the model.
- The view is responsible for presenting some facet of the model to the human user. For example, it might be the Publications held by the Library that are available for loan. In the case study the view is represented by the LibraryFrame class.
- The view and the controller(s) are combined into a view/controller. As well as displaying information about the model the view presents a button for each use-case. There is a controller (action object) attached to each button so that when there is a mouse click on it the controller stimulates the model in the appropriate manner. For example, there is a button labelled Display publications on loan. Attached to it is a controller of the class DisplayPublicationsOnLoanAction. As its name suggests, its role is to send the message displayPublicationsOnLoan to the Library. In the case study each controller is an instance of an inner class declared in the LibraryFrame.
- The view/controller also has classes to elicit information from the human user. For example, there is a class BookDialog that gets the various details of a Book to be added to the Library.
- As noted earlier the classes that make up the model specialize classes from a framework. There are no architectural relationships shown between them. It is the responsibility of the framework to establish and maintain the necessary relationships.
- There is a class Main whose role is to act as the main point of entry for the Java run-time environment. It creates a LibraryFrame which, having created the model, the controllers and the GUI starts up the application.

The problem with the current version is nothing to do with its functionality. It is that the use of the MVC and a framework is not evident from the design and ensuing code.

Figure 9.1 shows its top-level package/class structure (omitting relationships for clarity).

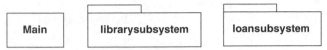

Figure 9.1   *The top-level package/class structure of the current version of the case study*

The role of Main is clear. Its single method is trivial and so we can safely ignore it. The classes in the loansubsystem package constitute a framework and so are not expected to be changed during normal maintenance. However, we should check that they are properly documented as part of this iteration. Note that extensions to the UML have been proposed (Fontoura 2001) to help document frameworks.

When it comes to the interfaces and classes located in the librarysubsystem it is a different matter. As is shown in figure 9.2 there is no indication that the designers had the MVC architecture and the use of a framework in mind.

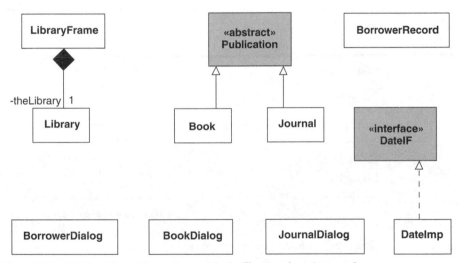

Figure 9.2   *The interfaces/classes located in the librarysubsystem package*

We can help remedy this by introducing a guisubsystem package for the view/ controller and use the librarysubsystem package for the model. Appropriate stereotypes on the three packages that result are also useful. Figure 9.3 illustrates the revised package structure.

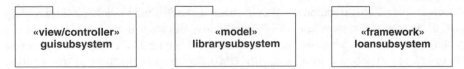

Figure 9.3   *A revised package structure*

Now we can locate those interfaces and classes that pertain to the model in the librarysubsystem package. They are shown in figure 9.4

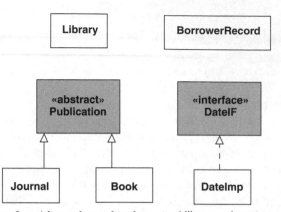

Figure 9.4   *The interfaces/classes located in the revised librarysubsystem package*

Finally the classes that pertain to the view/controller are located in the guisubsystem package as shown in figure 9.5.

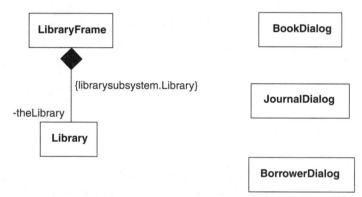

Figure 9.5   *The classes located in the guisubsystem package*

Note that the Library is given its qualified name. This makes it clear that the Library class referred to is located in the librarysubsystem and not the guisubsystem.

Having introduced these changes the dependencies between classes must be updated and the resulting code recompiled. This is a minor task as only a few classes, e.g. LibraryFrame and BorrowerDialog, are involved. To convince ourselves that all is well the tests from iteration 3 of chapter 6 are carried out. Happily the same behaviours are observed.

Before leaving this iteration we should check that all of the various UML diagrams that document the system are still accurate. All we have done is to introduce a package and relocate some of the code therefore no updates are required. However, interested readers should investigate the UML deployment diagram in this context (Pooley 1999).

Although it is a somewhat subjective judgement, we believe that the overall design intent behind the case study is now more obvious. Therefore the aim of this iteration has been achieved. File Lib9_1.uml is the completed model.

9.3   Iteration 2

One important reason for refactoring a design is to remove duplicate code. Apart from making the code size larger than it need be, duplicate code is a barrier to effective maintenance. The main problem is that any changes must also be duplicated. There is a risk of not being aware that duplicate changes are required or making different changes to achieve the same effect. Ideally a change should be made only in one place. The aim of this iteration is to remove duplicate code.

If we focus on the three dialog classes in the guisubsystem package of figure 9.5 then it seems likely that we will find some duplicate code. The reason is that they have similar responsibilities but are presently unrelated. It is not surprising to find that:

- they have the same superclass
- they have methods and attributes in common
- each has an inner class to handle the selection of the Ok and Cancel button on a dialog box.

From our discussion in chapter 5, it seems sensible to create a superclass with the common features of the three dialog classes. By doing this we should be able to remove significant amounts of duplicate code. Therefore we introduce an abstract superclass AbstractDialog with the attributes and methods common to the BorrowerDialog, BookDialog and JournalDialog classes. It also has the inner class OkCancelAction to handle selection of the Ok or Cancel button on a dialog box.

We then apply the same approach to the BookDialog and JournalDialog classes by introducing the abstract superclass PublicationDialog. It has the attributes and methods common to all Publications. Figure 9.6 shows the hierarchy of dialogs that results.

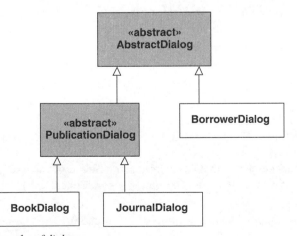

Figure 9.6   *A hierarchy of dialogs*

Now the Java code is significantly improved. As shown in the code that follows, the AbstractDialog class:

- uses its constructor to set up a basic dialog box with an action object for the Ok and Cancel buttons in place
- has an inner class to which the action object belongs
- has the method getUserAction used by a client, i.e. LibraryFrame to discover if Ok or Cancel has been clicked.

```java
// class AbstractDialog
package guisubsystem;

import javax.swing.*;
import java.awt.*;
import java.awt.event.*;

public abstract class AbstractDialog extends javax.swing.JDialog {
    protected    AbstractDialog(javax.swing.JFrame frame, String title) {
        super(frame, title, true);
        this.setSize(500, 200);

        Container contentPane = this.getContentPane();
        contentPane.setLayout( new BorderLayout() );

        //
        // Note that theDetailsPanel is now protected - subclasses will populate it
        contentPane.add(theDetailsPanel, BorderLayout.CENTER);
        contentPane.add(theButtonsPanel, BorderLayout.SOUTH);

        setResizable( false );
        setLocationRelativeTo( frame );
    } // method: AbstractDialog

    public int final getUserAction() {
        return theUserAction;
    } // method: getUserAction

    protected javax.swing.Box    theDetailsPanel = Box.createVerticalBox();
    {
        theDetailsPanel.add( Box.createVerticalGlue() );
    }

    private javax.swing.Jpanel    theButtonsPanel = new JPanel();
    {
        theButtonsPanel.add(theOkButton);
        theButtonsPanel.add(theCancelButton);
    }

    private OkCancelAction    theOkCancelAction = new OkCancelAction();
    private javax.swing.Jbutton    theOkButton = new JButton("Ok");
    {
        theOkButton.addActionListener(theOkCancelAction);
    }
```

```
private int    theUserAction;
protected static final java.awt.Dimension    PANELSIZE = new Dimension(400, 40);
private javax.swing.JButton    theCancelButton = new JButton( "Cancel" );
{
   theCancelButton.addActionListener(theOkCancelAction);
}
private class OkCancelAction implements java.awt.event.ActionListener {
   public void    actionPerformed(java.awt.event.ActionEvent event) {
     AbstractDialog.this.setVisible(false);
     if(event.getActionCommand().equals("Ok") )
        AbstractDialog.this.theUserAction = JOptionPane.OK_OPTION;
     else
        AbstractDialog.this.theUserAction = JOptionPane.CANCEL_OPTION;
   } // method: actionPerformed
} // class: OkCancelAction
} // class: AbstractDialog
```

The introduction of the AbstractDialog class makes the BorrowerDialog class straightforward. The code that follows shows that all it has to do is to:

- use its superclass to set up a basic dialog box
- amend the dialog box to be suitable for getting a borrower's details.

```
// class BorrowerDialog
package guisubsystem;

import javax.swing.*;
import java.awt.*;
import java.awt.event.*;

public class BorrowerDialog extends AbstractDialog {
   public    BorrowerDialog(javax.swing.JFrame frame) {
       super(frame, "Borrower details");
   } // method: BorrowerDialog

   public String    getName() {
     return theNameTextField.getText();
   } // method: getName
   private javax.swing.JLabel    theNameLabel = new JLabel("Name") ;
   private javax.swing.JTextField    theNameTextField = new JTextField(30);
   private javax.swing.JPanel    theNamePanel
                                  = new JPanel(new FlowLayout(FlowLayout.LEFT));
   {
     theNamePanel.add(theNameLabel);
     theNamePanel.add(theNameTextField);
     //
```

```
    theNamePanel.setMaximumSize(PANELSIZE);
    theDetailsPanel.add(theNamePanel);
  }
} // class: BorrowerDialog
```

The PublicationDialog is very similar as the only difference is that it:

• amends the dialog box to be suitable for getting a Publication title and catalogue number.

Its coding is as follows:

```
// class PublicationDialog
package guisubsystem;

import javax.swing.*;
import java.awt.*;
import java.awt.event.*;

abstract class PublicationDialog extends AbstractDialog {
  protected    PublicationDialog(javax.swing.JFrame frame, String title) {
    super(frame, title);
  } // method: PublicationDialog
  public int    getCatalogueNumber() {
    return Integer.parseInt( theCatalogueNumberTextField.getText() );
  } // method: getCatalogueNumber
  public String    getTitle() {
    return theTitleTextField.getText();
  } // method: getTitle
  private javax.swing.JLabel    theCatalogueNumberLabel = new JLabel("Num");
  private javax.swing.JTextField    theCatalogueNumberTextField = new JTextField(30);
  private javax.swing.JPanel    theCatalogueNumberPanel = new JPanel
    (new FlowLayout(FlowLayout.LEFT));
  {
    theCatalogueNumberPanel.add(theCatalogueNumberLabel);
    theCatalogueNumberPanel.add(theCatalogueNumberTextField);
    theCatalogueNumberPanel.setMaximumSize(PANELSIZE);
  }
  private javax.swing.JLabel    theTitleLabel = new JLabel("Title");
  private javax.swing.JTextField    theTitleTextField = new JTextField(30);
  private javax.swing.JPanel    theTitlePanel
                      = new JPanel(new FlowLayout(FlowLayout.LEFT));
  {
    theTitlePanel.add(theTitleLabel);
    theTitlePanel.add(theTitleTextField);
    theTitlePanel.setMaximumSize(PANELSIZE);
    //
```

```
        theDetailsPanel.add(theCatalogueNumberPanel);
        theDetailsPanel.add(theTitlePanel);
   }

} // class: PublicationDialog
```

As expected when it comes to the BookDialog and JournalDialog classes they are also straightforward. They inherit from PublicationDialog and through it AbstractDialog. Therefore they need only:

- use the superclass to set up a basic dialog box suitable for getting a Publication title and catalogue number
- amend the dialog box to be suitable for getting additional Book or Journal details.

The BookDialog code that follows illustrates.

```
// class BookDialog
package guisubsystem;

import javax.swing.*;
import java.awt.*;
import java.awt.event.*;

public class BookDialog extends PublicationDialog {

    public    BookDialog(javax.swing.JFrame frame) {
        super(frame, "Book details");
    } // method: BookDialog

    public String    getAuthor() {
        return theAuthorTextField.getText();
    } // method: getAuthor

    private javax.swing.JLabel    theAuthorLabel = new JLabel("Author");
    private javax.swing.JTextField    theAuthorTextField = new JTextField(30);
    private javax.swing.JPanel    theAuthorPanel
                        = new JPanel(new FlowLayout(FlowLayout.LEFT)) ;

    {
        theAuthorPanel.add(theAuthorLabel);
        theAuthorPanel.add(theAuthorTextField);
        theAuthorPanel.setMaximumSize(PANELSIZE);
        //
        theDetailsPanel.add(theAuthorPanel);
    }

} // class: BookDialog
```

Even though we have introduced significant changes by the introduction of a hierarchy of dialogs there is no impact on the rest of the system or its documentation. This is because a client's view of pre-existing classes is unaffected. They continue to "see" the same classes with the same features. Execution of our tests confirms this. Therefore having removed significant amounts of duplicate code we consider the second iteration to be successful. The model is given in the file Lib9_2.uml.

9.4   Iteration 3

A major benefit from the study of design patterns in chapter 8 is that we can learn from the experience of others. Therefore we can improve the case study design by reflecting on design patterns. For example, it would simplify the overall design if there was a single Library object with a common point of access. It would mean that we do not have to pass the Library as a parameter and that future changes could not introduce multiple Library objects by mistake. Modification of the Library class to be a singleton is an obvious strategy.

The Library is implemented as a singleton by providing a **public static** method getLibrary. It returns a reference to a single Library object held in a **private static** field. The important point we need to understand is that getLibrary uses a private constructor to create the Library object it references. This happens only once. As the Library constructor is not available to clients they cannot create a library and must use getLibrary instead. The following code illustrates the approach taken in more detail:

```
public class Library extends loansubsystem.LenderImp {
    private   Library(String aName) {
        super();
        theName = aName;
    } // method: Library

    public static librarysubsystem.Library getLibrary() {
        if(theLibrary == null) {
            Library.restore();    // See later
        }
        return theLibrary;
    } // method: getLibrary
    // ...
    private String    theName;
    private static librarysubsystem.Library    theLibrary = null;
} // class: Library
```

This implementation rather than the one discussed in section 8.5 has been chosen because a more subtle behaviour is required of the Library class. It originates from the fact that it implements the Serializable interface (as specified in its superclass). The problem we face is that a **static** field is associated with its class and not with an object belonging to the class. Therefore on object serialization a **static** field is ignored and is set to **null** on deserialization.

Three important consequences that affect the implementation of the Library as a serializable singleton are as follows:

- serialization of a Library object does not store the value of the **static** field theLibrary
- the **static** field theLibrary must be updated to reference the deserialized Library object otherwise it will have a **null** value

- each client must update its reference to the Library following serialization/ deserialization otherwise it could easily reference the Library object before serialization/deserialization.

We also discover that the LibraryFrame is responsible for the serialization/ deserialization of the Library. Clearly it is preferable that the Library itself should take on this responsibility.

Taking the first two consequences into account and reusing code from the LibraryFrame the Library class now supports persistence using Java's serialization/deserialization mechanism as follows:

```
public class Library extends loansubsystem.LenderImp {
    // ...
    private static void    restore() {
        // Attempt to restore the persistent application objects
        File file = new File(PERSISTENT_FILENAME);
        if(file.exists()) {
            try {
                FileInputStream fis = new FileInputStream(PERSISTENT_FILENAME);
                ObjectInputStream ois = new ObjectInputStream(fis);
                theLibrary = (Library)ois.readObject();
                ois.close();
            } catch(IOException ex) {
                JOptionPane.showMessageDialog(null, "Error reading persistent store",
                                                    "Library", JOptionPane.ERROR_MESSAGE);
                theLibrary = new Library("Napier");
            } catch(ClassNotFoundException ex) {
                JOptionPane.showMessageDialog(null, "Error restoring application objects",
                                                    "Library", JOptionPane.ERROR_MESSAGE);
                theLibrary = new Library("Napier");
            }
        } else {
            theLibrary = new Library("Napier");
        }
    } // method: restore

    public static void save() {
        try {
            FileOutputStream fos = new FileOutputStream(PERSISTENT_FILENAME);
            ObjectOutputStream oos = new ObjectOutputStream(fos);
            oos.writeObject(theLibrary);
            oos.close();
        } catch(IOException ex) {
            JOptionPane.showMessageDialog(null, "Cannot open persistent file",
                                                "Library", JOptionPane.ERROR_MESSAGE);
            System.exit(1);
        }
```

```
        System.exit(0);
    } // method: save
    // ...
    private static final String    PERSISTENT_FILENAME = "library.ser";
} // class: Library
```

The third conclusion leads us to ensure that each action class in the LibraryFrame requests a Library reference in its actionPerformed method, i.e. as close to the point of use as we can arrange. It is typical of design patterns that they can often appear straightforward to implement but are in fact rather subtle. The following outline code from the inner LibraryFrame class, AddBookAction, illustrates:

```
// class AddBookAction
public void actionPerformed(java.awt.event.ActionEvent event) {
    // Get the Library singleton
    Library library = Library.getLibrary();
    // ...
} // method: actionPerformed
```

Having made these changes to the Library and the LibraryFrame the resulting code is significantly improved. Inappropriate Library serialization/deserialization code in the LibraryFrame constructor has been moved to its rightful location in the Library class. In addition we have removed the risk of the existence of multiple libraries by making the Library a singleton.

Changes required to the documentation include:

- update the Library class diagram with the stereotype <<singleton>> and include getMethod, restore and save
- make the Library constructor **private**
- remove Library serialization/deserialization code from the LibraryFrame constructor
- update each action object in LibraryFrame to request a Library singleton reference.

Recompilation and successful execution of our tests confirm that all is well. Therefore we consider this iteration to be complete. The model is given in the file Lib9_3.uml.

Before leaving this iteration a previously undetected bug becomes evident. Sadly, despite our best efforts, this is a relatively common occurrence with software systems. The problem is that when adding a Book or Journal to the Library we expect a dialog box with the title Book details or Journal details. In fact neither is present (see figure 7.23).

Closer examination shows the PublicationDialog class has an operation getTitle that is its source. The purpose of getTitle is to deliver the title of the Book or Journal keyed into a text field by a human user. Unknown to us there is an operation with the same signature inherited from the Swing framework. Its purpose is to deliver the title of a dialog box when it is displayed. We have inadvertently redefined it to return an empty String! Hence the lack of a title on the dialog box.

Although we should not make any changes to the functionality of the system when refactoring it seems safe to do so in this case. Therefore we rename getTitle in the PublicationDialog class as getPublicationTitle. For consistency (and some peace of mind) we also rename similar operations in the AbstractDialog hierarchy as follows:

```
// class BorrowerDialog
getBorrowerName
```

```
// class BookDialog
getBookAuthor
```

```
// class JournalDialog
getJournalEditor
getJournalDateOfPublication
```

All that remains is to modify the actionPerformed method in the addBookAction, addPublicationAction and registerBorrowerAction classes accordingly. Subsequent testing shows that all is well.

An important point to learn from this experience is that a method should be **final** if it is intended that it should be invariant over specialization (see section 5.4). In this case it seems clear that getTitle inherited from the swing framework should have been **final**. This would have prevented us from inadvertently redefining it.

9.5   Iteration 4

Consider the screen dump from the ROME case tool shown in figure 9.7.

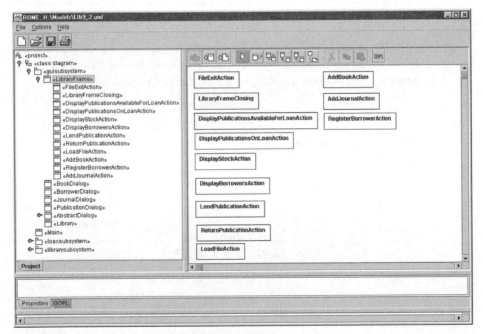

Figure 9.7   *An overview of the LibraryFrame class*

It indicates that the LibraryFrame class with 12 inner classes is probably too large and complex. Our suspicion is that as more and more functionality is required, more inner classes will be added.

The difficulty with inner classes is that they can be hard to maintain. For example, the size of the enclosing class can be daunting and a refactoring that tries to establish a hierarchy of inner classes can be technically challenging (in our opinion anonymous inner classes are even worse). We also suspect that the LibraryFrame has duplicate code, as some of the inner classes appear to have a similar role. Therefore the aim of this iteration is to rationalize and simplify the LibraryFrame class.

Closer examination shows that the LibraryFrame:

- establishes the GUI
- detects human user selections
- actions each selection and reports on the outcome.

Recall that the LibraryFrame is the view/controller in the MVC architecture. Therefore it is reasonable that it should establish the GUI and have responsibility for detecting/actioning human user selections. The real problem is how to manage the complexity that inevitably arises. One approach is to make a clearer separation of its two concerns, i.e. establishing the GUI and detecting/actioning human user selections.

Normally an inner class is used for encapsulation purposes. Since an inner class has direct access to the private features of its enclosing class it can also simplify the coding. However, we can also achieve an element of encapsulation by the introduction of an inner package to relocate the inner classes. The benefit for us is that the enclosing class is significantly reduced in size and complexity. If the inner classes are not heavily dependent on the private features of the enclosing class (as is the case here) then this alternative is even more attractive.

Weighing up the advantages and disadvantages of each approach we abandon the use of inner classes for action objects not directly associated with the LibraryFrame. As shown in figure 9.8, this leads to a revised structure for the guisubsystem package.

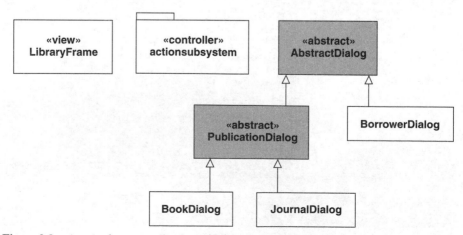

Figure 9.8 *A revised structure for the guisubsystem package*

Note that the stereotype <<view>> helps clarify the LibraryFrame's role. Similarly, the stereotype <<controller>> clarifies the role of the classes in the actionsubsystem package.

As the inner class LibraryFrameClosing is directly concerned with the LibraryFrame it is not relocated. However, the remaining eleven interact with the model (the Library) and are relocated to the actionsubsystem package. Of course while doing so we refactor them.

Using a similar approach to that taken in iteration 2, we introduce an AbstractLibraryAction class as shown in figure 9.9.

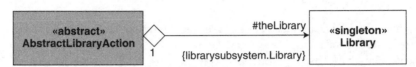

Figure 9.9   *The AbstractLibraryAction class*

It has three main responsibilities. The first is to initialize its superclass AbstractAction. The second is to initialize a **protected** LibraryFrame reference intended for use by all subclasses. This is necessary as the LibraryFrame is no longer directly visible. As normal the constructor discharges both of these responsibilities.

The third responsibility is unusual in that it gives subclasses access to the most recent Library singleton. This is achieved by setting the reference to it in its actionPerformed method. Recall that in the previous iteration each of the LibraryFrame inner classes had this responsibility. Subclasses can now use the superclass actionPerformed method to ensure that the Library singleton reference is up to date.

The following code illustrates.

```
// class AbstractLibraryAction
package guisubsystem.actionsubsystem;

import librarysubsystem.Library;

public abstract class AbstractLibraryAction extends javax.swing.AbstractAction {

    protected    AbstractLibraryAction(String label, guisubsystem.LibraryFrame aLibraryFrame) {
        super(label);
        theLibraryFrame = aLibraryFrame;
    } // method: AbstractLibraryAction

    public void    actionPerformed(java.awt.event.ActionEvent event) {
        // Get the Library singleton
        theLibrary = Library.getLibrary();
    } // method: actionPerformed

    protected guisubsystem.LibraryFrame    theLibraryFrame;
    protected librarysubsystem.Library    theLibrary;
} // class: AbstractLibraryAction
```

Before proceeding we should consider the fact that the LibraryFrame's private attributes are no longer visible to the action classes. Therefore we add two new operations appendToViewingArea and setStatusField to the LibraryFrame. They allow clients to display information on a scrollable text area and to update a status field. They are coded as follows:

```
// class LibraryFrame
    public void    appendToViewingArea(String text) {
        theViewingArea.append(text);
    } // method: appendToViewingArea

    public void    setStatusField(String aMessage) {
        theStatusField.setText(aMessage);
} // method: setStatusField
```

It is not unusual when refactoring that the original design must be modified in some manner to accommodate it.

The subclasses of AbstractLibraryAction:

- RegisterBorrowerAction
- LendPublicationAction
- ReturnPublicationAction
- AddBookAction
- AddJournalAction
- LoadFileAction and
- FileExitAction

are now relatively straightforward. Each constructor is given a LibraryFrame object and a check is made to ensure that the redefined actionPerformed method updates the Library singleton reference (if necessary). Finally the LibraryFrame methods appendToViewingArea and setStatusField are used as appropriate.

The following code for the RegisterBorrowerAction class illustrates:

```
package guisubsystem.actionsubsystem;

import javax.swing.*;
import guisubsystem.*;

public class RegisterBorrowerAction extends AbstractLibraryAction {
    // ----- Operations ----------
    public    RegisterBorrowerAction(java.lang.String label,
            guisubsystem.LibraryFrame aLibraryFrame) {
        super(label, aLibraryFrame);
    } // method: RegisterBorrowerAction

    public final void    actionPerformed(java.awt.event.ActionEvent event) {
        super.actionPerformed(event);
        //
        BorrowerDialog dialog = new BorrowerDialog(theLibraryFrame);
        dialog.setVisible(true);
        //
        if(dialog.getUserAction() == JOptionPane.OK_OPTION) {
```

```
        String borrowerName = dialog.getBorrowerName();
        //
        // Register the borrower with the library
        theLibrary.registerOneBorrower(borrowerName);
        //
        // Display the outcome
        theLibraryFrame.setStatusField(theLibrary.getStatus() );
      } else
        theLibraryFrame.setStatusField("No Borrower registered");

   } // method: actionPerformed

} // class: RegisterBorrowerAction
```

The AbstractLibraryAction hierarchy is now as shown in figure 9.10.

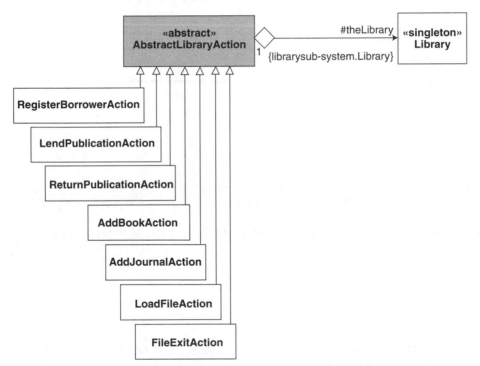

Figure 9.10   *The updated* AbstractLibraryAction *hierarchy*

All that remains are the classes:

- DisplayStockAction
- DisplayBorrowersAction
- DisplayPublicationsOnLoanAction
- DisplayPublicationsAvailableForLoanAction

Based on previous experience, it is obvious that we should introduce a superclass DisplayAction with the common features all four display classes. The final AbstractLibraryAction hierarchy is now as shown in figure 9.11.

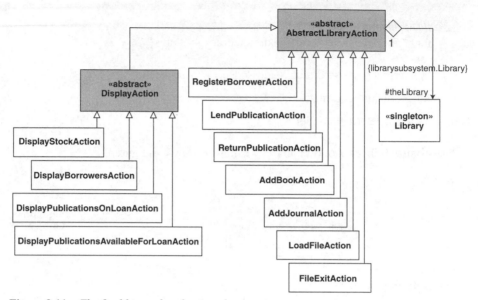

Figure 9.11   *The final hierarchy of action classes*

The DisplayAction class is quite straightforward. All that is required is that:

- the constructor must properly initialize its superclass AbstractLibraryAction
- the actionPerformed method ensures that it has the up-to-date Library singleton and
- the Library is displayed on the LibraryFrame's scrollable text area.

Its detailed coding is:

```
// class DisplayAction
package guisubsystem.actionsubsystem;

public abstract class DisplayAction extends AbstractLibraryAction {

    protected   DisplayAction(String label, guisubsystem.LibraryFrame aLibraryFrame) {
        super(label, aLibraryFrame);
    } // method: DisplayAction

    public void   actionPerformed(java.awt.event.ActionEvent event) {
        super.actionPerformed(event);
        // Display information about the Library
        theLibraryFrame.appendToViewingArea("\n" + "Library:" + theLibrary);
    } // method: actionPerformed

} // class: DisplayAction
```

Happily the four subclasses are also straightforward as they require only the expected modifications. The code for the DisplayStockAction class serves as an illustration.

```
// class DisplayStockAction
package guisubsystem.actionsubsystem;

import java.util.*;
import librarysubsystem.*;

public class DisplayStockAction extends DisplayAction {
    public   DisplayStockAction(String label, guisubsystem.LibraryFrame aLibraryFrame) {
        super(label, aLibraryFrame);
    } // method: DisplayStockAction

    public void   actionPerformed(java.awt.event.ActionEvent event) {
        super.actionPerformed(event);
        //
        // Display information about the publications
        theLibraryFrame.appendToViewingArea("\n\t" + "Publication stock");
        boolean publicationFound = false;
        Iterator iter = theLibrary.getLoanItemsIterator();
        //
        while( iter.hasNext() ) {
            Publication pub = (Publication) iter.next();
            theLibraryFrame.appendToViewingArea("\n\t\t" + pub);
            publicationFound = true;
        }
        if( publicationFound == false )
            theLibraryFrame.appendToViewingArea("\n\t\t" + "None");
            //
            // Display the outcome
            theLibraryFrame.setStatusField("");
    } // method: actionPerformed

} // class: DisplayStockAction
```

Recompilation and execution of our tests confirm that all is well. Therefore we consider this final iteration to be successful. All that remains is to make a final check on the documentation and this review of the case study is complete. The model is given in the file Lib9_4.uml.

9.6   Summary

1. Refactoring is often necessary as part of a final review. Although refactoring depends on experience, the subject has been well documented and a vocabulary exists to describe a sequence of refactorings that might be applied to a system.
2. Each refactoring should make a relatively small change. It should be accompanied with extensive testing to guarantee the same behaviour from the software.

3. Redistribution of classes in stereotyped packages clarifies their role and eases the maintenance burden.
4. Code duplication is a major cause for refactoring.
5. Exposure to design patterns can help identify useful refactorings.

9.7   And finally

This book has been concerned with Object-Oriented Design, the Unified Modelling Language and the Java Programming Language. Throughout we have stressed the importance of conducting an OOD prior to the implementation of a system. We have used the UML to enhance our understanding of the system by documenting different views of it. In our opinion this brings a significant measure of quality to the final program code.

From chapter 2 we have argued that system development should be architecture-centric, employ an iterative approach and be use-case led. A lightweight lifecycle was introduced to bring discipline to development activities without being encumbered by needless bureaucracy.

The expressive power of the UML has been crucial to the success of our designs. Its various diagrams gave important insights into the final system. For example, a sequence diagram reveals how message propagation through a collection of objects implements some part of its functionality. Often the diagram can be mapped directly into code.

Of all of the UML diagrams available, the class diagram has been the most important. It is central to the modelling activity as it defines the architecture we realize in code. In effect it drives our implementation development. We seek to forward engineer the program code from the class diagram.

The class diagram reveals association and aggregation relations that exist between classes. This led to the use of the Java Collections Framework. As well as acting as an exemplar of OOD their use markedly simplified our application code.

Specialization hierarchies are also documented in a class diagram. They helped us make use of the polymorphic effect and to aspire to design to an interface wherever possible. Ultimately this led to the development of a framework.

The more sophisticated applications of polymorphic substitution gave rise to advanced design patterns. They offer elegant solutions to many common design and programming problems. The vocabulary they introduce elevates the level of abstraction we can achieve in our designs above that of an ordinary class diagram.

Our discussion of design patterns led naturally to refactoring our designs. Jointly, refactoring and design patterns represent leading edge developments in object orientation.

Finally, we have made ROME available. We hope that this case tool has stimulated the reader to value the modelling capabilities of the UML and to confirm the merits of design before coding.

9.8   Exercises

1. Exercise 1 in chapter 8 noted that object creation is one of the most expensive operations for a Java program to perform. You were asked to use the singleton design

pattern to guarantee that only one BookDialog object is ever created. Now apply this strategy to other GUI objects in the design using Lib9_4.uml as your starting point.

You may like to investigate the use of a profiling tool as part of this exercise. It can be used to identify classes that are suitable for refactoring. There are several on offer as downloads from the World Wide Web.

2. Figure 8.12 in exercise 2 of chapter 8 presented a simplified class diagram for the class diagrammer used in ROME. Construct a similar class diagram for the collaboration diagrammer used in ROME.

Identify classes that are common to the class diagrams for the two diagrammer tools. Now produce a single unified class diagram. Can we consider this as the basis for a refactoring strategy?

3. Figure 9.12 is an extract of the relationship classes from figure 8.12. Additionally we have shown that a relation comprises one or more LineSegment objects representing the horizontal and vertical strips used to draw a relation.

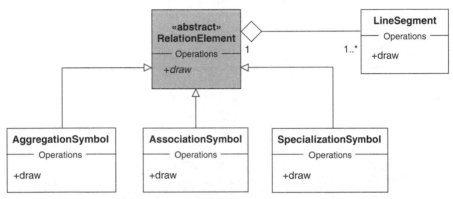

Figure 9.12   *Extract of the relationship classes*

The draw method in class AggregationSymbol renders the LineSegments then places a diamond decoration at one end. The draw method in class SpecializationSymbol renders the LineSegments then places an arrow tip at one end. The class AssociationSymbol simply renders the LineSegments.

Refactor this architecture by making the concrete draw method in the class RelationElement. What common behaviour does it take responsibility for? How do we deal with the particular decorations required by the concrete subclasses? What design pattern are we using?

4. During the development of iteration 3, it was noted that there was the possibility of having several Library singletons! The root cause was the serialization/deserialization the Library. Our solution was to ensure that each client of the Library updated its Library reference before sending it a message. Although this approach does work it has significant drawbacks. For example, there is no guarantee that someone maintaining our code will remember to update a Library reference before using it. It could also be argued that it is inefficient if the update is not required.

An alternative solution is give the Library responsibility for informing its clients that it has been deserialized. It could also supply each client with an updated reference to itself. Each client could then store this reference for its own use. The point is that this reference is guaranteed to refer to the most recently deserialized Library object.

(a) Suggest a design pattern that could be used to help us refactor iteration 3 using our new approach.

(b) Implement your ideas as iteration 5.

Bibliography

Beck 1999   Kent Beck, Martin Fowler
                 *Extreme Programming Explained: Embrace Change, and
                 Planning Extreme Programming*
                 Addison-Wesley 1999

Booch 1991   Grady Booch
                 Object Oriented Design with Applications
                 Benjamin/Cummings 1991

Booch 1999   Grady Booch, James Rumbaugh, Ivar Jacobson
                 The Unified Modeling Language User Guide
                 Addison-Wesley 1999

Cockburn 2001   Alistair Cockburn
                 Agile Software Development
                 Addison-Wesley 2001

Dietel 2003   H Dietel and P Dietel
                 Java: How to Program
                 Prentice Hall 2003

Eckel 2002   Bruce Eckel
                 Thinking in Java
                 Prentice Hall 2002

Elliott 2002   James Elliott et al.
                 Java Swing
                 O'Reilly 2002

Fontoura 2001   M Fontoura, W Pree, B Rumpe
                 The UML Profile for Framework Architectures
                 Addison-Wesley 2001

Fowler 1999   Martin Fowler
                 Refactoring: Improving the Design of Existing Code
                 Addison-Wesley 1999

Gamma 1994   Eric Gamma, Ralph Helm, Richard Johnson, John Vlissides
                 Design Patterns
                 Addison-Wesley 1994

Jacobsen 1999   Ivar Jacobsen, Grady Booch, James Rumbaugh
                 The Unified Software Development Process
                 Addison-Wesley 1999

Larman 2002   Craig Larman
                 Applying UML and Patterns
                 Prentice Hall 2002

McGregor 2001   J D McGregor, D A Sykes
                   A Practical Guide to Testing Object-Oriented Software
                   Addison-Wesley 2001

Pooley 1999   Rob Pooley, Perdita Stevens
                   Using UML
                   Addison-Wesley 1999

Priestley 2000   Mark Priestley
                   Practical Object-Oriented Design with UML
                   McGraw-Hill 2000

Rumbaugh 1991   James Rumbaugh et al.
                   Object Oriented Modeling and Design
                   Prentice Hall 1991

Topley 1999   Kim Topley
                   CORE Swing: Advanced Programming
                   Prentice Hall 1999

Watt 2001   D Watt, D F Brown
                   Java Collections
                   Wiley 2001

Appendix A

Setting up the Environment

The examples and exercises in this book were prepared with the ROME modelling tool. The ROME tool and the Java programs produced by it use the Java Software Development Kit (SDK) version 1.4. In this appendix we present how to install the SDK and the ROME tool, and how to configure the environment to run ROME and the Java programs. These notes assume a Microsoft Windows installation. Other installations follow similar procedures.

A.1   Installing the Java Software Development Kit

ROME is a pure Java application. In order to use it you must install a Java run-time environment. However, since ROME generates Java programs that must be compiled and run, then we must download and install the Java 2 platform, Standard Edition. You can download the Java SDK for Windows, Linux and Solaris environments from the java.sun.com website. See the ROME website (http://www.dcs.napier. ac.uk/~kab/jeRome/jeRome.html) for further details.

Installation of the SDK is relatively straightforward. Currently, on the main page (http://java.sun.com/j2se), follow the *releases* link to the download page. For a Microsoft Windows installation, download Windows (all languages including English) by selecting the Windows DOWNLOAD SDK link. Read and accept the terms of conditions, then start the download of the self-extracting compressed file (currently *j2sdk-1_4_1_03-windows-i586.exe*).

Run the self-extracting compressed file by double-clicking the file icon in the file manager tool. The decompression begins and the SDK is extracted into the default directory c:\j2sdk1.4.1_03. We shall refer to this as the *JDK home directory*. Now ensure that the JAVA_HOME environment variable is set to the installation JDK home directory, and that the Java bin directory is included in the search path. On a Windows system you set these by entering the following commands in a command prompt window:

`C:\> set JAVA_HOME = c:\j2sdk1.4.1_03`
`C:\> set PATH = %JAVA_HOME%\bin;%PATH%`

To test this set-up procedure, now run the command:

`C:\> java –version`

which will reply with details of the installed Java run-time environment.

Successfully setting a number of environment variables is dependent on there being sufficient environment space. It may be necessary to open the command window by selecting Run ... from the Start menu and entering the name of the program as:

command /e:4096

This will open a command window with a 4Kbyte environment space that is sufficient for our needs.

Later in this appendix we will show how we can prepare these environment settings in a batch (*.bat) file and run this file at the start of a ROME and Java session.

It is also worth downloading the Java documentation from the same site. You can download the documentation for the current version from http://java.sun.com/j2se. This documentation describes the Java tools such as a Java compiler and the Java language APIs.

A.2   Installing the ROME modelling tool

Create the empty directory C:\jeRome5 on your Windows system. This we will refer to as the *ROME home directory*. You can download the ROME modelling tool from the ROME website at http://www.dcs.napier.ac.uk/~kab/jeRome/jeRome.html (case sensitive URL). The *download* link brings you to a page from which to start the actual download procedure. Save the *jeRome5.Distribution.zip* file into the ROME home directory, then unpack this compressed file with a suitable zip tool such as WinZip (http://www.winzip.com/). Choose the ROME home directory as the output directory for this file decompression. Alternatively, you can unpack it by changing to the ROME home directory, then run the Java archive tool (jar) with the command:

C:\> **cd jeRome5**
C:\jeRome5> **jar xvf jeRome5.Distribution.zip**

Either method creates a number of subdirectories within the ROME home directory. The subdirectories include, among others, examples and jerome. The ROME home directory also includes a number of batch files such as setjava.bat, setrome.bat, jc.bat and jr.bat (discussed later). Like most software packages the ROME home directory contains a file entitled README.txt. Software distributions change and the README.txt should always be consulted for the most current installation instructions.

The examples directory includes the models and programs presented in the book. The models are in folders corresponding to the chapters. Hence in the examples directory there is a subdirectory chapter03 with the two models files Prog3_1.uml and Prog3_2.uml. The examples directory also includes subfolders AppendixF and AppendixG. They, in turn, have subfolders such as ProgramF01 and ProgramG01 that contain the Java files for the samples from appendices F and G.

The jerome directory has the start-up code for the ROME tool. Following our discussion of batch files (next) we show how to launch ROME.

A.3   Setting the compilation environment

We noted in section A.1 the two commands to enter in a command window:

C:\> **set JAVA_HOME = c:\j2sdk1.4.1_03**
C:\> **set PATH = %JAVA_HOME%\bin;%PATH%**

These and other environment settings have been prepared in the batch file
setjava.bat. This is a text file that can be edited with a simple text editor such as
Notepad. The reader must consult the README.txt file to discover how to make the
small number of revisions to the batch file to reflect the user's set-up.

When the revisions have been made we can execute the commands in the batch file with:

C:\> **c:\jeRome5\setjava**

We can now confirm that the settings are correct by compiling and running any of the
sample programs from appendices F and G. For example, if we change to the directory
ProgramF02:

C:\jeRome5\examples\AppendixF\ProgramF02>

Then we can use the batch file jc.bat to compile Java program files and the batch file
jr.bat to run a Java program. In the directory programf02 we have the program files
Main.java and Point.java. First we compile all the program files with the command:

C:\jeRome5\examples\AppendixF\ProgramF02> **jc *.java**

Now we can run the program with:

C:\jeRome5\examples\AppendixF\ProgramF02> **jr Main**

A.4   Setting the ROME environment

The batch file setrome.bat sets the environment to execute the ROME program.
As noted in the previous section this too is a text file that may require editing to correctly
reflect the user's configurations. Again, the README.txt file gives the necessary installation instructions.

Once setrome.bat has been modified, open a second command window. Notice we
have two command windows opened. The first was opened in section A.1 and is used to
compile and run the Java code produced by ROME. This second command window is
used to start ROME. Now run this batch file with the command:

C:\> **c:\jeRome5\setrome**

Now we can use the batch file rr.bat to start ROME:

C:\> **rr jerome.Rome**

When ROME starts the following message appears in the command window:

PROGRAM START... (do not close this window)

This reminds the user not to close the command window (though it can be iconized). Closing the command window will also close down the ROME application.

Section B.5 describes how to set the output directory using the Preferences Editor dialog. This sets the directory in which classes generated from a class diagram are placed (say, C:\demonstration for illustration). In the first command window opened (see section A.1), move to this output directory then compile and run the ROME generated code following the instructions from the previous section.

C:\demonstration> **jc *.java**
C:\demonstration> **jr Main**

Appendix B

ROME

ROME is a lower-case object-modelling environment. It is a UML compliant tool with support for activity, class, collaboration, object, sequence and use-case diagrams. Additionally, ROME provides support for generating and compiling Java code from a class diagram.

ROME is a lightweight tool with sharply focused functionality. The aim was to produce a modelling environment that is productive, uniform and consistent, has a low learning curve and has a small footprint. Such features make ROME a readily accessible tool for novices new to object-oriented analysis and design. Developers can, in time, consider progressing to the feature-rich professional tools that are available.

Unburdened by too many features, ROME proves to be a highly productive environment. For example, the class diagram supports visually preparing a static model of the object classes and their relationships. Simultaneously, the developer is able to see the emerging Java code. Further, any class can be compiled from within ROME and any errors are reported back into ROME.

All the diagrams are given a uniform and consistent behaviour. Most UML diagrams comprise graphs of nodes and relations that connect the nodes. Within any diagram a node and a relation have the same behaviours. When a node in a class diagram (such as a class symbol) is dragged then the effect is much the same when an instance symbol is dragged in a collaboration diagram. Equally, a relation in an object diagram (such as an aggregate link) operates in the same manner as a transition symbol in an activity diagram.

The class diagram is a key UML diagram. Under ROME the class diagram can include nested class and package symbols, and architectural relations (aggregation, association and specialization). Dialogs permit ROME to decorate these modelling elements. These decorations are then used in the generation of the Java code for the diagram. Working jointly with a compilation system, the developer can create and revise a UML class diagram, generate the source code, then compile and execute that code. The cycle can then be repeated, incrementally developing the application.

In this appendix we simply illustrate using ROME's class diagrammer. Since the other diagrams operate in a similar manner, the user will obtain sufficient familiarity with this one example. The reader can also consult the online help provide by ROME. The user can also load any of the supplied models (class diagrams), set the preferences (see section B.5), and press the OOPL button to code generate. The user is encouraged to follow the example presented while simultaneously using ROME.

This current implementation of ROME (version 5) was developed as a Java application, using version 1.3 of the Java Development Kit. The graphical elements of ROME are assembled as Swing components. All development has taken place under a variety of MS Windows operating environments. Further updates will be provided on the website and in the ROME online help.

B.1   User interface

ROME has the look and feel of any standard graphical application. Upon start-up, a ROME session appears as shown in figure B.1. The principal elements of the ROME application are typical of those found in most graphical products. The Window has a caption bar across the top bearing the title ROME: Untitled. In the upper left of the window is the usual control menu box. At the top right are the minimize and maximize buttons.

Below the caption is the application menu bar and toolbar. These provide the common services irrespective of which diagram the user is editing. The remainder of the ROME window is partitioned into four panels, respectively the project panel, the model panel, the properties panel and the message panel. Each panel is separated by moveable partitions. By adjusting the partitions each sub-panel can be enlarged or reduced.

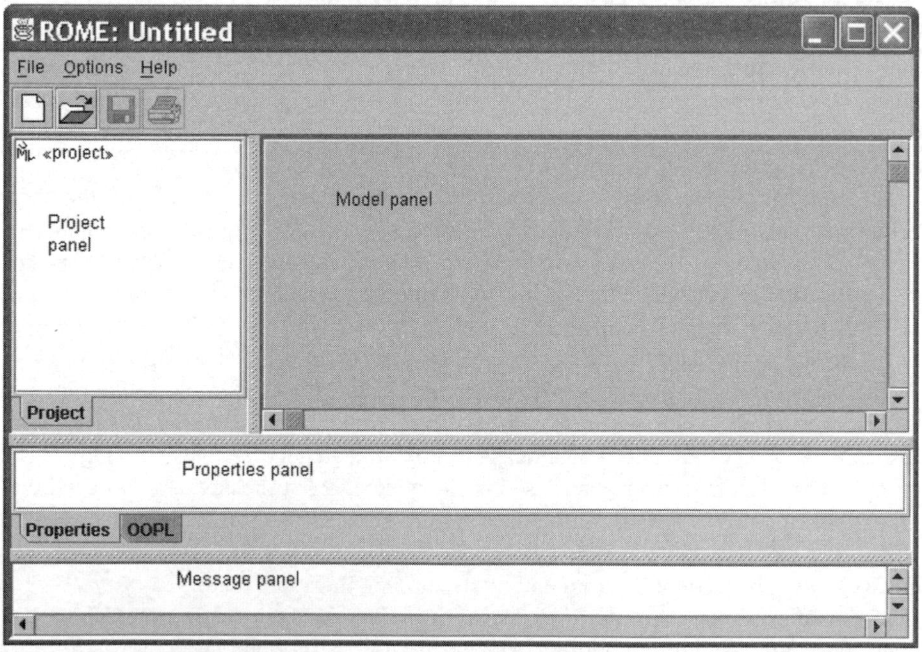

Figure B.1   *ROME start-up*

Below the caption is the menu bar including the menus File, Options and Help. We select a menu option and choose a command from the list of menu items to carry out some action. The File menu offers the choice shown in figure B.2. As we might expect New, Open..., Save and Save As are concerned with saving and retrieving models. Print prepares a hard copy of the diagram the user is editing and Exit terminates ROME.

Below the E<u>x</u>it menu item is a history list of the four most recent models the user had been editing. Selecting one of these will open the model into ROME.

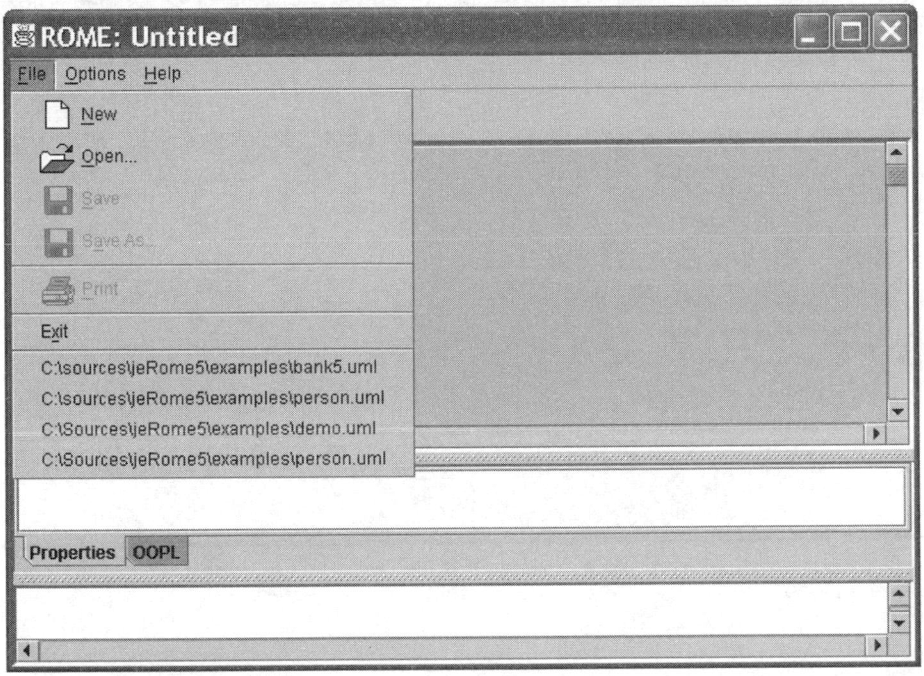

Figure B.2 *File menu*

Below the menu bar is a toolbar. The toolbar presents a list of iconic buttons providing fast and convenient access to the more common operations supported by ROME. For example, we may start a new design by selecting the command <u>N</u>ew from the <u>F</u>ile menu, or by simply clicking the leftmost button on the toolbar.

The four subpanels have been identified as the project panel, the model panel, the properties panel and the message panel. The model panel is where the user creates and edits a UML diagram. Where ROME is able to identify errors in the model, they are reported in the message panel. Error messages are also accompanied by an audible beep to alert the user. The properties panel is used to present information about the selected item in the model panel. For example, when editing a class in a class diagram, the properties panel will present some details about that class. The project panel partitions the project into one or more diagrams and permits the user to edit a selected diagram or to add further diagrams to the project.

B.2   Preparing UML diagrams

The project panel is an index into the diagrams and subdiagrams that comprise a UML file in ROME. A project can consist of one or more UML diagrams. The diagrams are presented as branches from the <<project>> root. The user selects one of the diagrams in the project panel and the model panel shows the UML diagram.

The user clicks the right mouse button while over the <<project>> root to reveal the floating menu shown in figure B.3. From this menu the user can select and add a new diagram into the project. Repeated application of this will populate the project with a number of separate diagrams.

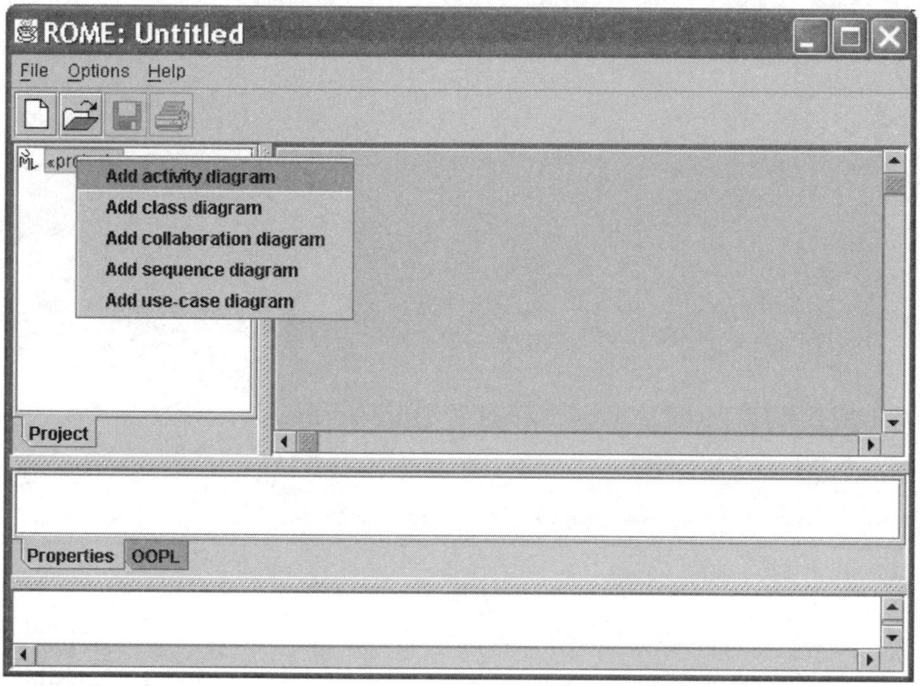

Figure B.3 *Project menu*

Figure B.4 shows the project panel when a class and a collaboration diagram have been added to the project. Notice how in all these illustrations that the model panel has remained greyed, indicating that it is inactive.

The following sections illustrate how to use the class diagrammer. Most of the features for preparing and editing this type of diagram are common to the other diagrams. The reader is encouraged to experiment with the other diagrams in ROME.

B.3   Preparing class diagrams

If in figure B.4 we now select the «class diagram» node in the project tree, then the model panel is opened to reveal a drawing surface on which the user assembles the diagram, and a context sensitive diagram toolbar. Since this is a class diagram, there are toolbar buttons for including the various elements associated with class diagrams. This is shown in figure B.5.

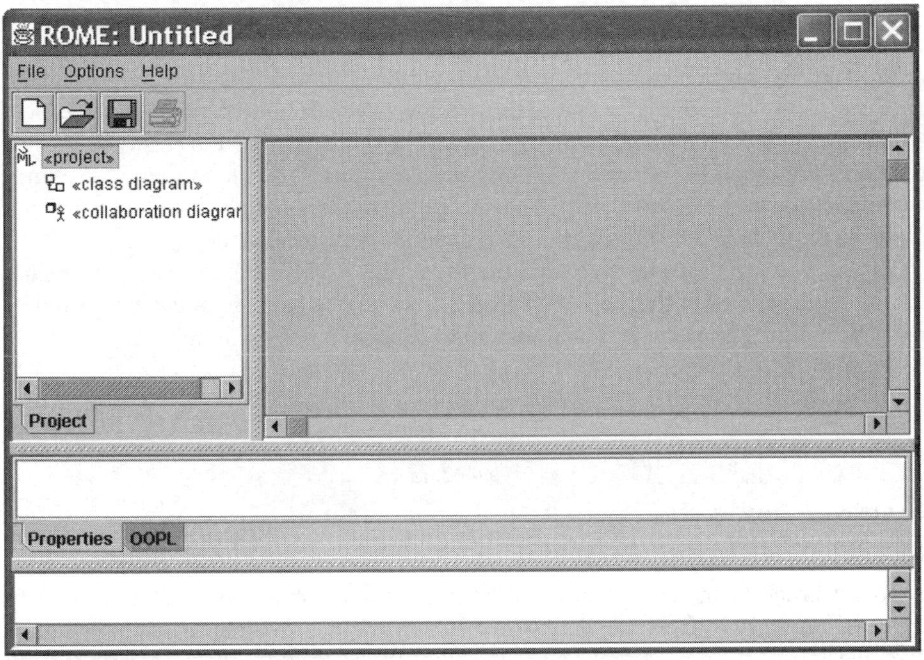

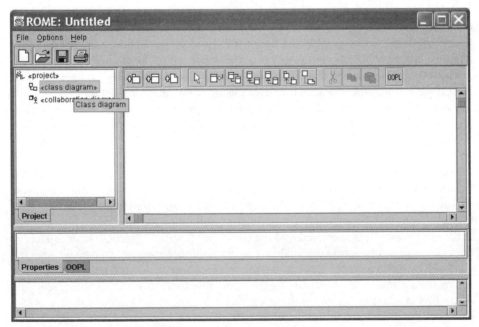

Figure B.4   *Project comprising a class and a collaboration diagram*

Figure B.5   *Class diagrammer*

Observe how the toolbar buttons are arranged into logical groups. The first group of three buttons is used to include, respectively, package, class and note into the class diagram. The next seven buttons are the relation building buttons. They are toggle buttons with only one ever active. The first of these is the selection button and is used to select an element in a diagram and, perhaps, drag it. The others are used to form the various relations. For example, the second of these seven buttons is used to introduce an aggregation relation between two classes. The next group of three buttons is for cut, copy and paste. Finally, the OOPL button is used to generate code (see later).

To describe the stages involved and the use of the ROME tool, we shall reconsider the model developed in chapter 3. The model was that of a banking organization and its many accounts. The example model we aim to develop is shown in figure B.6.

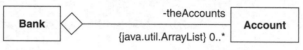

Figure B.6   *Target model*

To construct the model we must first introduce the Bank and Account classes into the diagram. In ROME we can do this in a variety of ways. With the mouse positioned over the model pane we simply click the right mouse button, then a floating menu appears and if we select the first option Add class, we get an empty class rectangle included in the class diagram at the mouse point.

The same can be achieved by pressing the add class toolbar button (second from left of toolbar). Note that when we position the mouse cursor above this toolbar button the tool tips reminds us that this button is for adding a class to the model. When we click this Add class toolbar button with the left mouse button a class symbol appears in the upper left of the model as shown in figure B.7. Here we have clicked the button twice and the class symbols have been stacked at the upper left of the diagram.

From figure B.7 we see that the newly entered class is obscuring an existing class box. If we position the mouse cursor over this new box, then press and hold down the left mouse button a rectangular outline for the selected class appears. While still holding the left mouse button we may drag the selected symbol to the desired location. When we are satisfied with the new position for the symbol we simply release the left mouse button.

We now name one of these new classes as Account. This we do with the Class Editor dialog shown in figure B.8. The dialog appears by first positioning the mouse cursor above the unnamed class, clicking the right mouse button and selecting Edit Class... from the floating menu. The primary information we give through this dialog is the class name. This is entered into the edit control labelled Name. Other aspects of the class can also be set by this dialog. For example, commentary associated with the class can be given by entering a Java style comment into the Comment field.

We need now to establish the aggregation relationship between the Bank and Account classes. This we do using the aggregate button on the toolbar. Select this tool (the tool tip indicates that this is an Aggregation tool for establishing aggregation relationships) then move the mouse cursor to the class representing the part in the

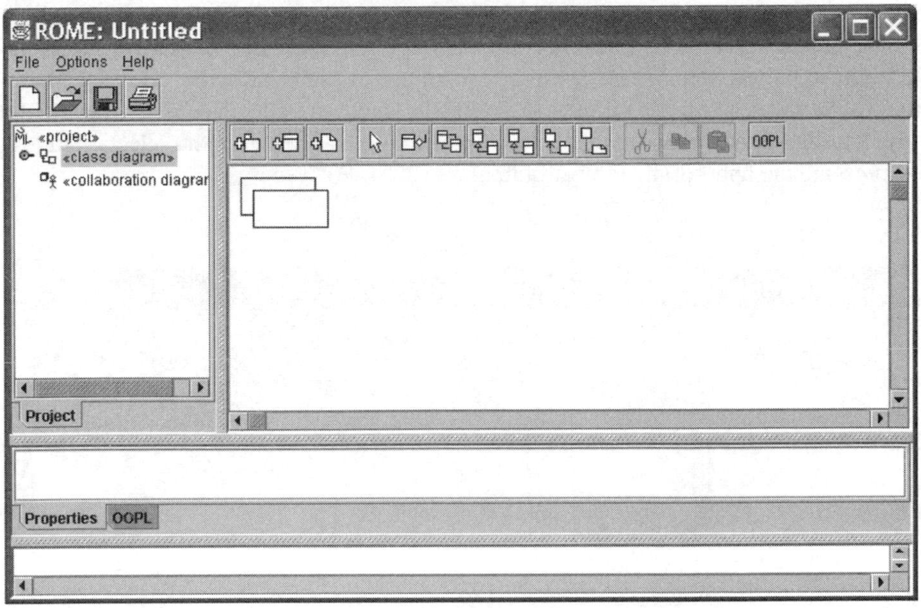

Figure B.7   *The effect of adding two new classes*

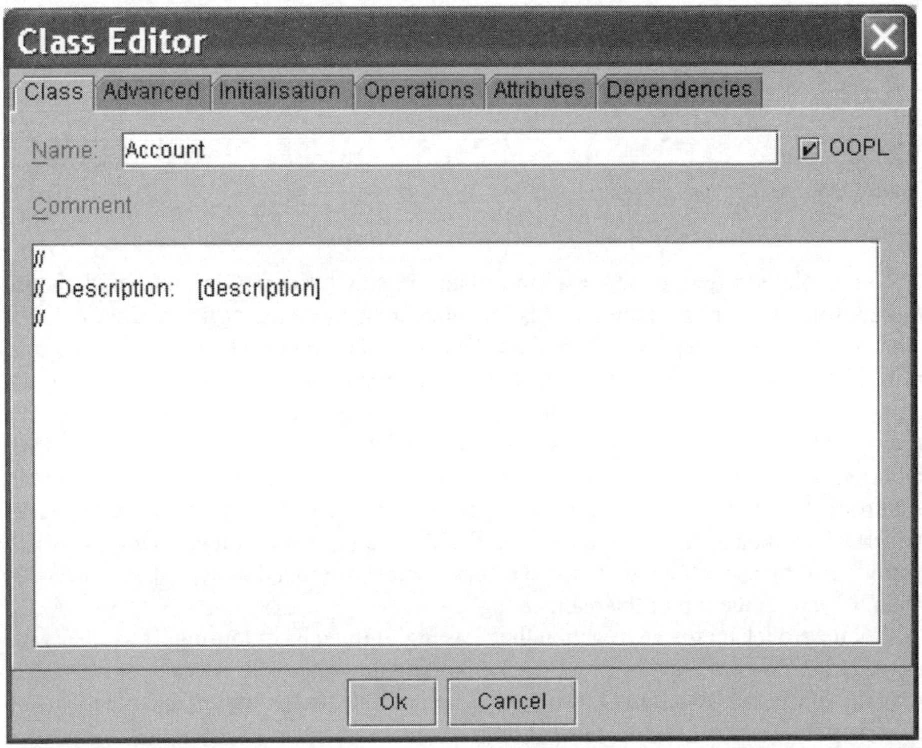

Figure B.8   *Class Editor dialog*

whole/part relationship (here, the class Account). Note that ROME has changed the mouse cursor to a crosshair to reflect the kind of operation that is underway. Press the left mouse button at this class and drag the mouse to the class acting as the whole in the whole/part relationship (Bank). Releasing the mouse results in ROME drawing the aggregation relation between these classes (see figure B.9). Restore the selection mouse by pressing the appropriate toolbar button.

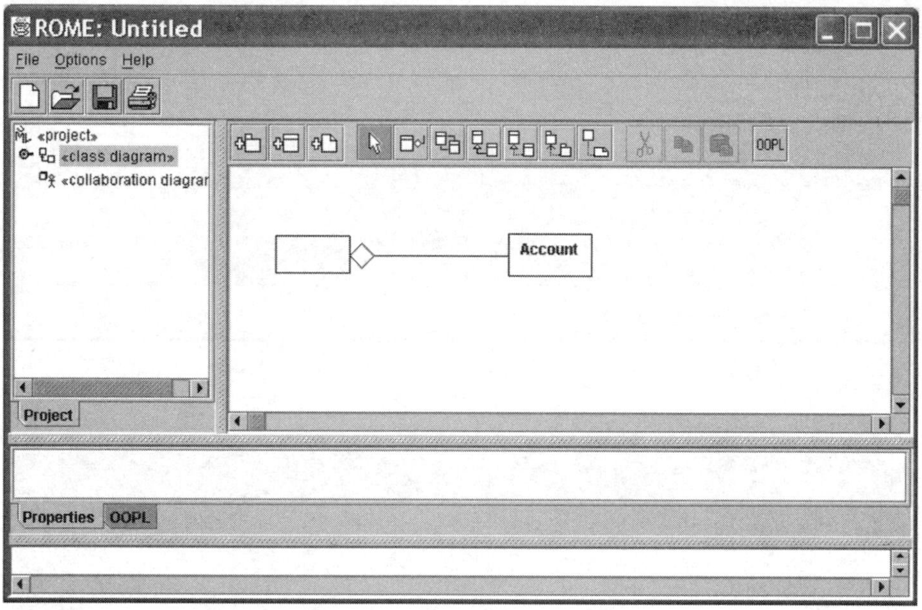

Figure B.9   *Adding an aggregation relation*

To complete this aggregation we must label it, attach the multiplicities and choose the appropriate implementation. This we enter with the Aggregation Editor dialog shown in figures B.10a and B.10b. First right click the mouse while over the aggregation relation and select Edit aggregation... from the context sensitive menu. The role name for this end of the aggregation is entered into the edit control labelled Role. The multiplicity is chosen from the list under Multiplicity, which is a fixed list of UML labels such as 0..* for many. The choice of implementation (here an ArrayList) is chosen from the list under the Implementation, again a fixed list of the standard Java containers. This dialog, like so many other ROME dialogs, is a tabbed dialog in which various editing aspects are distributed across a series of related dialogs. Each is chosen from the tabs at the top of the main dialog.

When we click on the aggregation line, it is highlighted by colouring. If we select the aggregation this way then by pressing and holding down the left mouse button we may drag the line either upward or downward. When we release the button the effect may be as shown in figure B.11. Note how ROME continues to retain the relationship between the two classes by extending the relationship in an appropriate manner.

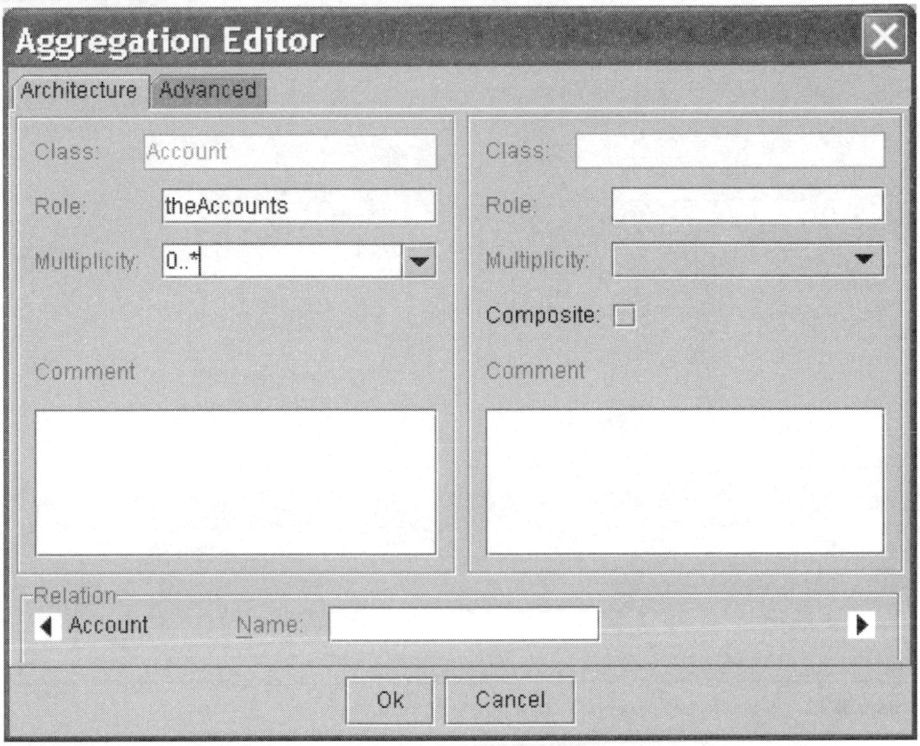

Aggregation Editor

Architecture | Advanced

Class: Account
Role: theAccounts
Multiplicity: 0..*

Comment

Class:
Role:
Multiplicity:
Composite: ☐

Comment

Relation
◀ Account   Name: _____   ▶

Ok   Cancel

Figure B.10a *Aggregation Editor dialog (Architecture tab)*

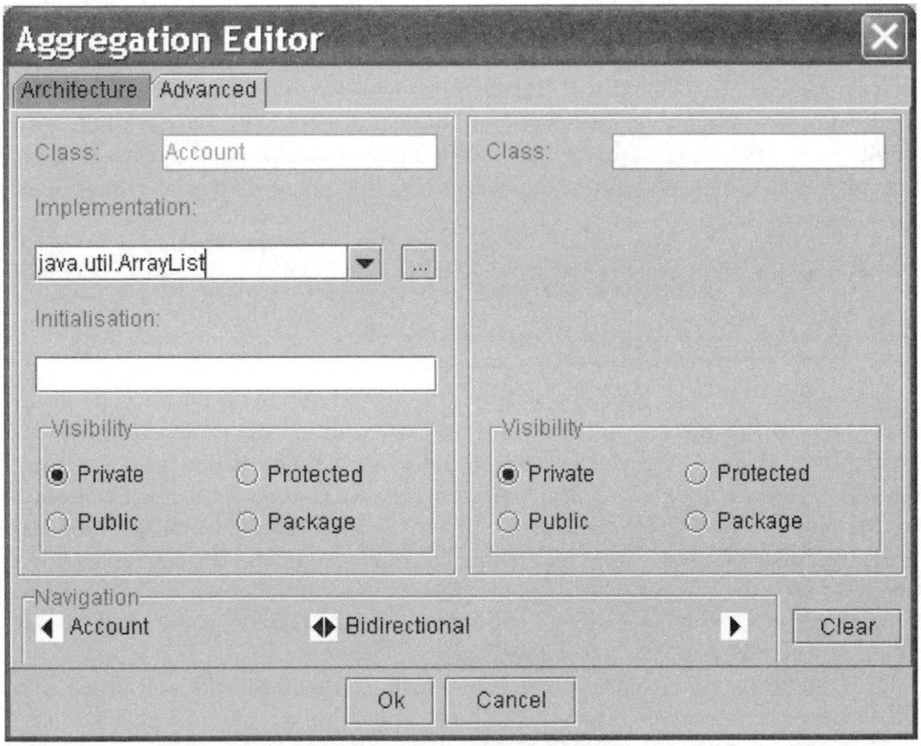

Aggregation Editor

Architecture | Advanced

Class: Account
Implementation:
java.util.ArrayList   ...
Initialisation:

Visibility
◉ Private   ○ Protected
○ Public   ○ Package

Class:

Visibility
◉ Private   ○ Protected
○ Public   ○ Package

Navigation
◀ Account   ◀▶ Bidirectional   ▶   Clear

Ok   Cancel

Figure B.10b *Aggregation Editor dialog (Advanced tab)*

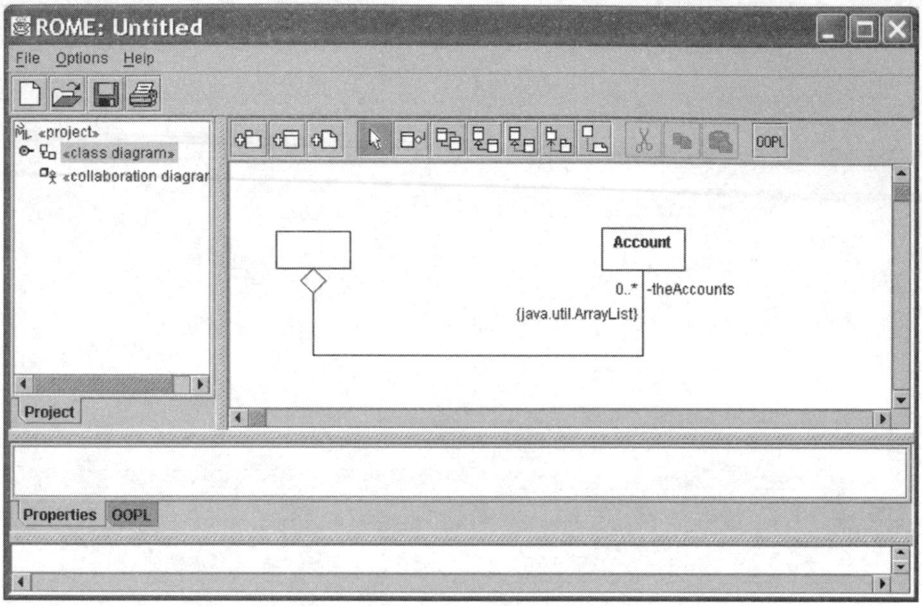

Figure B.11 *Moving and stretching the aggregation relation*

Aggregation and association relations can be recursive (see chapter 2). We achieve this in ROME by combining the corresponding tool with the CTRL button. For example, first select the association tool from the class diagram toolbar. Now with the CTRL key held down press and release the mouse left button above the class symbol. When the mouse is released ROME will put a recursive association into the class diagram. The relationship links the class symbol with itself. The relationship can be dragged and edited like any other.

B.4   Class attributes and operations

Now we shall enter the attributes and operations into the Account class. From the discussions of chapter 3, we know that the two attributes are theNumber and theBalance. To enter these details we invoke the Class Editor dialog we saw in figure B.8 and repeated in figure B.12. This time, however, we activate the dialog by positioning the mouse cursor above the Account class, clicking the right mouse button, and selecting Edit attributes... from the floating menu. Note this time the central list box in which we view the class attributes and the Add... button at the lower left of the dialog. Initially, of course, the list is empty. The button is used to obtain a second dialog into which we enter the new attribute theBalance.

Press the Add... button on the Class Editor dialog. This pops up the Attribute Editor dialog shown in figure B.13. Into the Name edit control we enter the attribute name theBalance and its type into the Type field. The type can be entered in a number of

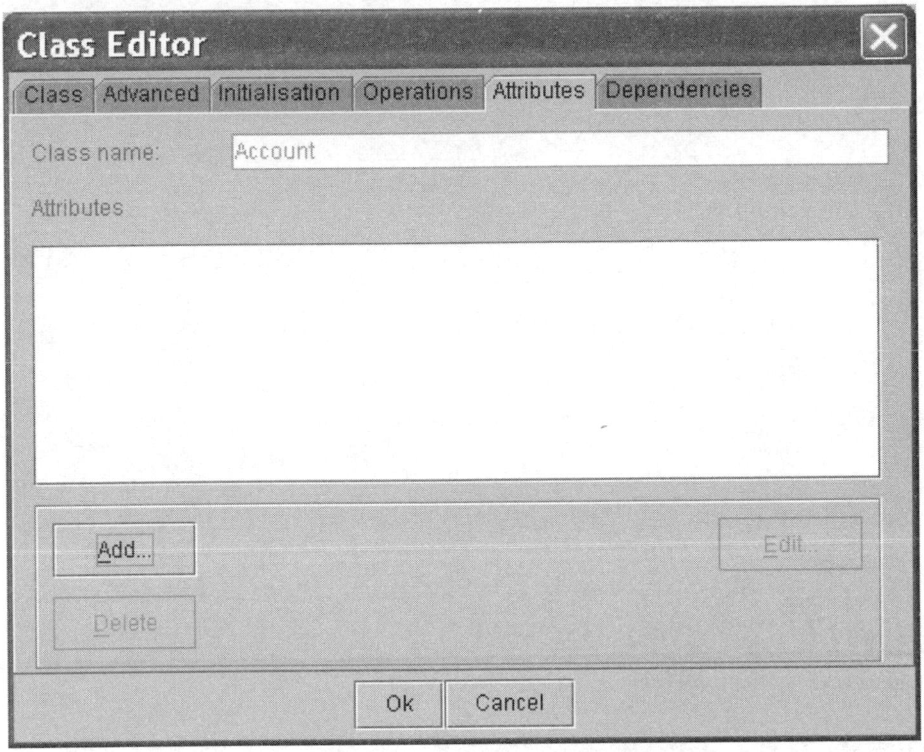

Figure B.12 *Class Editor dialog with empty attribute list and Add... button*

ways. We can enter the type name int directly into the field or we can select from the drop list of fundamental Java types.

We continue by now adding theNumber attribute. Again, press the Add... button from the Class Editor dialog (see figure B.12) and enter theNumber into the Name field. This time we wish this attribute to have the String type. Again, we can enter this directly into the field. We can also select from a list of Java types using the Class types dialog. Press the button labelled with the ellipsis (...) alongside the Type field then work your way through the packages to locate the correct class name. Figure B.14 shows the packages partially expanded. The String class is part of the lang package and we must open this and select the class name.

We can also give an attribute an initial value. Enter the Java expression into the Initial field on the attribute editor (figure B.13). Only the expression needs to be entered. ROME will complete the generated code by placing an assignment (=) between the attribute and the expression and a terminating semicolon (;) after the initialized declaration.

Also associated with an attribute is an initialization block. This comprises one or more Java statements entered into the field marked Initialisation block. Again ROME will code generate the necessary block markers (brace symbols {and}) around the statements. See chapter 7 for an example of this.

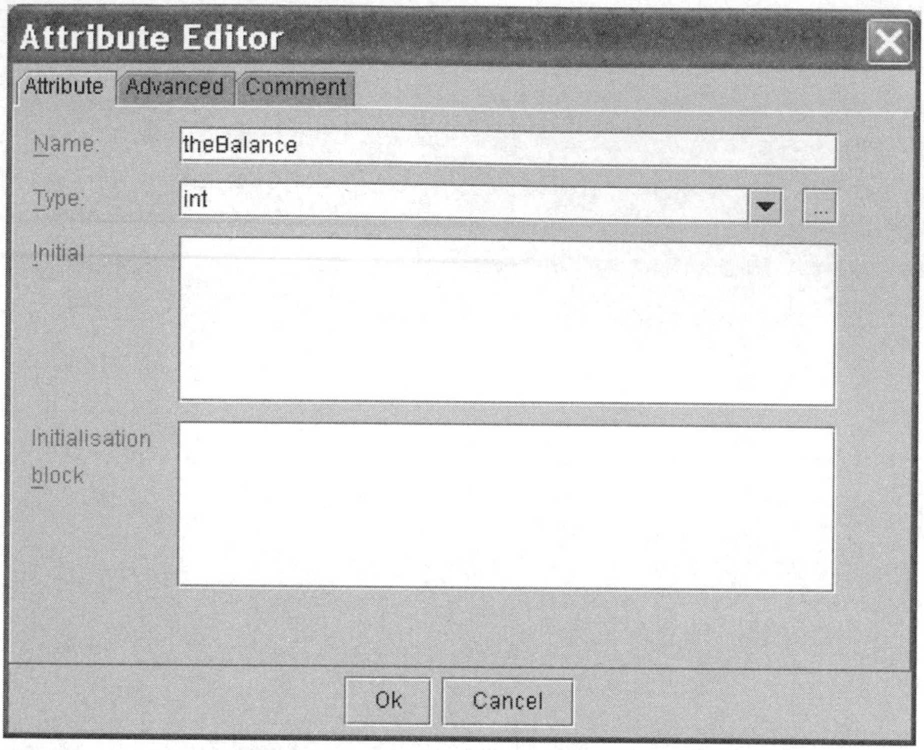

Figure B.13 *Attribute Editor dialog*

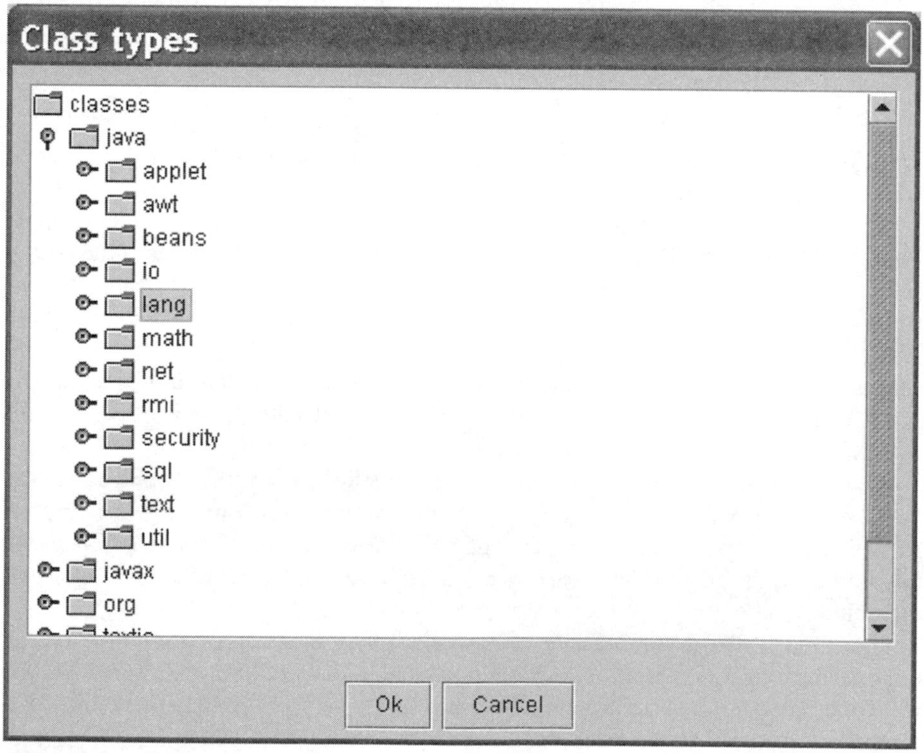

Figure B.14 *Class types dialog*

Having entered the representational attributes, the **Class Editor** dialog now appears as shown in figure B.15. The minus symbol prefix attached to each entry reminds us that the attribute is private (a plus symbol means public and an equal symbol means protected).

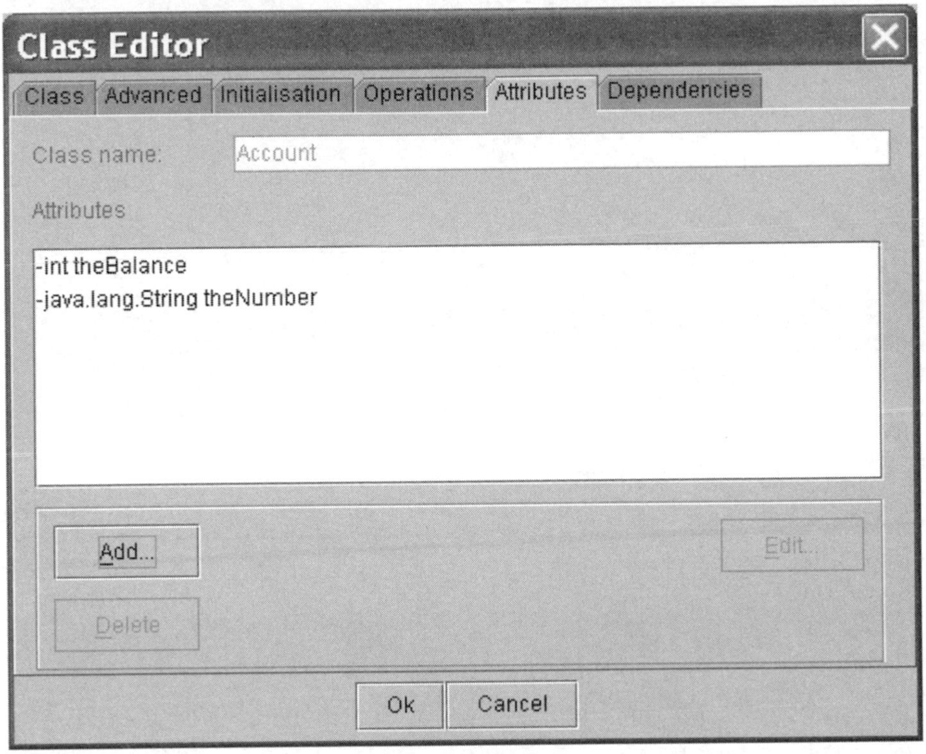

Figure B.15   *Class Editor dialog and attribute list*

We do much the same for the operations of the class **Account**. We first select the item **Edit Operations...** from the floating menu. The **Add** button brings up the **Operation Editor** dialog as shown by figure B.16. Into the Name text field enter the name for the operation. Enter the type of return value in the **Return type**. Like the type for an attribute we can set this value in a number of ways. We can enter the type name directly, or we can choose one of the fundamental types from the drop list. Finally, we can select a class name with the **Class types** dialog by following the ellipsis button alongside the **Return type** field.

Operations with parameters are introduced as shown in figure B.17. The centre of the **Operation Editor** dialog comprises a scroll list for the operation parameters. Each row in this list has a field for the **Name** and the **Type** of the parameter. Into the **Name** field provide the parameter name. The **Type** field is completed either by entering the type name directly, choosing a fundamental type from the drop list (appears automatically when the field is clicked), or by obtaining a class name from the **Class types** dialog by using the ellipsis button.

When editing an existing list of parameters, be sure to press the **RETURN** key when having completed editing either a **Name** or **Type** field. This is required to ensure that the correction is properly registered.

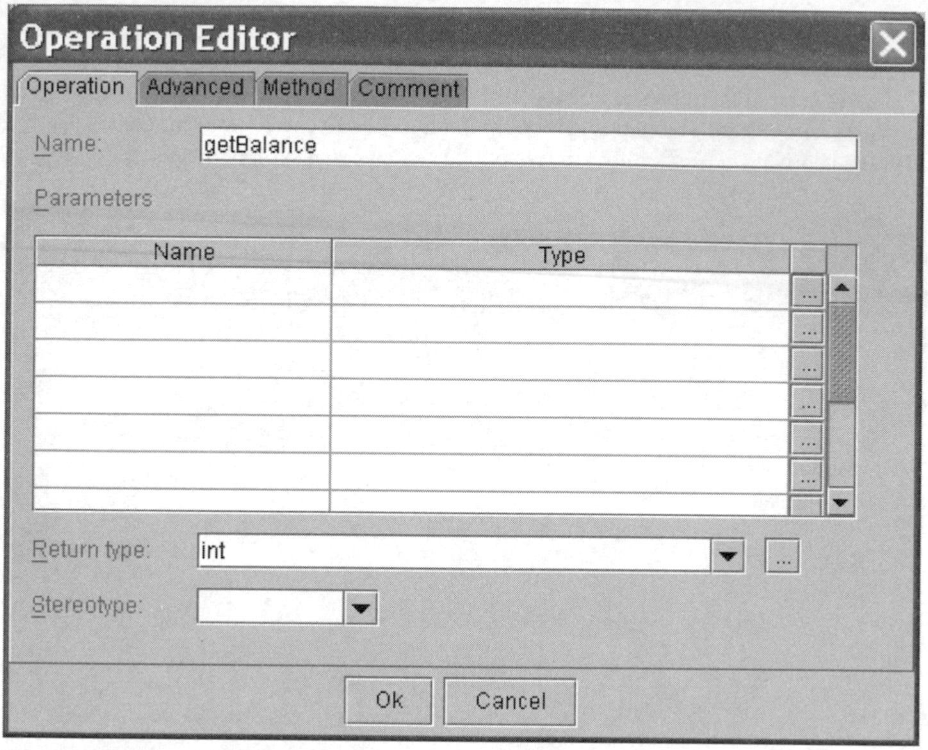

Figure B.16 *Operation Editor dialog*

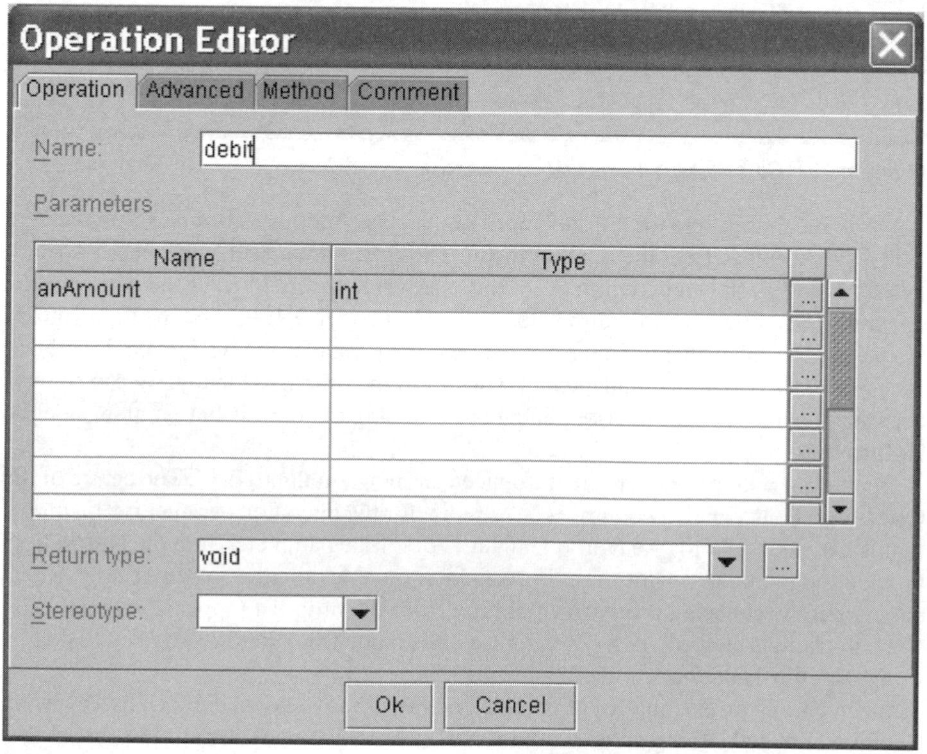

Figure B.17 *Operation parameters*

Note also the **Stereotype** combo box on these **Operation Editor** dialogs. You must choose **constructor** from this drop-list when introducing a constructor method into a class. Otherwise, the **ROME** code generator will handle constructor methods the same as all other methods.

The **Comment** tab on the **Operation Editor** dialog presents a template for documentation for use with the **javadoc** tool. The user replaces one or more templates enclosed with [and] with appropriate documentation. This can be used to name and describe the method, and to name and describe the method parameters and return value.

For any selected class in the model panel, the properties panel presents a list of all the features (attributes and operations) for that class (see figure B.18). This is useful for obtaining some details of the class without having to open the class dialog.

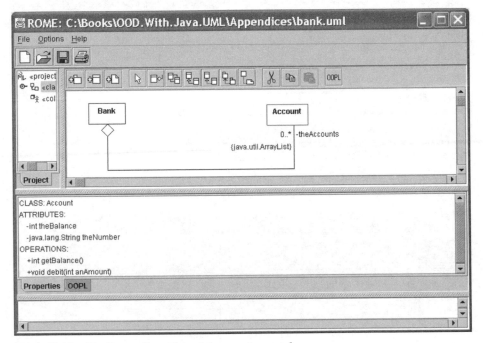

Figure B.18   *The Properties tab in the properties panel*

Notice how in figure B.18 the **Properties** tab in the properties panel is selected. If we select the **OOPL** (Object-Oriented Programming Language) tab, then for the selected class we obtain an on-the-fly presentation of the Java code for the class (see figure B.19). Observe that this generated code is not saved, it simply permits the developer to review the emerging Java code.

We can also see what features are associated with a class by requesting the class to show these details on the model. Place the mouse cursor over the **Account** class, click the right mouse button and select, say, **Show operations** from the floating menu. This would list the operations for that class. Additionally, if we repeat this and also select **Show signatures**, then further details for the operations are revealed. The effect is shown in figure B.20. Note how the effect is done on a per-class basis. Note also how the expanded class symbol fills the model and disrupts the arrangement of the model. Usually, some repositioning is needed to make sense of the diagram.

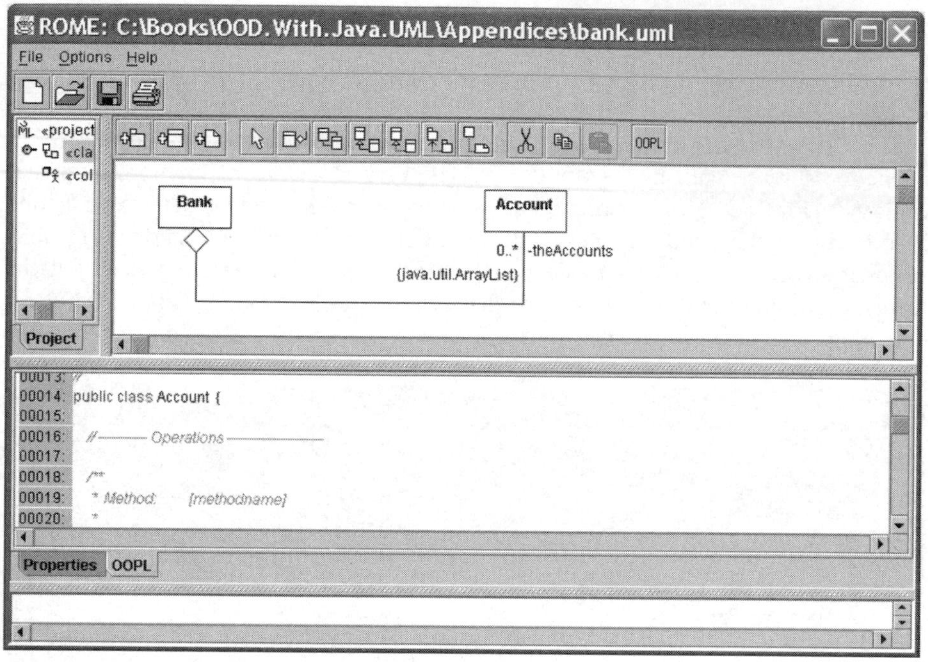

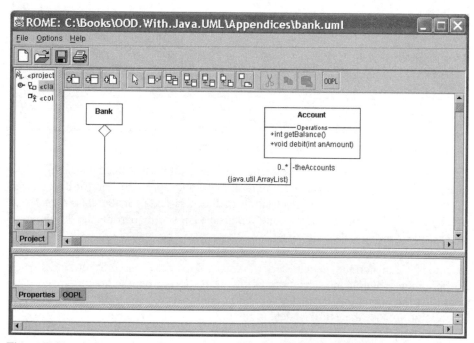

Figure B.19   *The OOPL tab in the properties panel*

Figure B.20   *A class with its Show operations and its Show signatures set*

The final work we have to do is fill the method bodies for the operations we have attached to the class. Right click the Account class and choose Edit operations... from the floating menu. From the list of classes select the one to edit and press the Edit button. This brings up the Operation Editor dialog as first shown in figure B.17. Now select the Method tab and complete the method body as shown in figure B.21. Observe how the operation name is repeated in a non-editable field at the top of the tab.

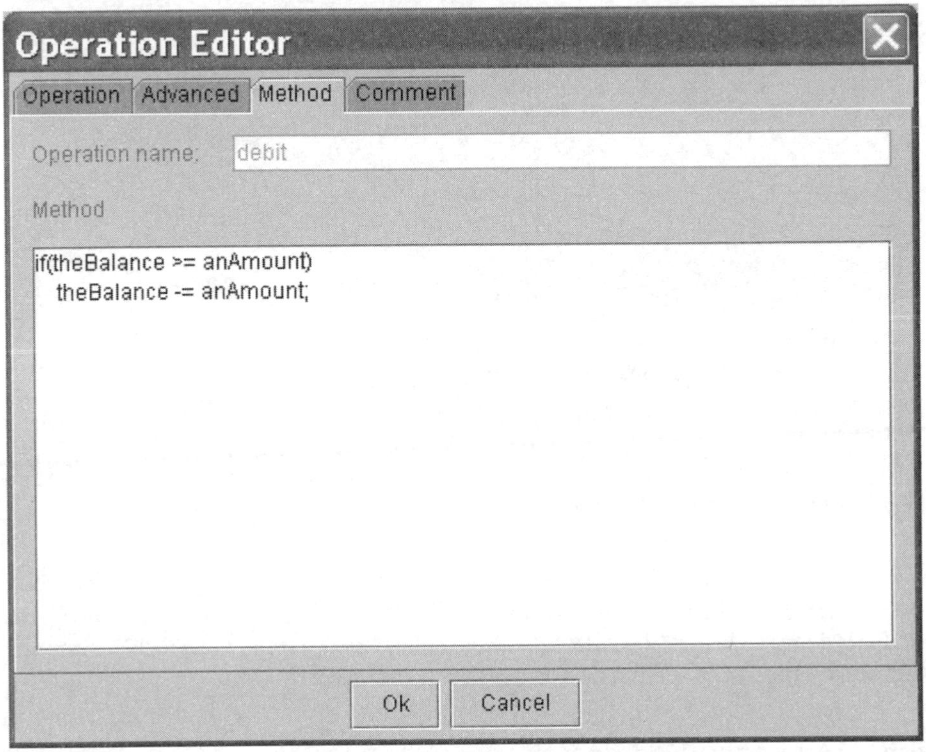

Operation Editor

Operation | Advanced | Method | Comment

Operation name: debit

Method

```
if(theBalance >= anAmount)
    theBalance -= anAmount;
```

Ok   Cancel

Figure B.21 *Method tab*

B.5   Code generation

When we have completed the class diagram, ROME offers us a code generation facility. We can have ROME produce the generated Java code. The button on the toolbar labelled OOPL activates this service. The service involves producing one or more Java files for each class in the model. To use this facility, we must first nominate the directory in which they will be placed. Before we use these buttons we set the directory to the current working directory with the Preferences... command on the Options menu. The result is the Preferences Editor dialog shown in figure B.22.

The important field on this dialog is the Output Directory and specifies where all files are placed. Typically we set this as the same directory in which the model resides but it can be placed anywhere (here the directory is c:\demonstrator). We may edit this field directly, or use the Browse... button to select the desired directory.

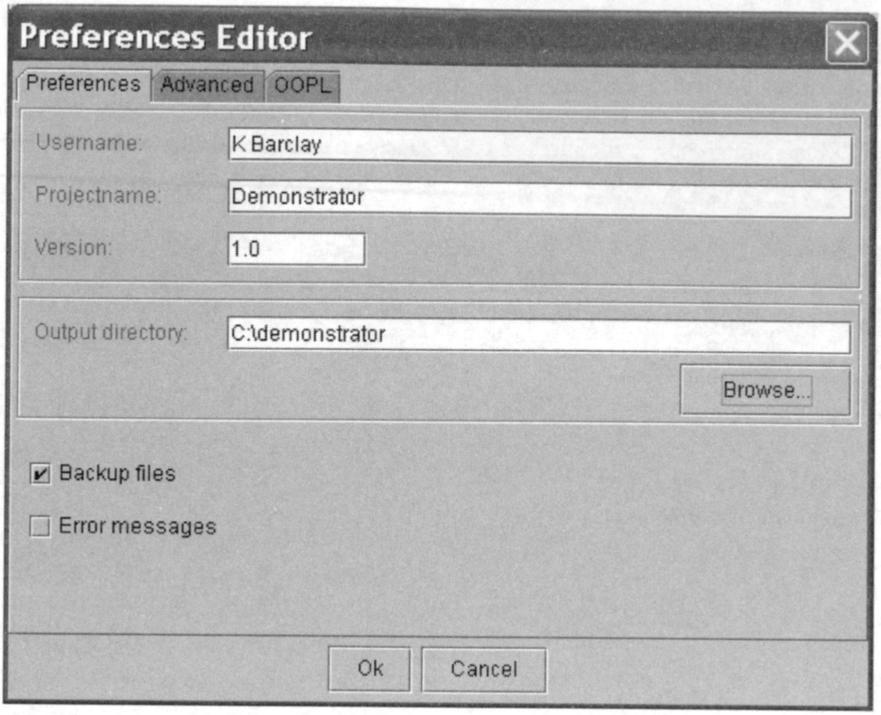

Figure B.22   *Preferences Editor dialog*

If we press the OOPL button on the toolbar, then ROME creates Java code files for each class in the model. Thus we would have the files Bank.java, Account.java etc. ROME also produces a Java code file called Main.java containing the Java main method that creates an Application object and sends it the run message (see chapter 3).

B.6   Packages and nested classes

Class diagrams can include packages as we saw in the final iteration of the library case study from chapter 4. Classes with nested inner classes were also introduced in chapter 7 when we discussed event handlers. Both of these are treated in a similar manner in ROME.

In figure B.23 a package has been introduced into a new class diagram and decorated with the name librarysubsystem. The package is given this name using the Package Editor, much like a class name is introduced.

Importantly, observe how the «class diagram» entry in the project panel has an expand/collapse button to its immediate left. The button has a horizontal tail indicating that this part of the project tree is currently collapsed. If we press this button and expand the tree we obtain figure B.24.

By selecting the «librarysubsystem» entry in the project tree we effectively open that package. As we see in figure B.25 a package in a class diagram can consist of the usual elements that comprise a class diagram. The toolbar reveals that a package can be an assembly of classes with relations and even other nested packages.

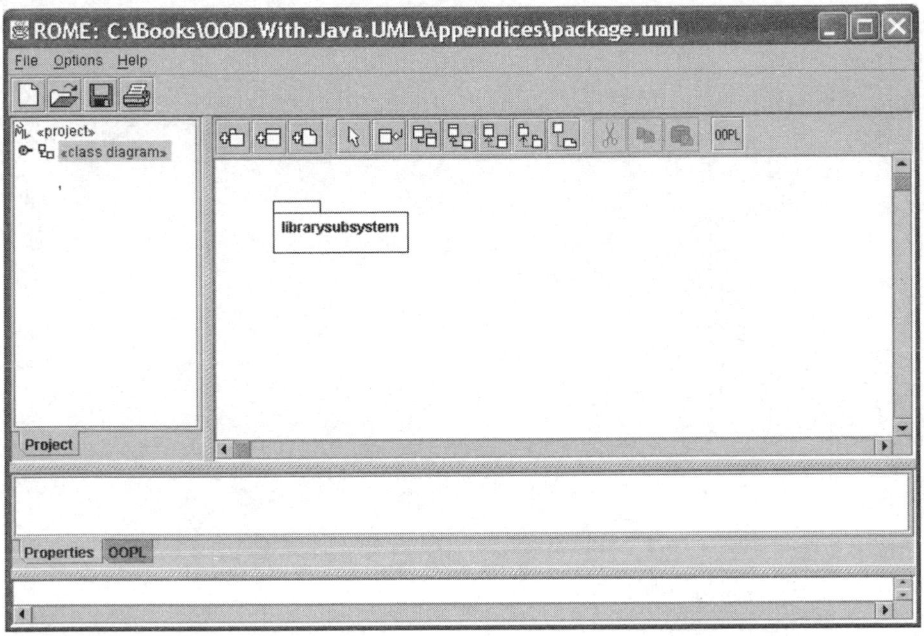

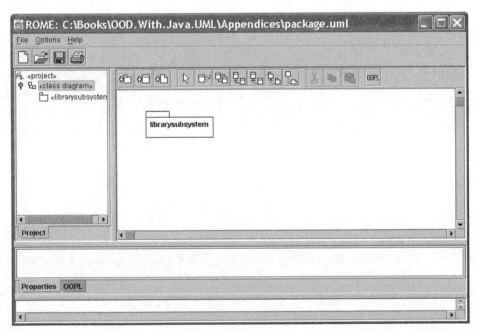

Figure B.23 *A package*

Figure B.24 *Expanded class diagram*

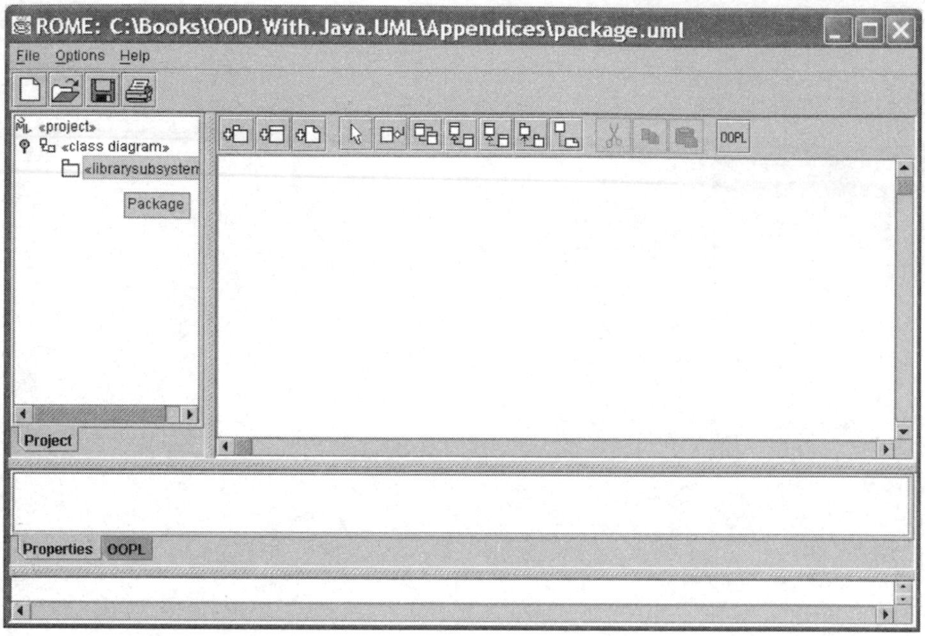

Figure B.25   *An opened package*

We have a similar arrangement when we include a class in a class diagram. In figure B.26 we have a class LibraryFrame in the model panel with a corresponding entry in the project tree.

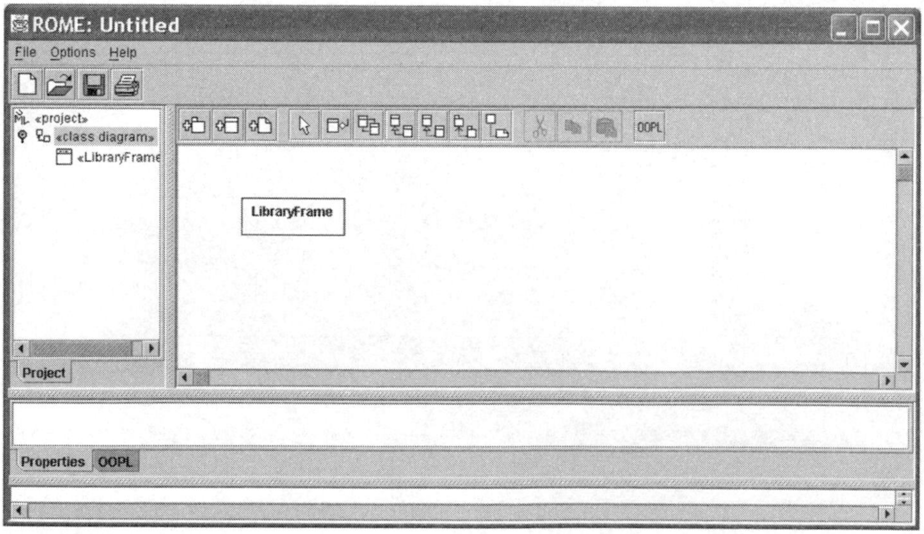

Figure B.26   *A class symbol*

Again, when we select the LibraryFrame entry in the project tree, we open the class as shown in figure B.27. Observe how in this figure ROME has disabled adding a package to a class, compared to figure B.25 in which a package can be a nested element for another package. In figure B.27 the toolbar button to add an inner class, such as WindowClosingAction, remains enabled.

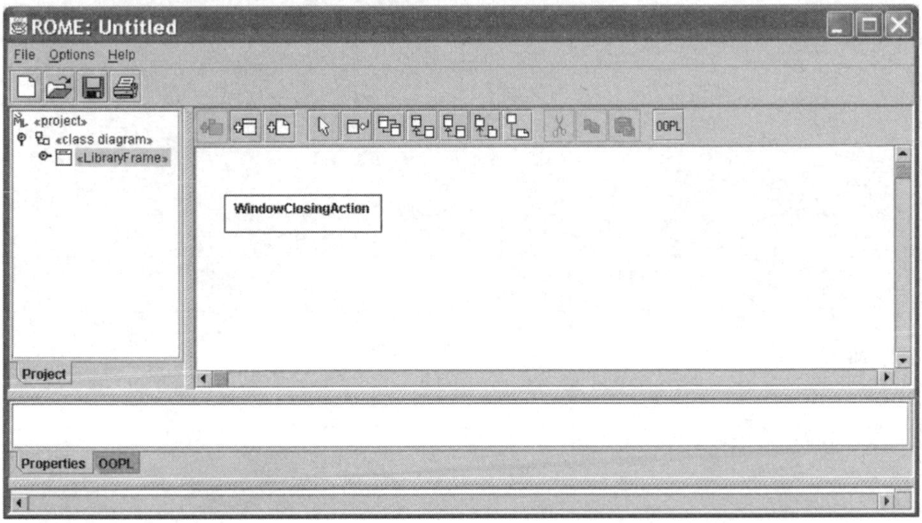

Figure B.27   *An opened class*

B.7   Other features

All ROME diagrams have a group of buttons to cut, copy and paste elements from a diagram. One or more model elements are first selected. The elements can then be cut or copied into the ROME clipboard using these toolbar buttons. The items in the clipboard can then be pasted into the same diagram or into another diagram. When we are pasting into another diagram, ROME will only permit this where the same diagram types are involved.

An individual model element is selected in the normal way. A number of model elements from a diagram can be selected using the selection tool by pressing and holding the left mouse button over an unoccupied part of a diagram and by dragging the mouse to a new part of the diagram. As the mouse is dragged an elastic rectangle grows and surrounds the items that are selected.

We can also cut, copy and paste text in any of the text fields such as a class name or a method body. All or parts of the text are selected with the mouse. The highlighted text can then be cut with CTRL-X and copied with CTRL-C. This text, now in the clipboard, can be pasted into another text field with CTRL-V.

ROME is also capable of compiling any class in a class diagram. To do so we must first specify the necessary directories using the OOPL tab from the Preferences Editor dialog.

In figure B.28 the three fields classpath, directory and sourcepath have been established. The classpath field is the search path used to locate program files during compilation. If we use the Browse... button repeatedly then each entry is appended to

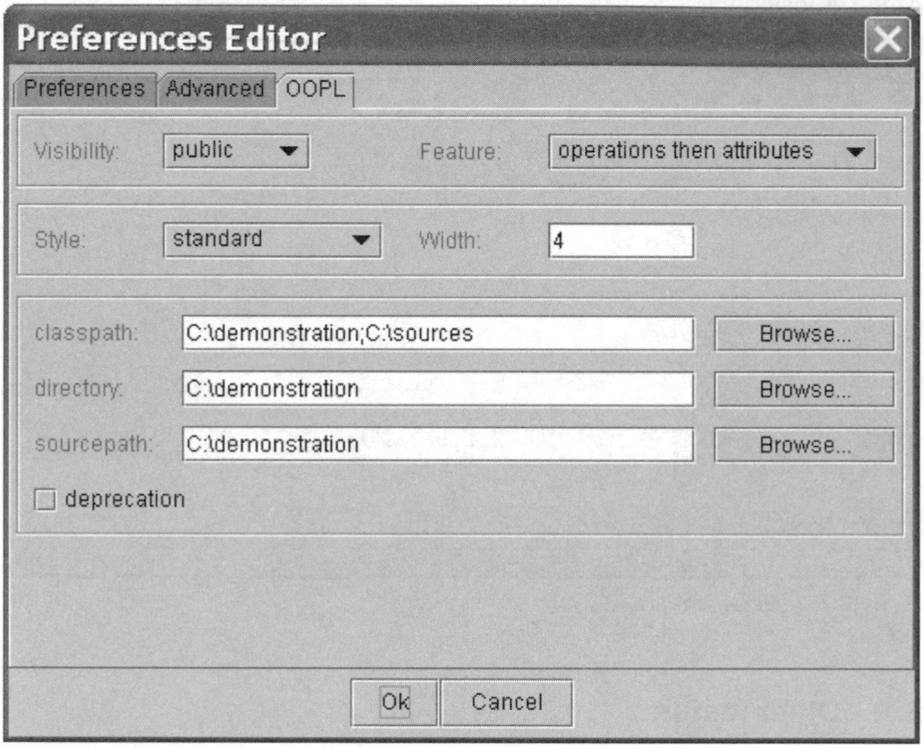

Figure B.28   *Setting the OOPL directories*

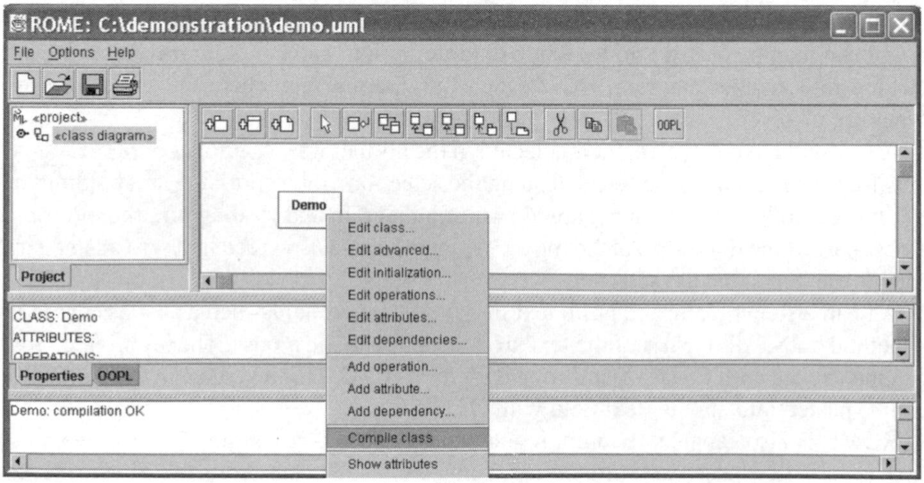

Figure B.29   *Compiling a class*

the existing entries with a semicolon separator. The directory field specifies the root
destination directory into which all compiled files are stored. This root directory will
contain other subdirectories if the class diagram contains packages. The sourcepath
field identifies the directory or directories in which the source code files reside. Again,
the Browse... button forms a list separated with semicolons.

To compile a particular class we select the menu item Compile from the context sensitive menu as shown in figure B.29. Any success message or any error messages appear
in the message panel. If necessary, we can remove the entries in the message panel by
issuing a right mouse click over this panel and selecting Clear from the floating menu.

Appendix C

Package textio

The input/output (IO) facilities of Java are provided by classes located in the java.io package. This is an especially sophisticated suite of classes assembled using a number of advanced design patterns. Unfortunately, the advanced designs are not suitable for the novice. To this end we have developed a relatively simple package, textio, suitable for console and file IO. The interfaces and classes in this package are described by the class diagram shown in figure C.1.

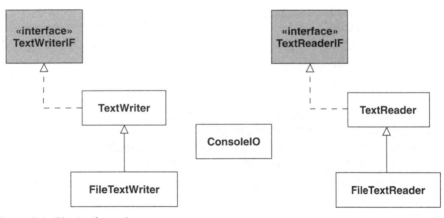

Figure C.1 *The textio package*

C.1 Output

The Java interface is an indispensable tool when designing software systems. It allows us to specify the behaviour of an object without worrying about how it is achieved. Another way of stating this is that the interface specifies a protocol for all classes that implement it.

For example, an analysis of our output requirements reveals that we need to be able to output language types such as **int**s, **boolean**s and Strings as character sequences. Our output requirement then leads us to a TextWriterIF interface as shown below in which there are overloaded print and println methods. Some have formatting parameters. We have the guarantee that any class that implements this interface must support them. Its declaration is as follows:

```
package textio;
public interface TextWriterIF {
    // ------ group #1 ------
    public abstract void    print(String text);
    public abstract void    print(char ch);
    public abstract void    print(int number);
    public abstract void    print(double number);
    public abstract void    print(boolean bool);

    // ------ group #2 ------
    public abstract void    println();
    public abstract void    println(String text);
    public abstract void    println(char ch);
    public abstract void    println(int number);
    public abstract void    println(double number);
    public abstract void    println(boolean bool);

    // ------ group #3 ------
    public abstract void    print(String text, int width);
    public abstract void    print(char ch, int width);
    public abstract void    print(int number, int width);
    public abstract void    print(double number, int width, int fraction);
    public abstract void    print(boolean bool, int width);

    // ------ group #4 ------
    public abstract void    println(String text, int width);
    public abstract void    println(char ch, int width);
    public abstract void    println(int number, int width);
    public abstract void    println(double number, int width, int fraction);
    public abstract void    println(boolean bool, int width);
} // interface: TextWriterIF
```

The print methods are categorized into groups 1 and 3 as shown in the listing. The first group displays the standard types unformatted, while those in the third group have additional formatting values. Those in the second and fourth groups mirror, respectively, those in groups 1 and 3, but additionally issue a newline symbol after having completed the output. The methods in groups 3 and 4 display the values formatted according to a field width. For example, for values of type **int**, the third method of group 3, displays the value right justified in a field width specified by the second parameter. For values of type **double** (see fourth method in group 3) we also specify the number of decimal places required.

The TextWriter class implements the interface TextWriterIF. Its outline class declaration is:

```
public class TextWriter implements TextWriterIF {
    // ------ Operations ------------
    public    TextWriter() { ... }
```

```
// Implementations for all of the operations in TextWriterIF
// ...
} // class: TextWriter
```

The class ConsoleIO includes a **static** member named out of the class TextWriter. It represents an output stream to the system console:

```
public class ConsoleIO {
    public static final TextWriter  out = new TextWriter();
    // ...
} // class: ConsoleIO
```

Assuming a suitable **import** statement we can deliver output to the command console with the following code samples:

```
int age = 21;
ConsoleIO.out.print("My age is:");
ConsoleIO.out.println(age);
```

This produces the following output (blanks in the output are emphasized with the symbol):

My age is:21

Formatted printing is achieved with:

```
double height = 1.6764;          // metres
ConsoleIO.out.print("My height is:");
ConsoleIO.out.println(height, 8, 2);     // total field width of 8, 2 decimals
```

producing:

My height is: 1.68

C.2   Input

An analysis of our input requirements reveals that we need to be able to input character sequences and form the various types such as Strings, **char**s, **int**s, **double**s and **boolean**s. One solution is to have a suitably named operation for each. For example, we might have the operation readInt to convert character input to an **int**. The TextReaderIF interface lets us impose a protocol based on this approach on any class that implements it. Its declaration is as follows:

```
public interface TextReaderIF {
    public abstract String          readString();
    public abstract String          readLine();
    public abstract char            readChar();
    public abstract int             readInt();
    public abstract double          readDouble();
    public abstract boolean         readBoolean();
} // interface: TextReaderIF
```

The method readString ignores any leading whitespace characters then reads a series of characters up to the next whitespace symbol. The methods readInt, readDouble and readBoolean ignore any leading whitespace then form the required input value from the subsequent character sequence. Any unexpected input, such as a letter when reading an **int** or **double**, stops further input and returns the value read. The method readChar reads the next available character, including whitespace characters. The method readLine reads a series of characters up to and including the newline symbol. The latter is not part of the returned String.

Input from the system console is provided by the static member in of the class ConsoleIO (see above). This TextReader class implements the interface TextReaderIF.

```
public class ConsoleIO {
    public static final TextReader  in = new TextReader();
    public static final TextWriter  out = new TextWriter();
} // class: ConsoleIO
```

Assuming a suitable **import** statement sample usage is:

```
ConsoleIO.out.print("Please enter your surname: ");
String surname = ConsoleIO.in.readString();
ConsoleIO.out.print("Your name is: ");
ConsoleIO.out.println(surname);
```

C.3 Files

Text files are human readable files suitable for text processing with utilities such as Microsoft NotePad. As shown in figure C.1, the textio package has two classes to support input and output from them. File handling uses the classes FileText Reader and FileTextWriter.

The FileTextReader is remarkably straightforward as shown in the following declaration:

```
import java.io.*;
```

```
public final class FileTextReader extends TextReader {

    // ------ Operations ------------
    public   FileTextReader(String fileName) throws IOException {...}
} // class: FileTextReader
```

As it extends TextReader then it inherits the various read operations. Therefore all that the constructor need do is any initialization it requires to read from a named file. Note that it can throw an exception belonging to the class IOException (a subclass of Exception in the package java.io). Therefore if we use this constructor then it must be in a **try** block.

Assuming that we have visibility of all of the classes in the textio package and the class IOException with:

```
import textio.*;
import java.io.*;
```

then a typical code fragment is:

```
try {
    String name = null;
    int age = 0;
    //
    FileTextReader data_in = new FileTextReader( "data_in.txt" );
    //
    while( data_in.isEOF() == false ) {
        name = data_in.readString();
        age = data_in.readInt();
        ConsoleIO.out.println("The age of " + name + " is " + age);
    }
}
catch(IOException e) {
    ConsoleIO.out.println("Cannot open data_in.txt");
}
```

Here we create a FileTextReader initialized to read from the file named data_in.txt. The method isEOF reports when the end of the file has been reached. Therefore the while loop displays the String-int pairs in the file until no more are left. In this simple example the exception handler just reports that the file cannot be opened. Specifically it does not use the IOException object passed to it.

As might be expected the FileTextWriter class is similar in that it inherits the functionality of the TextWriter, i.e. the various overloaded print and println operations. All its constructor need do is any initialization required to write to a named file. An outline class declaration is:

```
import java.io.*;

public final class FileTextWriter extends TextWriter {

    // ------ Operations -----------
    public    FileTextWriter(String fileName) throws IOException { ... }
} // class: FileTextWriter
```

Note that as with the FileTextReader, the constructor for the FileTextWriter class can throw an IOException. Therefore, it must be used in a try block. Again assuming we have the necessary import statements, an example of it in use is:

```
try {
    FileTextWriter data_out = new FileTextWriter( "data_out.txt" );
    String name = null;
    int age = 0;
    //
    name = ConsoleIO.in.readString();
    age = ConsoleIO.in.readInt();
    while( name.equals("XXX") == false ) {
        data_out.println("The age of " + name + " is " + age);
```

```
            name = ConsoleIO.in.readString();
            age = ConsoleIO.in.readInt();
        }
    }
    catch(IOException e) {
        ConsoleIO.out.println("Cannot open data_out.txt");
    }
```

The code accepts a String-int pair from the keyboard until XXX is input as the String. It copies them with some explanatory text to a file name data_out.txt.

Appendix D

UML Notation and
Java Bindings

In this appendix we catalogue the essential elements of the UML diagrams supported by ROME. We also provide a mapping from our designs into Java. We specify the bindings between our designs (expressed as UML class diagrams) and the corresponding Java class declarations.

D.1   UML notation

Class diagram

A UML *class diagram* is a static structure diagram of the application classes and the relationships formed between them. The application can be partitioned by recursively nested *packages*. Classes can also exhibit a nested structure with *inner classes*. The relations found in a class diagram include association, aggregation and specialization. Documentation in the form of a note can also be found in a class diagram.

A package is used to group model elements or subsystems. A package is a nested namespace, permitting elements with the same name to exist in different packages. A package is rendered as a tabbed folder with, optionally, the names of subordinate packages and classes shown within it, as depicted by figure D.1. The convention is to use only lower-case characters for the package name.

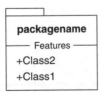

Figure D.1   *A package exploded to reveal its features*

A class defines the state and behaviours for objects of that class. An object's state is defined by the class attributes and an object's behaviour by the methods of the class. Attributes have a name and a type. Java supports class-level (static) attributes. A method has a name, none or more method parameters and a return type. Method

335

parameters also have a name and a type. Methods can be deferred (**abstract**) or class-level (**static**). Methods can also be tagged as **final**, indicating that they cannot be redefined in a subclass.

A class can be marked as abstract, containing none or more abstract methods. A class with only abstract methods and, possibly static attributes, is described as an interface. A class tagged as **final** cannot be subclassed. A sample class is shown in figure D.2.

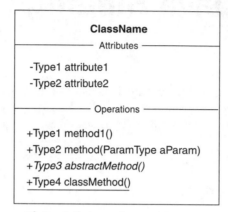

Figure D.2 *Sample class with its attributes and operations*

The permissible relations in a class diagram are association, aggregation and specialization. Association and aggregation relations are used to capture the architectural links between instances of the classes involved in the relationship. Two forms of aggregation are permitted: composite aggregation and shared aggregation. The former emphasizes a strong coupling between object instances of the whole and its parts. Shared aggregation weakens this coupling.

Specialization introduces the notion of subclass and superclass. A subclass inherits all the features (attributes and operations) of its superclass. The subclass may include additional features or redefine inherited operations. The term implementation is used where the superclass is an interface and the subclass is either abstract or concrete.

Figure D.3 shows two classes with a shared aggregation relationship. The relationship is that of a one-to-many, which is realized using an ArrayList collection.

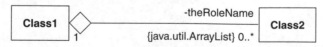

Figure D.3 *Shared aggregation between classes*

In figure D.4 the concrete class SubConcrete is related to the two classes SuperInterface and SuperConcrete. The class SubConcrete is shown to implement the interface SuperInterface and extend the concrete superclass SuperConcrete.

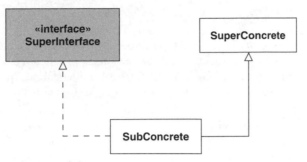

Figure D.4   *Specialization of classes*

Collaboration diagram

A UML *collaboration diagram* is used to present the dynamics of a part of a system. It comprises a set of object instances, the relations between them, and the messages that flow between the objects. The messages are numbered to reveal the order in which they occur. The association, composite or shared aggregation can be used as the relations. Each relation represents a link between objects. Each object is an instance of the class to which it belongs and will have the behaviours and attributes as defined by that class.

Figure D.5 presents a collaboration diagram with a single instance of a Bank object and three instances of Account objects. The Account objects are shown as aggregate sub-components of the Bank object. The three Account objects are distinguished by the identifiers ac1, ac2 and ac3. In the diagram the Account object ac3 has been exploded to reveal its state consisting of a bank account number ABC123 and a balance of 1200 monetary units. The diagram also shows the Bank object issuing the message getBalance, in turn, to each Account object.

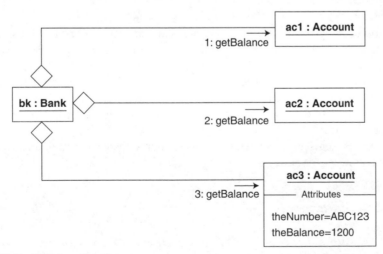

Figure D.5   *Collaboration diagram*

Sequence diagram

A UML *sequence diagram* is also used to present the dynamics of a part of a system. It comprises a set of object instances and the messages that flow between them. The objects are decorated with a lifeline in which time is understood to pass as we move down the diagram. The ordering of the messages is according to the passage of time as we read the diagram from top to bottom. Each object is an instance of the class to which it belongs and will have the behaviours and attributes as defined by that class.

Figure D.6 presents a sequence diagram with a single instance of a Bank object and three instances of Account objects. The diagram shows the Bank object issuing the message getBalance, in turn, to each Account object.

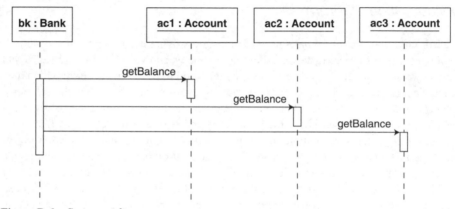

Figure D.6   *Sequence diagram*

Object diagram

A UML *object diagram* is used to present the static structure of a part of a system. It comprises a set of object instances and the relations between them. The association, composite or shared aggregation can be used as the relations. Each relation represents a link between objects. Each object is an instance of the class to which it belongs and will have the behaviours and attributes as defined by that class. In effect, an object diagram is a collaboration diagram without the messages. It is intended to reveal the architecture of the objects.

Figure D.7 presents an object diagram. It is similar to that shown in figure D.5 except there are no messages shown.

Use-case diagram

A UML *use-case diagram* is used to present an external view of the functionality required by a system. Each use-case will describe a single functional requirement of the system. All the use-cases describe the entire system.

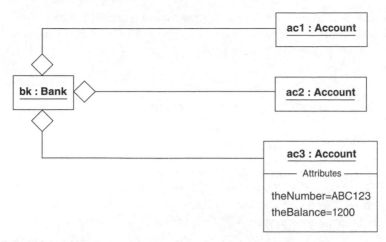

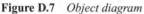

Figure D.7 *Object diagram*

A use-case diagram is shown in figure D.8. It consists of a use-case, an actor and an interaction. Figure D.8 shows a bank teller interacting with a computer system to open a new account for a customer.

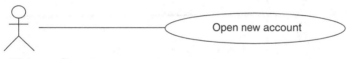

Figure D.8 *Use-case diagram*

Activity diagram

A UML *activity diagram* is often better known as a flow diagram or flowchart. It represents how some process is performed. It breaks the process into various steps or activities with, perhaps, decision points and loops. Figure D.9 shows an activity diagram in

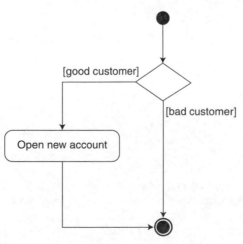

Figure D.9 *Activity diagram*

which a new account is only created for good customers. It comprises one activity (Open new account), one decision symbol (the diamond), a number of transitions as we proceed through the process, and start and stop symbols.

D.2   UML to Java class declarations

Concrete class

A *concrete class* represents a class from which object instances can be created. Concrete class features are its attributes (Java properties) and method definitions. Each feature is tagged with its visibility indication. Attributes can be initialized, finalized (constant) or static (class property). All operations of a concrete class have a method definition. Finalized methods cannot be redefined in a subclass. Static methods can only reference static features (attributes or methods). The class itself can be specified as **final**, indicating that it cannot be subclassed. The following example (and subsequent examples) illustrate class diagrams and associated Java code.

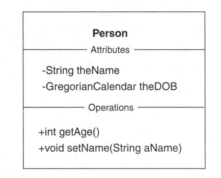

```
public class Person {

    public int                      getAge() { ... }
    public void                     setName(String aName) { ... }

    private String                  theName;
    private GregorianCalendar       theDOB;

} // class: Person
```

Abstract class

An *abstract class* is characterized by having deferred methods. A class may be abstract as a result of inheriting an abstract method from an abstract superclass (or an interface) that the class does not redefine. An abstract method cannot be **static** nor can it be **final**.

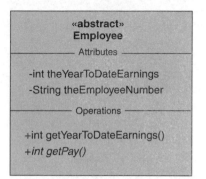

```
public abstract class Employee {
    public int                      getYearToDateEarnings() { ... }
    public abstract int             getPay();
    private int                     theYearToDateEarnings;
    private String                  theEmployeeNumber;
} // class: Employee
```

An interface

An *interface* is a class that specifies only **abstract** methods. An interface may include attributes but they must all be class variables (**static**). Like an abstract class we cannot create instances of an interface. By default, all features of an interface are **public** and all methods are **abstract**.

```
        «interface»
          DateIF
        Operations
    +int getDay()
    +int getMonth()
    +int getYear()
    +boolean isLeap()
```

```
public interface DateIF {
    public abstract int             getDay();
    public abstract int             getMonth();
    public abstract int             getYear();
    public abstract boolean         isLeap();
} // class: DateIF
```

One-to-one association (or aggregation)

An *association* (or *aggregation*) with multiplicity one is realized by a reference to an object of the associate (aggregate) class. This architectural feature can, like others, include a visibility tag.

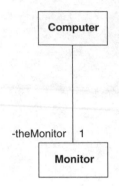

```
public class Computer {
    // ...
    private Monitor              theMonitor;
} // class: Computer
public class Monitor {
    // ...
} // class: Monitor
```

One-to-many *association (or* aggregation*)*

In a one-to-many *association* (or *aggregation*) an object of one class carries the references to many objects of the second class. Usually, the many object references are managed by a suitable collection.

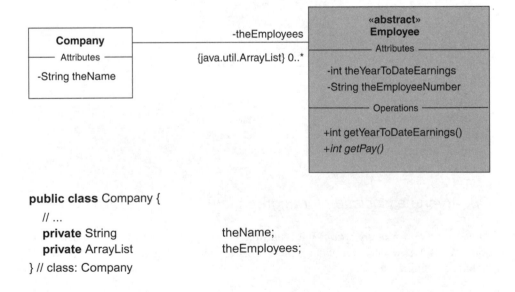

```
public class Company {
    // ...
    private String               theName;
    private ArrayList            theEmployees;
} // class: Company
```

```
public abstract class Employee {
    public int                       getYearToDateEarnings() { ... }
    public abstract int              getPay();

    private int                      theYearToDateEarnings;
    private String                   theEmployeeNumber;
} // class: Employee
```

Specialization from a concrete class

A concrete class can operate as the superclass for another concrete class or for an abstract class. The subclass is introduced as an *extension* of the superclass introducing additional features and/or redefinitions of inherited methods.

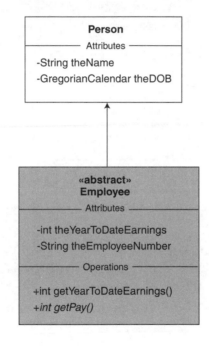

```
public class Person {
    // ...
    private String                    theName;
    private GregorianCalendar         theDOB;
} // class: Person
```
```
public abstract class Employee extends Person {
    public int                        getYearToDateEarnings() { ... }
    public abstract int               getPay();

    private int                       theYearToDateEarnings;
    private String                    theEmployeeNumber;
} // class: Employee
```

Specialization from an abstract class

An abstract class may act as the superclass for a concrete class or for another abstract class. A concrete subclass must provide definitions for all inherited abstract methods. An abstract subclass results from having its own abstract methods or from inheriting, without redefinition, abstract methods. Again, the subclass is said to *extend* the superclass.

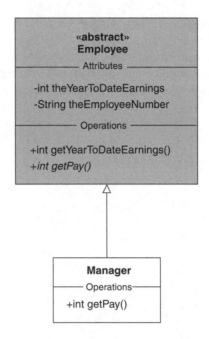

```
public abstract class Employee {
    public int                          getYearToDateEarnings() { ... }
    public abstract int                 getPay();

    private int                         theYearToDateEarnings;
    private String                      theEmployeeNumber;
} // class: Employee

public class Manager extends Employee {
    // ...
    public int                          getPay() { ... }
} // class: Manager
```

Specialization from an interface

An interface, an abstract class and a concrete class can be subclassed from an interface. An interface subclass is an extension of an interface superclass, simply adding further abstract methods to the interface. An abstract subclass will have one or more abstract methods either through inheritance from the interface or by the introduction of additional abstract methods. A concrete subclass will define all the abstract methods inherited from the interface. An abstract or concrete subclass is said to *implement* the interface.

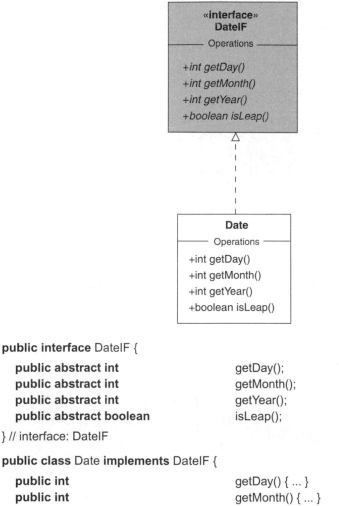

```
public interface DateIF {

    public abstract int            getDay();
    public abstract int            getMonth();
    public abstract int            getYear();
    public abstract boolean        isLeap();
} // interface: DateIF

public class Date implements DateIF {

    public int            getDay() { ... }
    public int            getMonth() { ... }
    public int            getYear() { ... }
    public boolean        isLeap() { ... }
} // class: Date
```

Specialization from multiple superclasses

Java does not support multiple inheritance. A subclass may have at most one superclass (either an abstract superclass or a concrete superclass). A class may implement any number of interfaces. An interface may also extend any number of other interfaces.

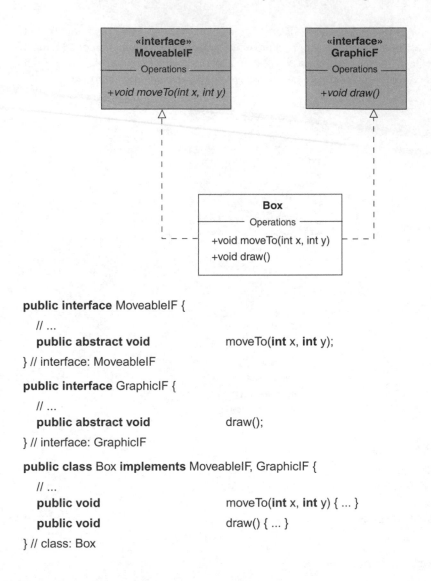

```
public interface MoveableIF {
    // ...
    public abstract void            moveTo(int x, int y);
} // interface: MoveableIF
```

```
public interface GraphicIF {
    // ...
    public abstract void            draw();
} // interface: GraphicIF
```

```
public class Box implements MoveableIF, GraphicIF {
    // ...
    public void                     moveTo(int x, int y) { ... }
    public void                     draw() { ... }
} // class: Box
```

Appendix E

The Java Collections Framework

This appendix demonstrates how the Java Collections Framework (JCF) is used to implement the one-to-many architectural relationships identified in an OOAD. It is not intended to be a definitive account. Interested readers should consult specialist sources such as http:/developer.java.sun.com/developer/online-Training/collections/ and Watt 2001.

We begin with a discussion of the design strategy used in the construction of the JCF. It should give the reader an insight into its architecture and so make it easier to deploy. It also serves as a good illustration of an industrial-strength framework that consolidates the material presented in chapter 6. Advice on the criteria for the selection of a collection follows. Finally we discuss some of the issues associated with using the JCF.

E.1   The architecture of the JCF

In object-oriented systems development there is a common requirement to have an object that acts as a *collection* of other objects. The intention is to treat the entire collection of those objects as a single entity. Typically it is used for their storage and retrieval. For example, we might use a collection to hold electronic mail messages or the details of each student studying OOAD.

The JCF is a unified architecture for representing and manipulating collections. Its design is based on the principle that an interface is partially implemented by an abstract class then fully implemented by a concrete class. It means that several concrete classes can benefit from having the same superclass. Figure E.1 illustrates that part of the JCF architecture that pertains to a particular kind of collection known as the set.

The Collection interface acts as the specification for any collection. As such it advertises methods common to all collections. The Set interface that specializes it acts as the specification for all collections that are sets. The AbstractCollection class implements those methods in the Collection interface applicable to all collections. Its subclass AbstractSet implements the remaining methods applicable to all sets.

Each concrete class now completes the implementation by implementing any methods that are specific to it. For example, the add method is implemented by the HashSet class so that its elements are added in no particular order. However, the TreeSet

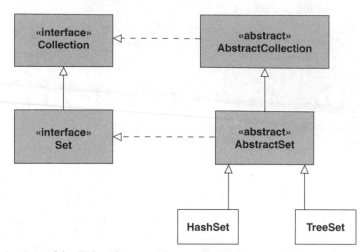

Figure E.1 *Part of the JFC architecture that pertains to the set collection*

provides a different implementation so that its elements are ordered. Both must also provide their respective constructors.

Notice how this elegant approach separates specification from implementation with an interface and a class hierarchy respectively. Many of the other interfaces, abstract classes and concrete classes that make up the JCF are structured similarly. They are all located in the java.util package.

It has been our experience that the advantages that come from using the JCF are enormous. It is no exaggeration to state that it radically changes the manner in which we develop our software. For example, collection objects make object-oriented systems easier and faster to develop. They also make them easier to understand. There is also the added bonus that the resulting software is more portable, reliable and efficient than would otherwise have been the case.

E.2   Choosing a Collection class

Although there are several different kinds of collection objects available, two are particularly useful. The first is the *set* and the second the *list*. The set is similar to that found in mathematics. It is a group of unique items or elements. The fact that the elements in the set must be unique is important because it forbids any duplicates. On the other hand the list is a sequence of elements that does allow duplicates. It also supports the positional indexing of its elements. This means that each element has its location in the list defined by a positive number known as its index. The purpose of the index is to allow access to an element that occupies a specific position in the list.

An important principle of object orientation is that one object should be independent of the internal workings of another. The essential idea is that a client should only rely on the specification of an operation as advertised by the receiver. To put it another way, the client should not depend on its method, i.e. its implementation. This supports the flexibility required to change a method's implementation with no unforeseen effects on clients.

In the case of a collection object, the choice of which one to use is not just determined by whether it is a set or a list but also how it is implemented. The reason is that a collection's implementation can often affect critical design criteria. For example, we might choose a collection with an implementation that minimizes the amount of memory used or one that retrieves an element faster than another. Note that there is no dependency on the collection's implementation. It is just that we choose one rather than the other because of our particular design requirements. This means that knowledge of a collection's implementation and its impact on the rest of the software is important.

The set supplied with the standard Java environment is implemented using either a *hash table* or a *tree*. Both are traditional data structures. Respectively, these are the Java classes HashSet and TreeSet. The hash table holds the location of each element of a HashSet to permit efficient access. Thankfully the details of how the hash table works need not concern us. In contrast the tree holds each element of a TreeSet in a particular sort order. Again we are not concerned with how this is actually achieved.

The list also has two implementations, namely the ArrayList and the LinkedList. The first uses an array and the second a linked list structure. Again, both are traditional data structures. The ArrayList holds its elements as a contiguous sequence, i.e. one follows the other in memory. Usually an ArrayList is the better choice when there is a need to access each element in the list at random locations. However, this is with the proviso that there are no (or at least a minimum of) insertions or removals from any place other than the end of the ArrayList. The reason for the first is that an element can be found rapidly with an index as it is mapped directly to a memory location. The reason for the second is that because each element in an array is adjacent in memory, an insertion or removal anywhere other than at the end must involve moving existing elements from one location to another. Obviously this can be time consuming.

With a LinkedList an element (often called a node) can be located anywhere in memory. This avoids rearranging elements but requires that each element is connected to the next by a link. A LinkedList is often the better choice when elements are added or removed from intermediate locations within the list. The reason is that there is no rearrangement of the elements required. An element's index is not directly mapped to a memory location. Therefore the LinkedList is not a good choice when the efficiency of random access is important.

Table E.1 summarizes the main attributes of the Java collection classes we commonly use. It can be used in determining the best design choice.

The UML default collection is the set. Therefore, we use the HashSet as our default Java implementation.

Table E.1   *A summary of the main attributes of the Java Collection classes*

Collection	JCF class	Description
set	HashSet	Unordered and no duplicates; fast insertion and retrieval
set	TreeSet	Ordered with no duplicates; fast insertion and retrieval
list	ArrayList	Positional ordering with duplicates; unsuitable for general insertions/deletions
list	LinkedList	Positional ordering with duplicates; suitable for general insertions/deletions

E.3   Using the JCF

There is no doubt that the JCF is a complex software artefact. However, so long as a few relatively simple rules are followed, it is not especially difficult to use. This section details how we employ the JCF in the implementation of one-to-many aggregations and associations identified in an OOAD.

E.3.1   Adding an element to a collection

Creating a collection object is the same as for any other object. For example, the declaration statement:

Collection accounts = **new** HashSet();

creates a HashSet object of a default size and

Collection accounts = **new** HashSet(10);

a HashSet initialized to hold ten elements.

The identifier accounts references the HashSet object created. Notice that accounts is declared as a Collection reference not a HashSet reference. This is good practice as it means accounts may reference any object whose class implements the Collection interface. We describe Collection as the *interface type* and HashSet as the *implementation type*.

Our four Java collection classes (HashSet, TreeSet, ArrayList and LinkedList) all support the method add. Its purpose is to allow a client to populate a collection with a reference to a particular object. For example, as the HashSet has no ordering of its elements the addition of an element is straightforward. We just add it to the HashSet with the operation:

boolean add(Object element)

As each collection is dynamically resizable we don't have to concern ourselves with memory allocation issues. It will just increase its size as necessary.

Each element in a collection is a reference to an object of the class Object. This is important as polymorphic substitution (see chapter 5) means that we can add a reference to any object of any descendant class of Object. As all Java classes are descended from Object, then it means that a reference to any Java object can be added to a collection. (This does not include the Java primitives such as **int** and **char**. In this case a wrapper class such as Integer and Character must be used.)

Notice that the add operation returns a **boolean** value. It is used to inform a client of the success or failure of the addition. For example, if we have:

```
public class Account {
    // ...
    }
```

and:

```
Collection  accounts = new HashSet();
Account    ac1 = new Account(...);
```

then we might have:

```
if(accounts.add(ac1) == true)
    // Take some action knowing that the addition was successful
else
    // Take some action knowing that the addition was not successful
```

We can also ignore the return value with:

```
// Ignore the boolean return value
accounts.add(ac);
```

With a LinkedList or ArrayList each element is part of a sequence. Therefore there is a positional index that can be used to locate an element. Although there are several (overloaded) add methods available, we normally use only the add method applicable to all collections, i.e. the one that does not involve a positional index. It just adds the element to the end of the list. This gives us the flexibility we require to change the implementation type with minimal impact on our code.

The only real difficulty is with the addition of an element to a TreeSet. The reason is that it must be added in some specified order. This means that the TreeSet must be able to compare the element to be added with those already present.

Focusing on the important question of the comparison of objects, if we have a collection of Integer objects then there is no problem. It is clear what the outcome of a comparison of one Integer with another should be. It is just that expected from arithmetic. Similarly for String values there is a natural ordering of English language Strings based on a *lexicographic comparison*. It relies on the fact that each character is represented in the computer as a numeric code.

When two Strings are compared it is actually a comparison of the numeric codes of the characters in the Strings. For example, John is less than Ken as the first character of John comes before the first character of Ken in the numeric coding sequence. Similarly john is greater than John as lower-case characters have higher numeric codes than upper-case ones. Finally Johnny is greater than John as it has two extra characters.

In fact all Objects that belong to standard Java classes have a natural ordering predefined. Our problem is that we need allow for the comparison of objects that belong to user-defined classes, e.g. the class Account. Clearly the Java development team could not anticipate what it means to compare objects of some arbitrary user-defined class. For example, we might compare two Accounts according to their account number or their balance or some other criteria. The point is that we must make the decision of what the basis for the comparison is.

Fortunately the situation is not quite so bleak as the previous discussion might suggest. The reason is that the Java development team designed the collections in such a way that the operation:

```
public int  compareTo(Object obj)
```

is used to compare two elements in a collection.

As with the add method its formal parameter is a reference to an Object thereby making it generally reusable. It returns a value of 0 if the object that receives the message compareTo is equal to the object referenced by the actual parameter, a negative value if it is less than it and a positive value if it is greater. For example, assuming the declaration statements:

```
String s1 = new String("John");
String s2 = new String("Ken");
String s3 = new String("Johnny");
```

and the three expressions:

```
s1.compareTo(s1)
s1.compareTo(s2)
s3.compareTo(s1)
```

The first returns 0, the second a value less than 0 and the third a value greater than 0. The collections can use this behaviour in selection statements such as:

```
if( s1.compareTo(s2) == 0 ) {
    // Take some action knowing that the Strings are equal
}
else if(s1.compareTo(s2) < 0 ) {
    // Take some action knowing that the first String is less than the second
}
else {
    // Take some action knowing that the first String is greater than the second
}
```

To add an object of a user-defined class to a TreeSet, all we have to do is to define a method for compareTo and the problem is solved. For example, we might have:

```
// ...
public class Account implements Comparable {
    // ------ Operations ----------
    public    Account(String aNumber, int aBalance) {
        theNumber = aNumber;
        theBalance = aBalance;
    }   // method: Account

    public String    getNumber() {
        return theNumber;
    }   // method: getNumber
    public int    compareTo(Object obj) {
        Account  acc = (Account) obj;
        String accountNumber = acc.getNumber();
        //
        return theNumber.compareTo(accountNumber);
    }   // method: compareTo
    // ...
```

```
// ----- Attributes ----------
   private final String   theNumber;
   private int   theBalance;
}   // class: Account
```

Notice that the Account class implements the interface Comparable (located in the java.lang package). As part of its sorting algorithm, a collection casts its elements to reference a Comparable object not to the actual class they belong to. The reason is that the collection only sends the compareTo message declared by the Comparable interface. Therefore it does not need to know the true class of each element.

In the Account class, the comparison is actually based on the operation compareTo defined for the class String. This is common practice as it limits the amount of work we have to do. Had the attribute used for the comparison been declared as **int** we might have had:

```
// class: Account
   public int   compareTo(Object obj) {
      Account  acc = (Account) obj;
      int accountNumber = acc.getNumber();
      int result;
      //
      if(theNumber < accountNumber) {
         result = -1;
      }
      else if(theNumber == accountNumber) {
         result = 0;
      }
      else {
         result = +1;
      }
      //
      return result;
}   // method: compareTo
```

or just:

```
// class: Account
   public int   compareTo(Object obj) {
      Account  acc = (Account) obj;
      int accountNumber = acc.getNumber();
      //
      return theNumber - accountNumber;
}   // method: compareTo
```

Also the attribute theNumber that is the basis for the comparison is qualified as **final** and is set by the constructor. This is intentional as it eliminates the risk that the attribute will be changed at a later date, perhaps invalidating the original comparison.

E.3.2   Mandatory profile considerations

Any object held by a TreeSet must have a method for compareTo otherwise a run-time error will result. Even though it is not required by other collections it is wise to insist that it is always available for any object held in a collection. The danger is that in a working system, if the collection used is changed to a TreeSet then an unexpected problem will result. It may be difficult to trace and fix.

It turns out that two other methods, namely equals and hashCode, are also used internally by the JCF. The first compares two elements for equality while the second delivers a number used for storage and retrieval of an element. Both are originally defined in the class Object. Again, they may not be strictly necessary for a particular collection but it is wise to include them.

Taking these considerations into account we insist that the class of any object to be held in a collection must have the following form:

```
public class ClassName implements Comparable {

    public int  compareTo(Object  obj) { ... }
    public boolean  equals(Object  obj) { ... }
    public int hashCode() { ... }

    // ...

} // class: ClassName
```

This is part of its mandatory profile. For example, for the Account class we have:

```
public class Account implements Comparable {
    public int    compareTo(Object obj) {
      // As in the previous section
    }   // method: compareTo

    public boolean    equals(Object obj) {
      boolean result  = this.compareTo(obj) == 0;
      //
      return result;
    }   // method: equals

    public int    hashCode() {
      return  theNumber.hashCode();
    }   // method: hashCode

    // ----- Attributes ----------
    private final String    theNumber;
    // ...

}   // class: Account
```

Notice that the method equals is defined in terms of compareTo. Although this might not always be the case, for our purposes it does bring consistency to our code. Therefore two Accounts that are equal occupy the same position when sorted. Similarly the method for hashCode uses the same **final** attribute as compareTo. This means that two equal Accounts have the same hash code.

E3.3   Traversing a collection

Clearly the Object references held in a collection are not directly available to a client. This poses a problem when there is a need to traverse a collection. Therefore the collections have a method iterator that delivers an Iterator (located in the java.util package) reference to a client. The idea is that an Iterator has privileged access to the internal workings of the collection. For example, it supports the methods:

// Return true if there is another element in a collection to visit otherwise return false
boolean hasNext()

and:

// Return the next Object in the collection
Object next()

Consider a bank that holds many accounts. We model it as shown in figure E.2.

Figure E.2   *A bank holding many accounts*

Outline code for the Bank class is:

```
import java.util.*;
import textio.*;

public class Bank {
   // ----- Operations ----------
   public    Bank(String aName) {
      theName = aName;
      //
      // Architecture initialisation.
      theAccounts = new TreeSet();
   }  // method: Bank
   //

   public void   openAccount(String aNumber, int aBalance) {
      Account acc = new Account(aNumber, aBalance);
      //
      // Establish the architecture between the bank and this new account.
      theAccounts.add(acc);
   }   // method: openAccount

   public void   displayAllAccounts() {
      ConsoleIO.out.println();
      ConsoleIO.out.println("All account details for: " + theName + "\n");
      //
      Iterator iter = theAccounts.iterator();
```

```java
      while(iter.hasNext()) {
         Account  acc = (Account) iter.next();
         acc.display();
      }
   }   // method: displayAllAccounts
   public void   displayAnAccount(String aNumber) {
      ConsoleIO.out.println();
      ConsoleIO.out.println("Single account details for: " + theName + "\n");
      //
      // Find the account with the given number and display it
      Iterator iter = theAccounts.iterator();
      while(iter.hasNext()) {
         Account acc = (Account)iter.next();
         String  number = acc.getNumber();
         if(aNumber.equals( number)) {
            acc.display();
            break;
            }
         }
   }   // method: displayAnAccount
   public int   getTotalAssets() {
      //   Form a running total of the balances for each account
      int totalAssets = 0;
      //
      Iterator iter = theAccounts.iterator();
      while(iter.hasNext()) {
         Account acc = (Account)iter.next();
         totalAssets + = acc.getBalance();
      }
      return totalAssets;
   }   // method: getTotalAssets
   // ..
   // ----- Attributes ----------
   private java.lang.String    theName;
   // ----- Relations ----------
   private java.util.Collection   theAccounts;   // of Account
   }   // class: Bank
```

Notice how the methods displayAllAccounts, displayAnAccount and getTotalAssets have the same form. Essentially it can be summarized by the following pseudo code (see appendix H):

```
FOREACH element IN theCollection DO
   get the next element
   use the element
ENDFOREACH
```

that maps to the Java code:

```
Iterator iter = theCollection.iterator();
while(iter.hasNext()) {
    ElementClass element = (ElementClass) iter.next();
    // send messages to element
}
```

For example, we have:

```
// class Bank
public void   displayAllAccounts() {
    ConsoleIO.out.println();
    ConsoleIO.out.println("All account details for: " + theName + "\n");
    //
    Iterator iter = theAccounts.iterator();
    while(iter.hasNext()) {
        Account  acc = (Account) iter.next();
        acc.display();
    }
}   // method: displayAllAccounts
```

Notice that the Iterator delivers an Object reference. Therefore it must be cast to reference an Account if a message such as display is sent. The reason is that display is not declared in the Object class. Also we have initialized the architectural attribute theAccounts to reference a TreeSet. Therefore the display prints each Account in its sort order, i.e. in order of increasing account number.

E.3.4   Removing an element from a collection

The Iterator object used to traverse a collection can also be used to remove an element. For example, in the Bank class we have:

```
// class Bank
    public void   removeAnAccount(String aNumber) {
        //   Find the account with the given number and remove it from the bank
        Iterator iter = theAccounts.iterator();
        while(iter.hasNext()) {
            Account acc = (Account)iter.next();
            String  number = acc.getNumber();
            if(aNumber.equals(number)) {
                iter.remove();
                break;
            }
        }
    }   // method: removeAnAccount
```

E3.5　Putting it all together

Program E.1 (ProgE_1.uml) in the software supplied demonstrates the use of JCF in the bank example. Complete listings are as follows.

Program E.1　Using the Java Collections Framework (model ProgE_1.uml)

```java
public class Main {

public static void    main(String[] args) {
   Application app = new Application();
   app.run();
      } // method: main

   }   // class: Main

public class Application  {

   // ----- Operations ----------
   public void    run() {
      // Open new bank
      Bank bk = new Bank("Object Bank");
      //
      //  Open three new accounts
      bk.openAccount("GHI789", 2000);
      bk.openAccount("ABC123", 1200);
      bk.openAccount("DEF456", 1000);
      //
      //  Display all accounts details
      bk.displayAllAccounts();
      //
      //  Display a single account's details
      bk.displayAnAccount("ABC123");
      //
      // Obtain total assets
      ConsoleIO.out.println("Total assets: " + bk.getTotalAssets());
      //
      // Remove an account
      bk.removeAnAccount("DEF456");
      //
      //  Display all accounts details
      bk.displayAllAccounts();
      //
      // Obtain total assets
      ConsoleIO.out.println("Total assets: " + bk.getTotalAssets());
      }   // method: run
   }    // class: Application

   import java.util.*;
   import textio.*;
```

```java
public class Bank  {
  // ----- Operations ----------
  public   Bank(String aName) {
    theName = aName;
    theAccounts = new TreeSet(); // a HashSet, ArrayList or LinkedList could also be used
  }   // method: Bank

  public void   openAccount(String aNumber, int aBalance) {
    Account acc = new Account(aNumber, aBalance);
    theAccounts.add(acc);
  }// method: openAccount

  public void   displayAllAccounts() {
    ConsoleIO.out.println();
    ConsoleIO.out.println("All account details for: " + theName + "\n");
    //
    Iterator iter = theAccounts.iterator();
    while(iter.hasNext()) {
      Account  acc = (Account) iter.next();
      acc.display();
    }
  }   // method: displayAllAccounts

  public void   displayAnAccount(String aNumber) {
    ConsoleIO.out.println();
    ConsoleIO.out.println("Single account details for: " + theName + "\n");
    //
    //  Find the account with the given number and display it
    Iterator iter = theAccounts.iterator();
    while(iter.hasNext()) {
      Account acc = (Account)iter.next();
      String  number = acc.getNumber();
      if(aNumber.equals(number)) {
        acc.display();
        break;
      }
    }
  }   // method: displayAnAccount

  public void   removeAnAccount(String aNumber) {
    //   Find the account with the given number and remove it from the bank
    Iterator iter = theAccounts.iterator();
    while(iter.hasNext()) {
      Account acc = (Account)iter.next();
      String  number = acc.getNumber();
      if(aNumber.equals(number)) {
        iter.remove();
```

```java
            break;
        }
      }
    }  // method: removeAnAccount

    public int   getTotalAssets() {
      //  Form a running total of the balances for each account.
      int totalAssets = 0;
      //
      Iterator iter = theAccounts.iterator();
      while(iter.hasNext()) {
        Account acc = (Account)iter.next();
        totalAssets += acc.getBalance();
      }
      return totalAssets;
    }  // method: getTotalAssets

    // ------ Attributes ----------
    private String   theName;

    // ----- Relations ----------
    private java.util.Collection   theAccounts;   // of Account
}   // class: Bank

import textio.*;

public class Account implements Comparable {
  // ----- Operations ----------
  public    Account(String aNumber, int aBalance) {
    theNumber = aNumber;
    theBalance = aBalance;
  }  // method: Account

  public final int   getBalance() {
    return theBalance;
  }  // method: getBalance

  public final void   display() {
    ConsoleIO.out.print("\t" + "Account");
    ConsoleIO.out.print("\t" + "Number: " + theNumber);
    ConsoleIO.out.println("\t" + "Balance: " + theBalance + "\n");
  }  // method: display

  public String   getNumber() {
    return theNumber;
  }  // method: getNumber

  public boolean   equals(Object obj) {
    boolean result   = this.compareTo(obj) == 0;
    return result;
  }  // method: equals
```

```
public int   compareTo(Object obj) {
  Account  acc = (Account) obj;
  String accountNumber = acc.getNumber();
  return theNumber.compareTo(accountNumber);
}   // method: compareTo

public int   hashCode() {
  return  theNumber.hashCode();
}   // method: hashCode

// ----- Attributes -----------
private final String    theNumber;
private int    theBalance;

}   // class: Account
```

The output produced is:

All account details for: Object Bank

Account Number: ABC123	Balance: 1200
Account Number: DEF456	Balance: 1000
Account Number: GHI789	Balance: 2000

Single account details for: Object Bank

Account Number: ABC123	Balance: 1200

Total assets: 4200

All account details for: Object Bank

Account Number: ABC123	Balance: 1200
Account Number: GHI789	Balance: 2000

Total assets: 3200

Appendix F

Programming with Java

This appendix visits some of the prerequisite Java programming knowledge required by this textbook. We do not intend this and the following appendix to be a detailed study of Java. Rather, we remind the reader of those language constructs that will figure prominently in our development of the study of object-oriented design. For a comprehensive introduction to the Java programming language see the specialist texts (Eckel 2002, Deitel 2003).

Here, we begin with an overview of some fundamental language constructs including classes, methods, constructors, parameters, objects, object initialization, and message passing. We review how, through object composition, one object is constructed from other objects. Where many composite objects appear, then we utilize the Java collection classes. Finally, we explore how an inner class is used as a helper class supporting and separating functionality from the enclosing class.

F.1 Objects and message passing in Java

Consider we have a class to represent geometric points in a two-dimensional cartesian coordinate space. Let this class be Point. In Java we introduce such an object with the statement:

Point p;

Here, p is the object's identifier and Point is the class to which it belongs. In Java, the identifier p is used to refer to the object and is known as the *object reference* or *object handle*. In this statement we have only introduced the object's reference and not the object itself. To create the object and associate a handle with it we use:

Point p = **new** Point(0, 0);

The keyword **new** specifies that a new Point object is created having the handle p. The two zero values represent the X and Y coordinates we wish for this newly created Point object. A second example is:

Point q = **new** Point(2, 4);

If the definition for the Point class includes an operation getX to interrogate such an object for its X coordinate, then we send such a message to the object referenced by p with:

int x = p.getX();

A message in Java specifies the identifier (reference) of the recipient object (p) and the operation (getX()), separated by the period (.) symbol. Since the operation requires no parameters (see later) then the parentheses are left empty. The value delivered by this message is stored in the primitive **int** x. Note that the primitive language types in Java represent themselves rather than object handles. We might now display this value with:

ConsoleIO.out.println("X coordinate is: " + x);

See appendix C for details of the **textio** package.

F.2   Classes in Java

The Java interpreter starts program execution from the method main that must be present in the class nominated for execution. Program F.1 simply prints two lines of text using two statements in method main, defined in the class Main. Here this class simply acts as a placeholder for the method main. It could equally have been embedded in some other class. For clarity, we choose to publish method main in its own class.

Program F.1

```
import textio.*;

public class Main {
    public static void  main(String[] args) {
        ConsoleIO.out.println("Hello.");
        ConsoleIO.out.println("My name is Ken.");
    }
}
```

The previous section introduced a Point class to represent a point in a two-dimensional space. Program F.2 introduces such a class. The class has three methods and two attributes. The private attributes comprise a pair of int values representing the X and Y coordinates. The *accessor* methods getX and getY provide operations to interrogate a Point object for its constituents. The first method listed is referred to as the *constructor method* (distinguished by having the class name), used to initialize a Point object. When, for example, we introduce the Point object p by:

Point p = **new** Point(3, 4);

the class constructor is called and uses its two actual arguments to initialize its internal attributes. We usually refer to this as a *parameterized constructor* since two parameter values are given to support the initialization.

Program F.2

```
import textio.*;

public class Main {
    public static void  main(String[] args) {
        Point p = new Point(3, 4);
```

```
      ConsoleIO.out.println("The X coordinate is: " + p.getX());
      ConsoleIO.out.println("The Y coordinate is: " + p.getY());
   }
}
```

and:

```
public class Point {
   public    Point(int x, int y) {
      theXCoordinate = x;
      theYCoordinate = y;
   }
   public int    getX() {
      return theXCoordinate;
   }
   public int  getY() {
      return theYCoordinate;
   }
   // ----- Attributes -----------------
   private int      theXCoordinate;
   private int      theYCoordinate;
}
```

The expression p.getX() in the program sends the message getX to the Point object referenced by p. When defining a method the code can reference local variables, the method parameters (if any), and the attribute values of the object. In method getX a copy of the private attribute theXCoordinate is returned to the sender object. In the parameterized constructor method the formal parameter x is used to initialize the value of theXCoordinate attribute.

Implicitly, any method body can also use the keyword **this**. Keyword **this** is the handle or reference to the object for which the method has been called. For example, **this** would be the handle for the Point object p when executing the method getX in the message call p.getX(). Class Point might also be defined as in program F.3. Observe how the constructor is defined. The statement:

this.x = x;

means assign to the attribute x of the recipient object the value of the parameter x, removing the ambiguity between the attribute and the formal parameter.

Program F.3

```
public class Point {
   public Point(int x, int y) {
      this.x = x;
      this.y = y;
   }
```

```
    public int  getX() {
       return this.x;
    }
    public int  getY() {
       return this.y;
    }
    // ----- Attributes ----------------
    private int       x;
    private int       y;
}
```

Class methods, including constructors, may be *overloaded* to offer various method definitions having the same name. Overloaded methods must be distinguished by the number and/or type of their parameters. This means that two or more methods may bear the same name provided the number and or type of the formal parameters differs. Program F.4 has overloaded the constructor, introducing a second with no arguments, called the *default constructor*. The method sets each attribute to have a default value.

Program F.4

```
import textio.*;
public class Main {
   public static void  main(String[] args) {
      Point p = new Point(3, 4);
      ConsoleIO.out.println("The X coordinate of P is:  " + p.getX());
      ConsoleIO.out.println("The Y coordinate of P is: " + p.getY());

      Point q = new Point();
      ConsoleIO.out.println("The X coordinate of Q is: " + q.getX());
      ConsoleIO.out.println("The Y coordinate of Q is: " + q.getY());
   }
}
```

and:

```
public class Point {
   public    Point(int x, int y) {
      theXCoordinate = x;
      theYCoordinate = y;
   }
   public    Point() {
      theXCoordinate = 0;
      theYCoordinate = 0;
   }
```

```
public int  getX() {
    return theXCoordinate;
}
public int  getY() {
    return theYCoordinate;
}
// ----- Attributes -----------------
private int   theXCoordinate;
private int   theYCoordinate;
}
```

It is useful to be clear about the order of initialization including a constructor call. We create an instance of a class with a statement of the form:

```
Point p = new Point(3, 4);
```

which calls the parameterized constructor with actual parameter values 3 and 4. Before the code for the constructor executes, the attributes of the object are first initialized. If no explicit initial values are specified, as in program F.4, then default "zero" values are used. Here, both theXCoordinate and theYCoordinate attributes of the new Point object p are first set to int zero. Then, the constructor method executes and initializes (strictly, resets) the attributes to the values of the constructor parameters.

This means that the default constructor need not have any statements to perform if the properties have been explicitly initialized as in the following code fragment. And, of course, where the default initializer for all attributes is zero, then we do not require the explicit initialization as shown. For the benefit of others that maintain our code there is an argument that we should always be explicit and unambiguous.

Further, observe that explicit or implicit attribute initialization occurs in the order of the declarations in the class. Thus an explicit initialization of the attribute theYCoordinate might involve the value of theXCoordinate.

```
public class Point {
    public    Point(int x, int y) {
        theXCoordinate = x;
        theYCoordinate = y;
    }
    public    Point() {
    }
    // ...

    // ----- Attributes --------------------------
    private int       theXCoordinate = 0;
    private int       theYCoordinate = 0;
}
```

The keyword **this** can also be used to define one constructor in terms of another. The default constructor in program F.5 has the single statement **this**(0, 0); which invokes the parameterized constructor method with a pair of zero **int** arguments. In a constructor method this construction is recognized as a call to another constructor. The actual constructor called is determined by matching the number and type of actual parameters to the number and type of the formal parameters.

Program F.5

```
public class Point {
    public          Point(int x, int y) {
        theXCoordinate = x;
        theYCoordinate = y;
    }
    public   Point() {
        this(0, 0);
    }
    // ...

    // ----- Attributes --------------------
    private int      theXCoordinate;
    private int      theYCoordinate;
}
```

Every object of a class has its own copies of its attributes. In program F.4, the two Point objects referenced by p and q both have the pair of attributes defined for the class. A **static** class member represents a single object shared by all instances of the class and by the application. In program F.6 class Point introduces the **static** member ZERO which happens to be initialized as a Point object having the values (0, 0). Further, the member is qualified as **final**, meaning that its value cannot be changed. We usually describe these as constants.

There is only one such object in an application, irrespective of how many Point objects (the class in which ZERO is defined) are created. Note then how the ZERO object is referenced in main. Since ZERO is defined as a **static** member in class Point we must qualify its usage with the class name as in Point.ZERO. To that Point object we send it the message getX with Point.ZERO.getX().

Program F.6

```
import textio.*;

public class Main {
    public static void      main(String[ ] args) {
        Point p = new Point(3, 4);
        ConsoleIO.out.println("The X coordinate of P is: " + p.getX());
        ConsoleIO.out.println("The Y coordinate of P is: " + p.getY());

        ConsoleIO.out.println("The X coordinate of ZERO is: " + Point.ZERO.getX());
        ConsoleIO.out.println("The Y coordinate of ZERO is: " + Point.ZERO.getY());
    }
}
```

and:

```
public class Point {
    public    Point(int x, int y) {
        theXCoordinate = x;
        theYCoordinate = y;
    }
    public    Point() {
        this(0, 0);
    }
    public int  getX() {
        return theXCoordinate;
    }
    public int  getY() {
        return theYCoordinate;
    }
    public static final Point      ZERO = new Point();

    // ----- Attributes -----------------
    private int   theXCoordinate;
    private int   theYCoordinate;
}
```

When we introduce **static** class members we must reconsider the order of initialization of objects. In a conventionally compiled language such as C++, when such an application starts execution all the program code is loaded into memory at the start-up and all **static** objects are created and initialized. With an interpreted language such as Java, a class is only loaded when we create an object of that class. When the class is loaded, any **static** objects defined by that class are first initialized.

The execution of program F.6 would begin by first loading the class Main. In method main a new Point object with handle p is created. The class Point is now loaded and before the constructor for the object p is invoked the **static** member ZERO is initialized. We have to be careful to fully understand this order of initialization that occurs in a Java program.

Had the class included two **static** declarations, then they are initialized in the order in which they occur in the class. Hence the second **static** object may have used the value of the first **static** object in its initialization.

Finally, had the **static** object ZERO been initialized to an object of some class other than Point, then that additional class would be loaded and any **static** objects defined there created and initialized. The constructor for this ZERO object is then executed before, finally, executing the constructor for the Point object reference by p that started this snowball effect.

F.3   Object composition

Our Point class consisted of two primitive attributes. Having defined such a class we can then use it as the type for attributes of another class giving rise to the notion of *object composition*: one object composed of other objects. For example, the class LineSegment might be introduced to represent a line in a two-dimensional coordinate space. One Point attribute represents the start of the line, the other Point the end of the segment. Figure F.1 shows the arrangement. Note how the coordinate space has the X coordinate increasing to the right and the Y coordinate increasing downward. This is in keeping with the coordinate system used by the Swing library (see chapter 7).

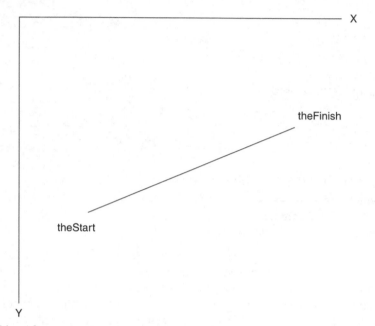

Figure F.1   *A line segment*

Class LineSegment is presented below. The class is realized with two Point attributes called theStart and theFinish. Two overloaded constructors are provided to initialize these values. The first has two Point parameters while the second uses four **int** values (two pairs of X, Y coordinates). The accessor methods getStart and getFinish give access to the attributes of a LineSegment object, while the enquiry methods isHorizontal and isVertical determine if a segment has these particular geometric qualities.

```
public class LineSegment {
    public   LineSegment(Point start, Point finish) {
        theStart = start;
        theFinish = finish;
    }
    public   LineSegment(int x1, int y1, int x2, int y2) {
        this(new Point(x1, y1), new Point(x2, y2));
    }
    public Point    getStart() {
        return theStart;
    }
```

```java
    public Point    getFinish() {
      return theFinish;
    }
    public boolean   isHorizontal() {
      return (theStart.getY() == theFinish.getY());
    }
    public boolean   isVertical() {
      return (theStart.getX() == theFinish.getX());
    }
    public double    getLength() {
      int xDiff = theStart.getX() - theFinish.getX();
      int xSquared = xDiff * xDiff;
      int yDiff = theStart.getY() - theFinish.getY();
      int ySquared = yDiff * yDiff;

      return Math.sqrt((double)(xSquared + ySquared));
    }
    // ----- Attributes -----------------
    private Point    theStart;
    private Point    theFinish;
}
```

Note the expression:

Math.sqrt((double)(xSquared + ySquared))

in the method getLength. Here we are executing the method sqrt (square root) from the Math class. This is a **static** method and hence is qualified with the class name.

Program F.7 shows this new class at work. Two segments are created. For the first we interrogate it for its start point, and for the second we find if it is vertical. We also obtain the length of the first line.

Program F.7

```java
import textio.*;
public class Main {
    public static void  main(String[] args) {
      Point p = new Point(3, 4);
      Point q = new Point(4, 5);
      LineSegment line1 = new LineSegment(p, q);
      LineSegment line2 = new LineSegment(q.getX(), q.getY(), q.getX(), 6);
      ConsoleIO.out.println("The start of line1 is: (" + line1.getStart().getX() +
                       "," + line1.getStart().getY() + ")");

      if(line2.isVertical())
        ConsoleIO.out.println("Line 2 is vertical");
```

```
    else
        ConsoleIO.out.println("Line 2 is not vertical");
        ConsoleIO.out.println("Line 1 length: " + line1.getLength());
    }
}
```

Continuing with this theme of object composition, let us introduce a class **Line** that describes a line comprising a series of connected line segments. Such a class will require a storage structure to collect together all the segments that form the line. The Java Foundation Classes (JFC) has a range of *collection classes* for this purpose (see appendix E). The simplest is the **ArrayList** that imitates a conventional array. However, this container is more sophisticated than the array in that it is an object that will dynamically grow to meet our requirements. If we include a new element into an **ArrayList** that has no free capacity, then it automatically resizes itself to make additional space for the incoming item. Here is this **Line** class:

```java
import java.util.*;

public class Line {
    public   Line(LineSegment first) {
        if(first != null)
            theLineSegments.add(first);
    }

    public   Line(LineSegment first, LineSegment second) {
        if(first != null)
            theLineSegments.add(first);
        if(second != null)
            theLineSegments.add(second);
    }

    public void   addSegment(LineSegment segment) {
        if(segment != null)
            theLineSegments.add(segment);
    }

    public double   getLength() {
        double totalLength = 0;
        int cardinality = theLineSegments.size();
        for(int k = 0; k < cardinality; k++) {
            LineSegment segment = (LineSegment)theLineSegments.get(k);
            totalLength += segment.getLength();
        }
        return totalLength;
    }

    // ----- Attributes ----------------
    private ArrayList        theLineSegments = new ArrayList(10);
}
```

Class ArrayList is defined in the utilities package java.util, hence the **import** statement at the top of the file. The Java classes are arranged into packages. A *package* is a collection of classes or sub-packages. The **import** statement requests that all the classes from the (sub-)package util of the java package be accessible to this program code file. The attribute theLineSegments is the container holding references to the segments comprising the line. This ArrayList object is initialized with start-up size of 10. The two class constructors for Line initialize the container with one or two segments. The method addSegment incorporates a further segment into the container using the method add of class ArrayList.

Method getLength computes the sum of the lengths of the individual segments. It does this by iterating across all the segments in the collection, getting their length and forming a running total. The number of elements in the container is found from the method size in class ArrayList, and a particular element is selected with the get message.

Note the statement:

LineSegment segment = (LineSegment)theLineSegments.get(k);

The ArrayList method get accesses the item in the container at some index position. The ArrayList class is a *generic container* capable of holding any type of object. Thus the method get simply returns a reference or handle to the object at the given index. We must *coerce* or *cast* this reference to the required type as shown. Since we know that the only items in the container are LineSegment objects then we use the cast prefix (LineSegment) to coerce the handle to the required reference type.

Program F.8 shows this new class in action. An empty Line object is first established then populated with two line segments. The length of the line is then determined.

Program F.8

```
import textio.*;
public class Main {
    public static void    main(String[] args) {
        Line polyLine = new Line(null);

        Point p = new Point(3, 4);
        Point q = new Point(4, 5);

        LineSegment l1 = new LineSegment(p, q);
        LineSegment l2 = new LineSegment(q.getX(), q.getY(), q.getX(), 6);

        polyLine.addSegment(l1);
        polyLine.addSegment(l2);

        ConsoleIO.out.println("The length of line is: " + polyLine.getLength());
    }
}
```

F.4   Inner classes

The Java language supports various styles of *inner classes*. This feature permits one class to be defined as a member of another class in a manner similar to that for class attributes and class methods. A variety of inner classes are supported by Java. Here, however, we are primarily interested in *inner member classes* that operate as helper classes to the enclosing class. A characteristic of these inner classes is that code within an inner member class can refer to any of the attributes and methods of the enclosing class, including those with **private** visibility.

Consider our Line class from the previous section. For some application we may have a need to cycle across the individual line segments and perform some action upon them. Of course, the class Line does not make visible its implementation, otherwise we would break the rules of *encapsulation* and *information hiding*. Hence we have no direct access to the line segments contained by the Line object.

What we require is another abstraction called an *iterator abstraction* or simply an *iterator*. An iterator permits us to visit the members of a collection without exposing the underlying storage mechanism of the collection object. We can achieve this using the mechanisms we have already introduced. A separate iterator class would usually have the collection it is to iterate across passed in the constructor and may also require various accessor methods introduced into the container object to support iteration.

An inner member class object, however, is always associated with an instance of the enclosing class. Inner class methods can directly reference the attributes of the instance of the enclosing class. These two features mean that we can forego some of the requirements noted at the end of the last paragraph, resulting in a more elegant solution.

Into class Line we introduce the inner class LineIterator with methods isExhausted, getSelection and advance. The class has a single property theIndex, used as the subscript into the ArrayList object theLineSegments of the enclosing class Line. Method isExhausted determines when all the segments have been visited. Method getSelection directly accesses the collection of segments and retrieves the reference to the LineSegment object at the index position. Method advance simply moves the index to reference the next segment.

```java
import java.util.ArrayList;
public class Line {
    public    Line(LineSegment first) {
      if(first != null)
         theLineSegments.add(first);
    }
    public    Line(LineSegment first, LineSegment second) {
      if(first != null)
         theLineSegments.add(first);
      if(second != null)
         theLineSegments.add(second);
    }
    public void    addSegment(LineSegment segment) {
      if(segment != null)
         theLineSegments.add(segment);
    }
    public double    getLength() {
      double totalLength = 0;
```

```
    int cardinality = theLineSegments.size();
    for(int k = 0; k < cardinality; k++) {
        LineSegment segment = (LineSegment)theLineSegments.get(k);
        totalLength += segment.getLength();
    }
    return totalLength;
}
// ----- Attributes -----------------
private ArrayList            theLineSegments = new ArrayList(10);

// ----- inner class ---------------------

public class LineIterator {
    public boolean           isExhausted() {
        return theIndex >= theLineSegments.size();
    }
    public LineSegment       getSelection() {
        return (LineSegment)theLineSegments.get(theIndex);
    }
    public void              advance() {
        theIndex++;
    }
    // ----- Attributes ---------------------
    private int              theIndex = 0;

}

}
```

Note how the inner class methods directly reference the private attributes of the enclos-ing class. For example, method isExhausted uses the code theLineSegments.size() to find how many segments there are. For clarification we may also use Line.**this**. theLineSegments.size() with the class qualifier Line and implicit attribute **this** prepended to signify the member theLineSegments of the instance of the enclosing class. This special syntax can be necessary to remove ambiguity where the inner class and the enclosing class have features with the same name. It is also useful documentation where the programmer wishes to highlight that the inner class methods are accessing outer class features.

Program F.9

```
import textio.*;
public class Main {
    public static void    main(String[] args) {
        Line polyLine = new Line(null);
        Point p = new Point(3, 4);
        Point q = new Point(4, 5);
        LineSegment l1 = new LineSegment(p, q);
        LineSegment l2 = new LineSegment(q.getX(), q.getY(), q.getX(), 6);
```

```
      polyLine.addSegment(l1);
      polyLine.addSegment(l2);
      int   verticalSegments = 0;
      Line.LineIterator iterator = polyLine.new LineIterator();
      while(iterator.isExhausted() == false) {
         LineSegment segment = iterator.getSelection();
         if(segment.isVertical())
            verticalSegments++;
         iterator.advance();
      }
      ConsoleIO.out.println("Number of vertical segments is: " + verticalSegments);
   }
}
```

Program F.9 demonstrates the usage. Here we iterate across the segments of a line and determine how many of these are vertical. The while loop utilizes the iterator methods to achieve the effect. Of particular interest is the statement:

Line.LineIterator iterator = polyLine.**new** LineIterator();

The LineIterator class must be given its fully qualified name (Line.LineIterator) since it is an inner class of a normal top-level class. The new LineIterator object is always associated with an instance of the enclosing class, here the instance polyLine, and the object creation is achieved with the code fragment polyLine.**new** LineIterator()

The iterator is an example of a *design pattern*, see chapter 8. The principal concept in this pattern is to assign responsibility for access and traversal of an aggregate object (such as class Line) to a separate iterator object (such as LineIterator). The iterator has a traversal interface for accessing the elements of the aggregate (namely, isExhausted, getSelection and advance).

A separator iterator class confers a number of advantages. A separate iterator class obviates the need for the traversal interface appearing in the aggregate class. Since an iterator object manages its own traversal state, then we can have more than one iterator object applied to the same aggregate. By subclassing (see next appendix) of the iterator class we can introduce variations on the traversal. For example, a specialized iterator may only select those elements that meet a particular requirement.

The relation between class Line and class LineIterator is shown by figure F.2.

Figure F.2   *LineIterator class diagram*

Note that the iterator and aggregate are coupled. To instantiate a LineIterator object we must provide a Line object to traverse. This is the essence of the code:

Line.LineIterator iterator = polyLine.**new** LineIterator();

The new LineIterator object (iterator) is created in association with the Line object (polyLine).

In this book's chapters we shall see that the Java collection classes are accompanied with iterator classes achieving the same effect we have illustrated (see appendix E). We shall use these in further code examples. Further, inner classes will play a significant role in our graphical applications (see chapter 7), acting as *event handlers* that process user requests.

F.5   Exercises

1. How are the attributes of a class introduced in Java? Why are they normally given private visibility? Under what circumstances can we remove this restriction?
2. What is the syntax for message sending in Java? Illustrate with messages with and without parameters and messages with and without return values.
3. Show how an object instance is created in Java. Which method in the object's class is involved in the creation process? Describe its purpose.
4. What do you understand by the term "overloading"? Give examples of where it might be used.
5. In this appendix we used the ArrayList collection class. Visit the Sun website: http://developer.java.sun.com/developer/onlineTraining/collections/Collection. html for a discussion of the other Java collection classes.
6. In this appendix we used inner member classes. Visit the Java website (http:// java.sun.com) and find what other types of inner classes are supported by Java. Identify the use for each type of inner class.
7. Develop the class Rational to represent a fractional value expressed as the quotient of two positive integers. Examples of rational numbers are 1/2, 3/4 and 8/5. A rational number comprises a pair of integers for the numerator and non-zero divisor. For this class implement suitable constructors and accessor operations getNumerator and getDenominator.

    Generally, a rational number is expressed in its simplest form. The rational number 8/6 reduces to 4/3. The reduced form is achieved by determining the highest common factor (hcf):

```
private static int   hcf(int num, int den) {
    if(den > num) {
        int temp = num;
        num = den;
        den = temp;
    }
    while(num % den != 0) {
        int temp = den;
        den = num % den;
        num = temp;
    }
    return den;
}
```

In this class introduce the operations to add and multiply rational numbers:

public void add(Rational rat) { ... }
public void mult(Rational rat) { ... }

The arithmetic of rationals is:

a / b + c / d = (a * d + b * c) / (b * d)
a / b * c / d = (a * c) / (b * d)

8. Develop the class Rectangle using the class Point. A rectangle is represented by a pair of Points for the upper left and lower right vertices. Include methods to determine the length of the perimeter and the area occupied by the figure. Augment this class with the methods contains and inflate:

public boolean   contains(Point p) { ... }
public void   inflate(**int** x, **int** y) { ... }

Method contains returns **boolean true** if the Point parameter is contained within the bounds of the rectangle (including the border).

Method inflate expands a rectangle by the given amounts. For a positive value for x, the rectangle should grow by that amount at both its left and right sides. For a positive value for y, the rectangle will grow by that amount at both its top and bottom. Negative values for either of these will shrink the rectangle.

9. Use the Point class to develop a new class Triangle declared as:

public class Triangle {

    public   Triangle(Point p, Point q, Point r) { ... }
    public double   perimeter() { ... }
    public double   area() { ... }
    public boolean   isRightTriangle() { ... }
    public boolean   isIsoscelesTriangle() { ... }

    private Point   p1;
    private Point   p2;
    private Point   p3;
}

A right triangle is one with a right-angled vertex. An isosceles triangle is one with two sides of equal length. If the length of the sides of a triangle are a, b and c, then the perimeter is:

a + b + c

and the area is given by:

sqrt(s * (s - a) * (s - b) * (s - c))

and where the semi-perimeter is given by:

s = (a + b + c) / 2

10. Review the system documentation for the class ArrayList. What operation is available to change the size of an ArrayList object dynamically? If we add one item more than the present capacity of an ArrayList what happens?

Appendix G

Object-Oriented Programming with Java

In this appendix we focus on the use and deployment of specialization. We illustrate Java's support for class specialization, substitution, method redefinition and the polymorphic effect. Further, we review abstract classes with deferred methods and interfaces.

Specialization permits a subclass to inherit features from a superclass, leveraging a measure of code reuse. Since a subclass object has all the features of a superclass object, then the former can substitute for where the latter is required. By carefully exploiting substitution we can make our software more extensible and more adaptable to change. Redefining methods in subclasses exploits the effect of dynamic binding, where a particular message adapts to the recipient's behaviour. These effects are a consequence of working to the interface described by a superclass.

One of the more distinctive features of object-oriented programming languages is their support for *specialization*. A specialized class is one that is developed out of an existing class by inheriting all the features of that class. Further, the specialized class can introduce additional attributes, additional methods, and may choose to redefine any of the inherited methods. Specialization can lead to significant amounts of code reuse where the specialized class only implements its differences with the existing class.

The specialized class is called the *subclass*. The class from which a specialized class is developed is referred to as the *superclass*. In Java the subclass is said to be an *extension* of the superclass. This is intended to connote the idea that the subclass has all the features of the superclass and possibly more. The subclass/superclass arrangement is usually presented in a UML class diagram as shown by figure G.1.

Specialization also gives rise to the notion of *isA*. An instance of a subclass is an instance of a superclass. This is correct since a subclass inherits all the features of its

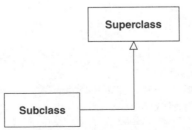

Figure G.1 *A class diagram with specialization*

superclass and hence an instance of the subclass can receive all the messages defined in the superclass. Further, this implies that where in our code an instance of a superclass is expected an instance of a subclass may be used. This is known as the *principle of substitutability*.

Substitution makes our software extensible since we may substitute an instance of a subclass that may be defined later in the lifetime of the software. Careful architecting of our systems is required to gain this leverage, but it can result in the introduction of instances of a new class without recourse to wholesale changes to the existing fabric of the software.

Substitutability combines with the *polymorphic effect* to have a profound effect on our programming style. Through polymorphism we leverage improved program code organization. Conventional procedural code is frequently concerned with obtaining an appropriate effect according to the type of some object. The polymorphic effect removes this consideration from our code. Where once we might have used a complex if or switch statement, now we have a single simple program statement. Frequently, these conditional statements would populate large parts of our code and make program maintenance a major concern.

Through dynamic binding one object sends a message to another, indifferent to its actual type. Substitution means the recipient object may be an instance of any appropriate subclass. The recipient object knows to which type or class it belongs and executes the correct method. If the recipient is an instance of a subclass in which the method has been redefined, then it is this redefined method that executes. Redefined methods of various subclasses will have different effects but the sending object is decoupled from this concern and for the need to select the appropriate behaviour.

G.1   Specialization

Consider a class that represents an Employee of an organization. From the previous chapter this class might appear as:

```java
public class Employee {
    public   Employee(String aName, String aJobTitle, int aSalary) {
        theName = aName;
        theJobTitle = aJobTitle;
        theSalary = aSalary;
    }

    public String   getName() {
        return theName;
    }

    public void      printDetails() {
        ConsoleIO.out.print("Employee: " +theName);
        ConsoleIO.out.print(", " +theJobTitle);
        ConsoleIO.out.println("," +theSalary);
    }
```

```
// ----- Attributes -----------------
protected String    theName;
protected String    theJobTitle;
protected int    theSalary;
}
```

The attributes of the class have protected visibility. Protected features behave as if they were private to other classes but public to subclasses. This means that any subclass of Employee will be able to refer to these values directly.

Consider now a manager. A manager is also an employee of an organization but with added responsibilities. A manager may head up one or more projects, or have responsibility for some budgetary amount. Let us introduce class Manager as a specialization of class Employee. A class diagram for this is given in figure G.2.

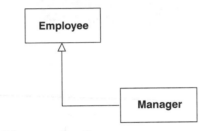

Figure G.2   *Employee/Manager class diagram*

The corresponding code for class Manager is:

```
public class Manager extends Employee {
    public   Manager(String aName, String aJobTitle, int aSalary, int aBudget) {
        super(aName, aJobTitle, aSalary);
        theBudget = aBudget;
    }

    public void    printDetails() {
        ConsoleIO.out.print("Manager: " +theName);
        ConsoleIO.out.print("," +theJobTitle);
        ConsoleIO.out.print("," +theSalary);
        ConsoleIO.out.println("," +theBudget);
    }
    // ----- Attributes -----------------
    private int       theBudget;
}
```

The keyword extends notifies that class Manager is an *extension* or a specialization of the class Employee. The Manager class inherits all the features of the class Employee. Since an instance of the class Employee can be sent the message getName, then so too can an instance of the class Manager.

Observe how the method printDetails is reintroduced in the subclass Manager. This is an example of *method redefinition*. The subclass is a more specialized version of the superclass with a budgetary responsibility, and requires a different presentation from this method. Note also how this method directly references the **protected** attributes theName, theJobTitle and theSalary from the superclass (but see also our discussions in chapter 5).

We have said that a subclass inherits all the features of the superclass. Strictly, a subclass inherits all but the superclass constructor. This is not unreasonable since a subclass will frequently have additional attributes that will require appropriate treatment under, for example, initialization. The subclass constructor will be responsible for initializing both its own and its inherited attributes. When creating an instance of the class Employee that class' constructor is employed to perform the correct initialization. In a similar manner the subclass Manager achieves proper initialization of its inherited attributes through the superclass constructor. The code is:

```
public    Manager(String aName, String aJobTitle, int aSalary, int aBudget) {
    super(aName, aJobTitle, aSalary);
    theBudget = aBudget;
}
```

The first statement is responsible for invoking the superclass constructor. The remaining statement initializes the additional subclass attribute in the usual way.

Program G.1 illustrates the two classes in use. An instance of the class Employee and an instance of the class Manager are first created. Both are then sent the message printDetails, producing the output:

```
Employee: Ken Barclay, Lecturer, 1000
Manager: John Savage, Senior Lecturer, 1200, 2000
```

Since class Manager has redefined this method then the new effect is exhibited by the second line of the output.

Program G.1

```
public class Main {
    public static void  main(String[ ] args) {
        Employee e1 = new Employee("Ken Barclay", "Lecturer", 1000);
        Manager m2 = new Manager("John Savage", "Senior Lecturer", 1200, 2000);
        e1.printDetails();
        m2.printDetails();
    }
}
```

This next example illustrates the principle of substitution and the polymorphic effect. The class diagram is shown in figure G.3. An organization has any number of people working for it. They are classified as plain employees or managers as described in the preceding example.

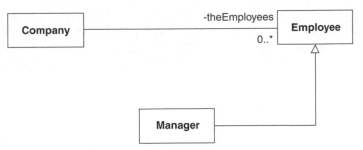

Figure G.3   *Company/Employee class diagram*

The one-to-many relationship between a Company object and its Employee objects is realized by an ArrayList object as part of the Company class. The method hire in the Company class is supplied with a handle to some Employee object that is added to the ArrayList container. In program G.2 the actual parameter objects given to method hire are instances of both the Employee class and the Manager class, illustrating the use of substitution.

```java
import java.util.*;
import textio.*;
public class Company {
    public   Company(String aName) {
        theName = aName;
    }
    public void   hire(Employee emp) {
        theEmployees.add(emp);
    }
    public void   printEmployeeDetails() {
        ConsoleIO.out.println("Company: " +theName);

        Iterator iter = theEmployees.iterator();
        while(iter.hasNext() == true) {
            Employee emp = (Employee)iter.next();
            emp.printDetails();
        }
    }
    // ----- Attributes -----------------
    private String       theName;
    private ArrayList    theEmployees = new ArrayList(10);
}
```

The method printEmployeeDetails from class Company cycles through the members of the collection and sends each the message printDetails. Here, we utilize the Iterator class from the Java Collections Framework classes. The various containers provide an Iterator object through the message iterator. The principal methods from this class are hasNext (which determines if there are more elements in the collection yet to visit) and next (which returns a handle to the next available item from the collection).

The collection is, of course, a mix of **Employee** and **Manager** objects. When we select an item from the collection we cast it to some kind of **Employee** object. The polymorphic effect of the operation printDetails is determined dynamically at run-time according to the type of the recipient object. This is exhibited by the program's output:

```
Company: Napier
Employee: Ken Barclay, Lecturer, 1000
Manager: John Savage, Senior Lecturer, 1200, 2000
Manager: Jessie Kennedy, Reader, 1200, 3000
```

According to the actual object's type the correct printDetails method is executed.

Program G.2

```
public class Main {

    public static void   main(String[] args) {
        Employee e1 = new Employee("Ken Barclay", "Lecturer", 1000);
        Manager m2 = new Manager("John Savage", "Senior Lecturer", 1200, 2000);
        Manager m3 = new Manager("Jessie Kennedy", "Reader", 1200, 3000);

        Company co = new Company("Napier");

        co.hire(e1);
        co.hire(m2);
        co.hire(m3);

        co.printEmployeeDetails();
    }
}
```

Substitution and polymorphism combine to make our designs extensible. The **Employee/Manager** class hierarchy of figure G.3 may be both broadened and deepened without incurring changes elsewhere. For example, we could introduce further kinds of employees or more specialized managers. No change would be necessary to the class **Company**. Through substitution, any of these specialized employees may be hired by the company. The company can produce a listing of all its employees through the polymorphism of the operation printDetails.

G.2   Abstract classes

An abstract class has one or more *abstract* (or *deferred*) *methods*. An abstract method does not include a method body, only a signature. The abstract methods may be introduced in the class itself or inherited from another abstract class. These abstract methods introduce a *protocol* that must be respected by all subclasses. Thus, a concrete subclass of an abstract superclass must present an implementation for all inherited abstract methods.

   Abstract classes are illustrated with the example of figure G.4. A **Company** has any number of employees working for it. The employees are either sales staff or sales managers.

All staff have a name, job title and salary. A SalesEmployee has a sales target set against them. A SalesManager has a budget and is a manager for any number of the sales employees.

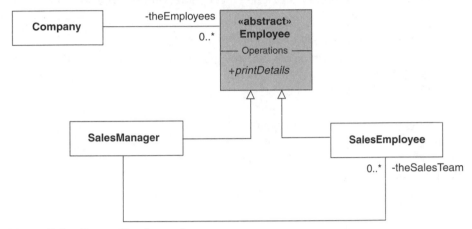

Figure G.4 *Abstract Employee class*

Class Employee is abstract with a deferred method printDetails. Both concrete subclasses SalesEmployee and SalesManager must define this method. For class SalesEmployee this method gives an employee summary listing their name, job title, salary and sales target. For class SalesManager the printDetails method again lists the manager's details but additionally gives the names of the sales staff in her team. Here is class Employee:

```
public abstract class Employee {
    public     Employee(String aName, String aJobTitle, int aSalary) {
        theName = aName;
        theJobTitle = aJobTitle;
        theSalary = aSalary;
    }
    public String   getName() {
        return theName;
    }
    public abstract void    printDetails();    // DEFERRED
    // ----- Attributes -----------------
    protected String    theName;
    protected String    theJobTitle;
    protected int        theSalary;
}
```

Note that the class is declared as **abstract**. No instances of an **abstract** class can be created. Observe also how the deferred operation printDetails is introduced. No method body is given, simply the operation signature giving the name, return type and formal parameters, if any. The operation is qualified as **abstract**.

Concrete subclass SalesManager and SalesEmployee must provide an implementation for printDetails. For class SalesManager we first print the details for the manager, then we cycle through the sales team members printing their name. Note also the method addSalesMember for adding a new SalesEmployee object to the team overseen by the manager.

```java
public class SalesManager extends Employee {
    public  SalesManager(String aName, String aJobTitle, int aSalary, int aBudget) {
        super(aName, aJobTitle, aSalary);
        theBudget = aBudget;
    }

    public void  addSalesMember(SalesEmployee emp) {
        theSalesTeam.add(emp);
    }

    public void  printDetails() {
        ConsoleIO.out.print("SalesManager: " + theName);
        ConsoleIO.out.print(", " +theJobTitle);
        ConsoleIO.out.print(", " +theSalary);
        ConsoleIO.out.println(", " +theBudget);

        Iterator iter = theSalesTeam.iterator();
        while(iter.hasNext() == true) {
            SalesEmployee emp = (SalesEmployee)iter.next();
            ConsoleIO.out.println("\tSales team member:" +emp.getName());
        }
    }
    // ----- Attributes ----------------
    private int         theBudget;
    private ArrayList   theSalesTeam = new ArrayList(10);
}
```

Class Company is modified from the previous version. Two hire methods are available. The first simply includes a new manager as part of the workforce. The second (overloaded) hire method includes the sales employee in the company workforce, but additionally includes her into the sales team managed by a particular team leader. To implement this we must find an existing employee that is the given manager. The code for this is:

```java
if(emp instanceof SalesManager && emp == man) { ... }
```

where the first part of the conditional checks the run-time type of the employee.

```java
public class Company {
    public  Company(String aName) {
        theName = aName;
    }
    public void  hire(SalesManager man) {
        theEmployees.add(man);
    }
```

```java
public void   hire(SalesEmployee sales, SalesManager man) {
   Iterator iter = theEmployees.iterator();
   while(iter.hasNext() == true) {
      Employee emp = (Employee)iter.next();
      if(emp instanceof SalesManager && emp == man) {
         SalesManager manager = (SalesManager)emp;
         manager.addSalesMember(sales);
      }
   }
   theEmployees.add(sales);
}

public void   printEmployeeDetails() {
   ConsoleIO.out.println("Company: " +theName);

   Iterator iter = theEmployees.iterator();
   while(iter.hasNext() == true) {
      Employee emp = (Employee)iter.next();
      emp.printDetails();
   }
}

// ----- Attributes ----------------
private String       theName;
private ArrayList    theEmployees = new ArrayList(10);
}
```

Finally, the application is shown in Program G.3. The program run produces the output shown below. The manager Jessie Kennedy presents both her details as well as listing the names of her sales team. The other employees simply list their details.

```
Company: Napier
Manager: Jessie Kennedy, Professor, 1200, Budget: 8000
   Sales member: Ken Barclay
   Sales member: John Savage
Sales employee: Ken Barclay, Lecturer, 1000, Sales target: 2000
Sales employee: John Savage, Lecturer, 1200, Sales target: 2000
```

Program G.3

```java
public class Main {
   public static void   main(String[] args) {
      SalesEmployee e1 = new SalesEmployee("Ken Barclay", "Lecturer", 1000, 5000);
      SalesEmployee e2 = new SalesEmployee("John Savage", "Senior Lecturer", 1200, 2000);
      SalesManager m3 = new SalesManager("Jessie Kennedy", "Professor", 1200, 10000);
      Company co = new Company("Napier");
      co.hire(m3);
```

```
        co.hire(e1, m3);
        co.hire(e2, m3);
        co.printEmployeeDetails();
    }
}
```

The UML collaboration diagram of figure G.5 shows the configuration of objects and the message flows between them. The application, represented by the actor, sends the message printEmployeeDetails to the Company object with identity co. This Company object then sends the message printDetails in turn to the three employee objects. When the message printDetails is received by the manager object it executes the redefined version for this method. The method printDetails from class SalesManager prints the managers details then gets the names of the two employees that are members of that manager's team.

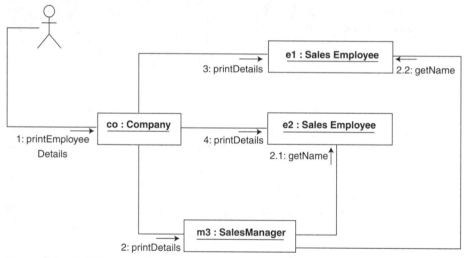

Figure G.5   *Collaboration diagram*

G.3   Interfaces

An *interface* is a particular kind of abstract class in which all methods are deferred and for which there are no (non **static**) attributes. In effect, an interface is a pure abstract class. A Java interface is implicitly **public** as are all its methods. Further, these methods are implicitly qualified as **abstract**. It is permissible in Java to make these qualifications explicit, rendering additional documentation for the reader.

A Java interface is permitted to contain class attributes but these must be **static** and **final**. Java interfaces are sometimes used to collect a number of related constants. Other classes can then refer to them using the class name qualifier or by subclassing to provide direct referencing.

Since an interface has no state we say that any specialized class *implements an interface*. An abstract class or a concrete class can implement an interface. A subclass that is itself another interface is said to *extend* the interface superclass.

Java supports *multiple inheritance* through interfaces. A class can be an extension of at most one abstract class or concrete class and implement any number of interfaces. The *isA* relation is still present, but now an instance of the subclass can be substituted for any of the superclasses, including the interfaces.

Consider any two-dimensional closed figure such as a rectangle or a triangle for which we wish to obtain the length of its perimeter or its area. The interface ClosedFigure introduces the protocols for obtaining these values, which every concrete subclass must implement. Further, we can develop code in terms of this interface without consideration for the actual type of closed figure with which we are interacting. That way, our code is made much more adaptable.

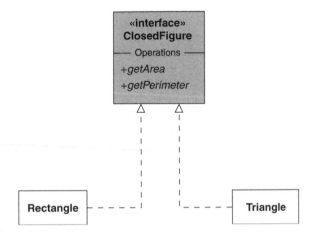

Figure G.6   *Specialization hierarchy*

Figure G.6 shows the class hierarchy we shall develop. The code for the interface ClosedFigure is:

```
public interface ClosedFigure {
    public abstract double   getPerimeter();
    public abstract double   getArea();
}
```

The code for the two concrete subclasses must include implementations for these two deferred methods. Below is the code for the Rectangle class using a variant of the Point class (using **double**s rather than **int**s). The Triangle class would be developed similarly.

```
public class Rectangle implements ClosedFigure {
    public   Rectangle(Point upperLeft, double width, double height) {
        theUpperLeft = upperLeft;
        theWidth = width;
        theHeight = height;
    }
```

```java
public   Rectangle(double upperLeftX, double upperLeftY, double width,
                        double height) {
   this(new Point(upperLeftX, upperLeftY), width, height);
}

public double   getPerimeter() {
   return 2 * (theWidth + theHeight);
}

public double   getArea() {
   return theWidth * theHeight;
}

// ----- Attributes -----------------
private Point        theUpperLeft;
private double       theWidth;
private double       theHeight;
}
```

The sample application code is given in program G.4. Here, we create instances of the Rectangle and Triangle classes and interrogate each for their perimeter and area values.

Program G.4

```java
public class Main {

   public static void    main(String[] args) {
      Rectangle rect = new Rectangle(0.0, 4.0, 4.0, 4.0);        // square 4 x 4
      Triangle tri = new Triangle(0.0, 0.0, 3.0, 0.0, 3.0, 4.0);   // 3-4-5 right triangle

      ConsoleIO.out.println("Rectangle: " + rect.getPerimeter() + ", " +rect.getArea());
      ConsoleIO.out.println("Triangle: " + tri.getPerimeter() + ", " +tri.getArea());
   }
}
```

G.4   Exercises

1. Carefully distinguish between the terms interface, abstract class and concrete class.
2. Carefully distinguish between the Java keywords **extends** and **implements**, ensuring that you are clear where they are used.
3. Carefully distinguish between the terms superclass and subclass. How may a subclass differ from its superclass? How much commonality is there? Describe what is meant by substitutability when discussing superclass and subclass.
4. What do you understand by the terms redefinition and the polymorphic effect? Explain their value to OO development.
5. What feature does a subclass constructor method use to initialize inherited attributes?

6. How is a redefined method introduced into a Java subclass?
7. The Java keyword **abstract** has two uses. Give examples of each and explain their usage.
8. Answer the following, giving explanations.
   (a) Can an interface extend another interface?
   (b) Can an interface extend more than one interface?
   (c) Can an interface subclass an abstract class?
   (d) Can an abstract class extend another abstract class?
   (e) Can a concrete class extend an interface?
   (f) Can a concrete class implement an interface?
   (g) Can a concrete class implement two interfaces and extend an abstract class?
   (h) Can a concrete class extend two abstract classes?
9. Develop a Person class with name and age properties. Now develop two specializations from class Person. Prepare the class Student as a subclass of Person with a matriculation number attribute. Revise the Employee class from this chapter with Person as its superclass.
10. Develop the class Employee with name and employee number attributes. Provide suitable methods for this class. Introduce into the class the abstract operation computeMonthlyPay that is redefined in the subclasses SalariedEmployee and HourlyEmployee. A salaried employee has a salary, which is paid out monthly. An hourly employee has a fixed pay rate and a number of hours worked per month. Implement the method computeMonthlyPay for these two subclasses.
11. Develop a class Account having a balance and account number attributes. Introduce the methods debit and credit that implement deposit and withdrawal transactions on a bank account. Form two specializations for the classes DepositAccount and CurrentAccount. A deposit account accrues interest and cannot be overdrawn while a current account may be overdrawn up to some overdraft limit.
12. Develop the concrete class Circle as a subclass of ClosedFigure. Test its behaviour using program G.4.
13. Develop the class Quadrilateral to represent any four-sided figure, then specialize this into the subclasses Rectangle and Square. Choose a suitable representation for the Quadrilateral and implement the operations getPerimeter and getArea. Redefine both these for the two subclasses.
14. Look up the Swing class Jframe in the Java documentation and present its superclass hierarchy.
15. Look up the Swing classes WindowListener and WindowAdapter in the Java documentation files. What is the class hierarchy of these two classes? Why is class WindowListener an interface? How is class WindowAdapter a concrete class?

Appendix H

Procedural Code in Java

Appendix D is primarily concerned with the mappings from the elements of a class diagram to Java classes. They were simply concerned with the major constructs including interfaces, classes, relations, class attributes and methods. In this appendix we examine the procedural code of class methods. In addition, we outline the pseudo-code that we use to describe program logic.

H.1 Procedural code

Java's procedural code is used to implement a method body. This code can also appear in initialization blocks as described in chapter 7. Procedural code is assembled from the three control structures of sequence, selection and iteration (repetition). These three control structures are prepared in any combination to describe the logic required. In Java the **if** and **switch** statements represent the language's two selection statements. The **while**, **do** and **for** are the repetition statements. Combining these structures in any manner means that an **if** statement can be nested within another **if** statement, or a **while** statement can be used in an **if** statement, and so on.

As we form complex control logic then it too will benefit from the application of some design. This we achieve by developing pseudo-code to describe the intended logic before converting it into actual Java statements. Pseudo-code, as its name suggests, is not Java code but bears some passing resemblance to it. Pseudo-code aims to describe the intention of some logic by using code-like constructs.

Pseudo-code is sometimes also known as a Program Design Language (PDL) or by the term Structured English, in which some restricted form of natural language is used to describe our intent.

The language used with pseudo-code is a mix of fixed constructs and paraphrased statements. The idea of fixed constructions is evident from the name Structured English. Some parts of the language used are constrained to some fixed structures. The remainder of the pseudo-code is composed of short phrases that describe the intent. We distinguish the latter by using italic font. The fixed structures are presented as emboldened text.

A simple illustration of pseudo-code with no structured elements is given below. It simply identifies that we need to obtain a date from the user and that we prompt the user when it is required. It represents an example of sequential code.

prompt the user for a date
read the date from the user

393

Observe how program code is totally absent. This is intentional, we simply wish to convey the actions required in getting some input from the user. If we then subsequently add some detail, then the following Java emerges:

```
ConsoleIO.out.print("Please enter date as DD MM YYYY: ");
int day = ConsoleIO.in.readInt();
int month = ConsoleIO.in.readInt();
int year = ConsoleIO.in.readInt();
```

The following pseudo-code presents some logic in which the user is prompted to provide a date that is then validated and an appropriate message displayed.

```
prompt the user for a date
read the date from the user
IF the date is valid THEN
    display "date is valid"
ELSE
    display "date is invalid"
ENDIF
```

Observe how the italicized text simply describes what should occur without detailing how it should be implemented. In particular note the phrase *the date is valid*, which seeks to determine if we have a valid date. No attempt is made to describe how this will be done. In fact, further pseudo-code can be introduced to outline the required logic for this as a separate activity.

Note also structured logic described by the **IF THEN ELSE ENDIF** control structure. This part of the pseudo-code identifies the need for a two-way selection to determine the output to be produced. The **THEN** and **ELSE** indicate choice, while **IF** and **ENDIF** delimit the extent of the selection.

The pseudo-code is readily translated into its Java implementation. Ignoring how the date is validated, and using the textio package from appendix C, we arrive at:

```
ConsoleIO.out.print("Please enter date as DD MM YYYY: ");
int day = ConsoleIO.in.readInt();
int month = ConsoleIO.in.readInt();
int year = ConsoleIO.in.readInt();
if(...)
    ConsoleIO.out.println("date is valid");
else
    ConsoleIO.out.println("date is invalid");
```

Iteration 2 of the case study from chapter 4 introduces a simple menu into the text-based user interface. The menu operates by first prompting the user with a set of numbered choices. The user selects from this list and the necessary service is provided. The menu is repeated and the process continues until the user selects numbered choice 0 to exit the system. The pseudo-code that would have first described this logic is given as:

```
present the menu to the user
read the selection made by the user
```

```
WHILE selection is not choice number 0 DO
    IF selection is choice number 1 THEN
        provide service for choice number 1
    ELSE IF selection is choice number 2 THEN
        provide service for choice number 2
    ELSE IF ... THEN
        ...
    ENDIF
    present the menu to the user
    read the selection made by the user
ENDWHILE
```

Again, this maps directly into Java as shown below. Note how presenting the menu and obtaining the user's selection is relegated to a support method called getSelection. The Structured English described by the **WHILE** and **IF** have direct counterparts in Java.

```java
String choice = this.getSelection();
while(choice.equals("0") == false) {
    if(choice.equals("1")) {
        // Get the borrower details from the human user
        ConsoleIO.out.print("\t" + "Enter the borrower name >>>");
        String borrowerName = ConsoleIO.in.readString();
        // Register the borrower with the Library
        library.registerOneBorrower(borrowerName);
    }
    else if( choice.equals("2") == true) {
        // ...
    }
        // ...
    else {
        ConsoleIO.out.println("\n\t\t" + "Unknown selection - try again" + "\n");
    }
    choice = this.getSelection();
}
```

Alternatively, we might have considered a **do** loop and nested **if-else** statement:

```java
String choice = "";
do {
    // Get the human user's choice.
    choice = this.getSelection();
    ConsoleIO.out.println();
    //
    // Action the human user's choice.
    if(choice.equals("0")) {
        ConsoleIO.out.println("\n\t\t" + "SYSTEM CLOSING" + "\n");
        ConsoleIO.out.println();
    }
```

```
else if(choice.equals("1") == true ) {
    // Get the borrower details from the human user
    ConsoleIO.out.print("\t" + "Enter the borrower name >>>");
    String borrowerName = ConsoleIO.in.readString();
    // Register the borrower with the Library
    library.registerOneBorrower(borrowerName);
}
else if(choice.equals("2") == true) {
    // ...
}
    // ...
else {
    ConsoleIO.out.println("\n\t\t" + "Unknown selection - try again" +"\n");
}
//
} while(choice.equals("0") == false);
```

Chapter 3 utilized a Java collection object to realize a one-to-many architecture. Appendix E gives details of the Java collection classes. These discussions introduced the notion of an iterator object that is bound to a collection object to provide a means to visit, in turn, all the items in the collection. In the Bank class of chapter 3 we prepared the creditAccount method:

```
// class Bank
public void creditAccount(java.lang.String aNumber, int anAmount) {
    Iterator iter = theAccounts.iterator();
    while(iter.hasNext() == true) {
        Account acc = (Account)iter.next();
        if(aNumber.equals(acc.getNumber())) {
            acc.credit(anAmount);
            break;
        }
    }
} // method: creditAccount
```

The method visits each Account object maintained by the collection theAccounts as part of the Bank class. The method identifies the Account object with the required account number and performs a credit operation against that account. This implementation might have been derived from the matching pseudo-code:

```
FOREACH acc:Account IN theAccounts DO
    IF account number of acc is that required THEN
        perform credit operation on acc
        break from this loop
    ENDIF
ENDFOREACH
```

The **FOREACH** clause introduces acc as a reference to one of the Account objects in the collection theAccounts. Within the **FOREACH** loop we use acc to represent the next Account object we are processing.

We can also use a variant of the **FOREACH** pseudo-code to indicate repetition a set number of occasions. We might, for example, use a **FOREACH** to show how the sum of the first ten input values is achieved:

initialize the total to zero
FOREACH *num* **IN** *0..9* **DO**
    prompt user for the next input value
    read the value
    add the value to the total
ENDFOREACH

Of course, the Java for this is:

```java
double total = 0.0;
for(int num = 0; num<10; num++) {
    ConsoleIO.out.print("Please enter next value: ");
    double value = ConsoleIO.in.readDouble();
    total += value;
}
```

Index